Technical Arts
in the
Han Histories

SUNY series in Chinese Philosophy and Culture

Roger T. Ames, editor

Technical Arts
in the
Han Histories

Tables and Treatises
in the *Shiji* and *Hanshu*

Edited by
Mark Csikszentmihalyi
and Michael Nylan

Cover: Bronze oil lamp from the tomb of Liu He, the Marquis of Haihun, ca. 64 BCE. On the side is inscribed: "Changyi Eunuch Imperial Messenger [regulation] ding-style lamp, weighing six jin ten liang, made in the second year." Photograph by Mark Csikszentmihalyi.

Published by State University of New York Press, Albany

© 2021 State University of New York

All rights reserved

Printed in the United States of America

No part of this book may be used or reproduced in any manner whatsoever without written permission. No part of this book may be stored in a retrieval system or transmitted in any form or by any means including electronic, electrostatic, magnetic tape, mechanical, photocopying, recording, or otherwise without the prior permission in writing of the publisher.

For information, contact State University of New York Press, Albany, NY
www.sunypress.edu

Library of Congress Cataloging-in-Publication Data

Names: Csikszentmihalyi, Mark, editor. | Nylan, Michael, editor.
Title: Technical arts in the Han histories : tables and treatises in the Shiji and Hanshu / Mark Csikszentmihalyi and Michael Nylan.
Description: Albany : State University of New York Press, [2021] | Series: SUNY series in Chinese Philosophy and Cultyure | Includes bibliographical references and index.
Identifiers: ISBN 9781438485430 (hardcover : alk. paper) | ISBN 9781438485447 (ebook) | ISBN 9781438485423 (pbk. : alk. paper)
Further information is available at the Library of Congress.

10 9 8 7 6 5 4 3 2 1

To Michael Loewe

by unanimous vote of the contributors,
with deep respect

Tamdiu discendum est quemadmodum vivas,
quamdiu vivas. — Seneca, *Letters*
As long as you live, keep learning how to live.

Contents

Acknowledgments ... ix

Introduction ... 1
Michael Nylan 戴梅可 *and Mark Csikszentmihalyi* 齊思敏

1 Land Tenure and the Decline of Imperial Government
  in Eastern Han ... 49
  Michael Loewe 魯惟一

2 Water Control and Policy-Making in the *Shiji* and *Hanshu* ... 101
  Luke Habberstad 何祿凱

3 The *Hanshu* Geographic Treatise on the Eastern Capital ... 135
  Lee Chi-hsiang 李紀祥

4 Celestial Signs in Three Historical Treatises ... 181
  Jesse J. Chapman 柴傑思

5 On *Hanshu* "Wuxing zhi" 五行志 and Ban Gu's Project ... 213
  Michael Nylan 戴梅可

6 Western Han Sacrifices to Taiyi ... 281
  Tian Tian 田天

7 Writing Abstractly in Mathematical Texts from
  Early Imperial China ... 307
  Karine Chemla 林力娜

8 Commentarial Episodes in Early Chinese Medicine:
   An Experiment in Decentering the Standard Histories   339
   Miranda Brown 董慕達

9 Narratives of Decline and Fragmentation, and the
   Hanshu Bibliographic Taxonomies of Technical Arts   367
   Mark Csikszentmihalyi 齊思敏 *and Zheng Yifan* 鄭伊凡

Contributors   407

Index   409

Acknowledgments

In writing a book, especially an edited book, one always incurs many debts, and it is a pleasure to acknowledge them in due time. This book came into being to address some of the needs of our graduate students and best undergraduates to show them how scholars young and old tackle the technical issues, partly by doing and partly by watching other experts in the field discuss their own research projects. We editors were aware that the fields of Classics and the history of science were doing innovative work on the technical arts in antiquity, pushing the boundaries in ways that remained largely closed to those of us working in Early China studies, so we decided to take the plunge.

Of course, we must begin with the chief dedication of the volume, supported unanimously by the contributors, all of whom were canvassed through emails along with other routine inquiries, until it became clear that we had to stop emailing the probable object of the dedication himself. For each and every person save Michael Loewe put Michael at the top of the list of people we wanted to see honored for contributions to the field of Han history in general, and the technical arts in particular. As it would take far too long to list Michael's virtues as a scholar, colleague, teacher, and human being, suffice it to say that his Chinese name says it all: he is 鲁惟一, "the only one," that is, "peerless."

All the contributors would like to thank Nathan Sivin, who commented on all the papers in various stages, sometimes more than once. Originally, the editors had planned to invite Professor Sivin to contribute his own chapter for the volume, which would describe where the history of the technical arts has been and is likely going in Chinese studies. Then, because of circumstances beyond everyone's control (but common, alas, in academia), a few chapters did not appear on time for the volume, at which point it seemed unfair to ask Professor Sivin to plough through every chapter all over again and produce his own in short order. That said, a number of contributors consulted him via email about their chapters, and we have predictably benefited from those exchanges, we believe.

As a student, Csikszentmihalyi's mentors in the technical arts were Sally Church, Robin Yates, Lothar von Falkenhausen, and Albert Dien. Since coming to Berkeley, he has been fortunate to learn from his students, Michael Nylan, and her students.

Nylan's work on the technical arts began with study of the *Taixuan jing* under Nathan Sivin, proceeded with Paul Serruys and linguistics, and took an unexpected turn with early optics, with medicine and with manuscript culture, rooted always in history undergirded by archaeology. In matters archaeological, she has often benefitted from the superb guidance of Robert Bagley, her teacher, friend, and colleague, as well as from Michael Loewe.

A great many contributors commented on each other's papers (as, for example, Luke Habberstad and Jesse Chapman read for Michael Nylan and each other). This has been an incredibly collegial bunch of people to work with: mostly on time with deadlines, endlessly patient in response to queries large and small, alert to the ways that they would like the papers to articulate with one another, despite the separate fields they largely represent. Besides this, we would like to thank the librarians at UC-Berkeley (Jianye He, Peter Zhou, Bruce Williams, and Deborah Rudolph especially) for their tireless work on behalf of faculty and students. Charles Aylmer, in the Cambridge University Library, hunted down materials for Michael Loewe.

At UC-Berkeley, the editors moreover had the help and support of the Taiwan–United States Alliance (TUSA), which funded our initial meeting for potential contributors to plan such a volume late in 2014. We owe thanks to the Institute for East Asian Studies (then under Martin Backstrom's leadership). Our respective departments (East Asian Languages and Cultures; History) gave the editors additional material and logistical support for the initial meeting with contributors. Finally, the Marjorie Meyer Eliaser Chair in International Studies provided support for the project at several key junctures.

Editors and contributors would like to thank Christopher Ahn for his encouragement early on with the project. We owe thanks to James Peltz and Roger T. Ames for their unfailing kindness and clear directions while shepherding the co-editors through the volume's final stages. We were fortunate enough to have the invaluable editorial assist from Vanessa Davies, an Egyptologist, whose steady intelligence, good cheer, and superb sense of the English language relieved the editors of considerable stress and gave us a "good read" on how best to convey some technical points to non-specialists. As Nylan was concurrently dealing with a less capable copy editor at another press, she was supremely mindful of what all academics owe to good editors at every stage of the complex process. The editors would also like to proffer warm thanks to Diane Ganeles at SUNY Press, and to Laura Tendler, for help during the final editing process.

Introduction

MICHAEL NYLAN 戴梅可 AND
MARK CSIKSZENTMIHALYI 齊思敏

> A good singer makes people follow his notes.
> A skilled builder inspires others to take up his work.
> Fitting the wheel of a cart requires a trained apprentice.
> Zhou's perfections were the artifact of Zhougong's efforts.
>
> 夫善歌者使人續其聲。
> 善作者使人紹其功。
> 推/椎車之蟬攫,負子之教也。
> 周道之成,周公之力也。
>
> *Yantie lun, juan 2, pian 7*

For more than a decade now, we two editors of this volume have been teaching our students at the University of California at Berkeley about the more technical aspects of the *Shiji* and *Hanshu*, not to mention how those Han histories relate in turn to the fifth-century *Hou Hanshu* and Liu Zhiji's assessment in the early eighth century, in an attempt to correlate our readings in the standard histories with the growing body of archaeological evidence at hand (much of it not yet adequately tabulated, let alone parsed). Too often, essays about history, religion, and literature still cleave to a handful of famous chapters in the two official histories, relying heavily, for example, on the assassin-retainer chapter of the *Shiji*, or the economic and legal treatises of the *Hanshu*, or the so-called "Letter to Ren An," where serviceable, if occasionally outdated, translations exist.[1]

Determined to push ourselves and our students out of the usual comfort zones, we decided to try to frame a set of more systematic responses to the technical arts for those interested in early China, in the twin beliefs, first, that responsible historians must delve into every aspect of the universe of discourse that the subjects of their research knew superbly well; and, second, that our counterparts in Classics and the history of science were pushing to find comparative materials they could "think with." Moreover, in a vital sense, everything worth in-depth study in early China was "technical" (see below), and the range of topics subsumed under these two rubrics of the "techniques" (*shu* 術) and "arts" (yi 藝) reflects the high value placed on forms of expertise that typically required a special vocabulary and set of disciplined practices. This is the broad sense in which the present volume is dedicated to the "technical arts," building upon an earlier volume titled *China's Early Empires: A Reappraisal* (2010), intended as a supplement to the somewhat outdated *Cambridge History of China, vol. 1*, and a recently published volume from Paris devoted to the technical monographs in the *Suishu*, which were modeled on the *Hanshu* treatises.

During the Han, a wide variety of *shu* spoke to the arts of governing in both theoretical and practical terms, whether the *shu* were used individually or combined as means to "govern the realm and achieve peace" 治國致平之術, "order the people" 治民之術, or "face south, as does the ruler of the people" 君人南面之術. And while the English term "technical arts" comes closest to the range of Han practices that sometimes take *shu* as the second graph in a binomial expression (terms such as *dao shu* 道術), we caution against a cross-cultural comparison that is already too widely in use, drawing a careless equivalence between sets of culturally-bound terms, leading some to mistakenly conclude that *shu* is an exact counterpart of the earlier Greek *techne* (τέχνη).[2]

A related sense of the term "technical arts" that we are resisting is the backward projection of the modern categories of science and technology. To see an example of this projection, we need go no further than the current ICS Concordance Series (CHANT) website, which sorts all Han and pre-Han writings into a series of rubrics, in which the "Science and Technology" (*Kexue jishu* 科學技術) category occupies a different segment of the website than those allotted to the Classics, Masterworks, Literary Studies, or Histories. Separating out early historical, philosophical, or literary texts from technical learning, we would argue, seriously impedes moderns' ability to understand the rhetorical patterns and bodies of specialized knowledge that informed them all. Similarly, lumping together pre-Han and Han materials in a single database, without offering at least some informed speculation about dating, may seem a minor design choice, but this feature comes with two unfortunate, momentous effects: (a) reifying the unwarranted assumption that dynastic rises and falls correlate neatly with cultural changes; and (b) lumping together Western and Eastern Han, when the two dynasties operated on starkly different sociopolitical, economic, and cultural bases. In consequence, the present website's classification system reinforces anachronistic and misleading divisions that many scholars had hoped to lay to rest decades ago.[3]

Surveying the breadth of materials, what is strikingly obvious is the sheer quantity of technical knowledge implicated in the attainment of high cultural literacy—the ability to recite, read, and compose elegant phrases—required of nearly all who aspired to serve the upper ranks of the two Han ruling houses.[4] Men and women of cultivation commanded the technical arts as part of their stock-in-trade, for, lacking such knowledge, they could not begin to grasp the arguments for and against specific innovations and alterations at court, in the provinces, or even on their estates. Simply to project a commanding presence was not enough. This integration of such facility into other areas of life that moderns tend to view as separate is one important dimension of the "technical arts" in the Western and Eastern Han.

The dynamic nature and changing status of classical and technical learning during the four centuries under review pose challenges to any simple generalization about the social applications of different fields of knowledge. As readers may recall, while the Western Han, especially through Xuandi's reign (74–48 BCE), aimed at a high degree of political centralization, the Eastern Han rulers from the beginning were, perforce, accustomed to oligarchic rule by a small group of leading families, which gave the capital elites and provincial magnates considerably more power. So far as we know, the Eastern Han never managed to carry out a successful empire-wide cadastral survey, although local surveys must have been conducted for tax purposes.[5] While the Western Han, until the reigns of Yuandi and Chengdi (r. 48–33, 33–7 BCE), paid less lip service to the authority of the Classics, its courts being understandably proud of their "new" form of rule, the weaker Eastern Han was determined to shore up the legitimacy of its "restored" dynasty by frequent references to its warm embrace of the "antique Way" of Confucius, the Duke of Zhou, and even the legendary sage-rulers Yao and Shun. The court also expended considerable energy managing the interpretation of omens and portents, linking the Classics to both an evolving symbolic vocabulary and to rearticulations of the theory of Tianming 天命 such as that of Ban Biao (d. 54) early in Eastern Han.⁶ And whereas the Xiongnu semi-nomadic groups were a significant worry for the Western Han courts, groups of them had been so far domesticated before Eastern Han, with their confederacies severed,⁷ that they served as the frontier forces for the "restored" Eastern Han dynasty, fighting against other newly powerful groups, such as the Qiang and Xianbei. One of the few "constants" that remained in place throughout the four hundred–odd years of the two Han dynasties, then, was the acceptance of the administrative structures inherited from Qin, as well as the dynastic pretensions to extend *jiaohua* 教化 (civilizing influences) to the local subject populations (as the Qin did before them)⁸ and to rule, whenever possible, by *wuwei* 無為 (nonintervention, delegation of authority). (In Han sources, *wuwei* is an administrative or social strategy, rarely if ever "nonaction" or "effortless action.")⁹

Both Western and Eastern Han, like Qin before it, offered court positions to men and women of learning, in the realization that neither the laws nor the usual administrative rules could suffice to answer every need.¹⁰ This structural continuity created the conditions for the development of a set of categories in the *Shiji* and *Hanshu*, providing us with a descriptive taxonomy of the technical arts in the Han administrations. It would have been possible, and valuable in a different way, to address this topic using redescriptive categories like those on the spines of the volumes of *Science and Civilization in China*. The present book instead draws attention to ways that Han authors tried to make sense of the various *shu* by linking their past legacies to their present deployment in a culture buffeted by sociopolitical ebbs and flows.

Given the complexity of these developments over time, as well as the paucity of work devoted to the *Hanshu*, the editors invited a group of well-respected Early China experts drawn from disparate communities in Europe, the United States, and the Chinese-speaking world to weigh in on the very sort of technical problems presented by the *Shiji* and *Hanshu* that frequently present major obstacles to modern readers. The result has been this interdisciplinary venture, in which each contributor explicates one very difficult topic for the edification of present and future scholars. Mindful of the extraordinary debt we owe to such earlier giants writing in English as Ho Peng Yoke and Joseph Needham, William Hung and Yang Lien-sheng, Nancy Lee Swann and A. F. P. Hulsewé, not to mention the savants of previous generations writing in Chinese, Japanese, French, and German (too numerous to mention here), this group of contributors in the present generation of scholars hopes, by their concerted efforts, to add to the fund of accumulated knowledge provided by the earlier generations.¹¹ Hence this dip into still largely uncharted waters.

Some of the topics explained here may, at first glance, strike readers as utterly abstruse, with the volume a virtual "Chinese encyclopedia" in the style of Borges. However, each chapter explores a major preoccupation of the courts during the early empires, such as landholdings or medical cures, sacrifices and omen-readings, the crucial relations pertaining between the heavens above and the earth below, predilections when reading and writing in manuscript culture, and the precise methods for accurate calculations. Throughout, the chapters in this volume ground these seemingly disparate topics firmly in a range of sources that describe them, attending especially to the standard histories of the *Shiji* and *Hanshu*, so as to analyze the different rhetorical patterns and cultural concerns that accompanied discussions of early technical knowledge.

Absent a better sense of these disparate pieces of highly crafted rhetoric, it is hard to believe that students of early China will not be misled when piecing together even a rudimentary picture of the early empires. Greg Anderson, in his bracing *The Realness of Things Past* (2018), has already shown how often the reigning experts on the early empires in classical Greece have trafficked, wittingly or unwittingly, in a panoply of shopworn concepts borrowed straight from the grand "systems" theories of the early modern eras.¹² Increasingly in other fields, those theories have drawn fire in cutting-edge studies, but sometimes students of early China still seem to be in their thrall. Due attention has therefore been paid by the editors to the underlying assumptions that inform the chapters in this book, and the precise vocabulary we contributors intend to introduce. But more simply, as the editors themselves have learned so much from reading this collection of chapters, they believe that other students of Chinese culture may profit from these pages, too, regardless of their respective stances on modern and postmodern theories.

I. On the "Technical Arts" (*shu* 術) and "Ordering" (*zhi* 治)

While *zhi* 治 most often means to govern or administer, it also is the main verb meaning "to heal" and "to make orderly," as when working farmland or healing a body.¹³ In a volume devoted the technical arts, its multivalence proves of immense significance, insofar as *zhi* is indisputably the general thrust of treatments of the "technical arts" in the chapters ahead, as it is in many chapters of the *Shiji* and *Hanshu*. Order is the aim of *shu* in these contexts in two precise senses: first, the technical arts all aim to induce or maintain order, where disorder threatens; and, second, the arts of governing are the crown of the various technical arts, each of whose contribution conduces to an orderly empire.[14] In consequence, the ruling metaphor *zhi* is as likely to intrigue generalists interested in the construction of the sociopolitical discourse in the early empires in China as it is good "to think with" for historians of science. The latter group will doubtless pick up on the narrower construction of "technical arts" supplied by the *Hanshu* bibliographic category, the "technical and computational arts" (*shu shu* 術數),[15] or the still looser definition of *shu* 術 as any skill that requires "know-how" to achieve the goal.[16] Notably the broader and narrower uses of the term "technical arts" often cannot be separated in the surviving literature from Zhanguo and Qin, Western and Eastern Han, where the phrase "techniques of the Way" (*dao shu* 道術)[17] implies cultivation of the person and also bringing civilization to the realm. To reiterate, the "technical arts" are paired with "governing" in root metaphors whose ramifications rippled throughout society, and the preoccupations of the members of the governing elite anxious to maintain their prerogatives and standing make this evident.

In the end, it was the degree to which the dynasty and its upper echelon of officials could legitimate themselves by their creation or maintenance of stable orders that became the standard used to assess the current occupant of the throne and his representatives. For that reason, ambitious would-be and actual officials sought mastery of one or more specialties subsumed in that narrower category of the technical arts, insofar as a special expertise would allow them relatively quick entry into the higher divisions of the administration via participation in the court conferences convened to devise solutions for current crises and to advance the court's "civilizing influence" inside and outside the capital. Most expected to practice only a subset of the technical arts (for example, omen prediction or jurisprudence), given how many technical fields there were, and it was only the highest-ranking ministers and chancellors who sought to model mastery of the entire range of the technical arts of governing, including "harmonizing the cosmic forces of yin and yang *qi*"[18] and arguing policy decisions from the standpoint of the "Great Way" or the "sages' intent."[19] As the images and patterns on which human society is modeled were there for all to see, the special ability of the good ruler or administrator depended upon his ability to "distinguish" (*bian* 辨) and "infer from categories" (*tui lei* 推類), sifting through the shifting phenomenal experience to ascertain the essential underlying significance. A sage, by definition, through long practice perceives the constant patterns underlying disparate phenomena (seemingly incommensurate, irregular, and ungovernable), so that he can act decisively and well in a bewildering array of situations. In the language of the day, he has attained "comprehensive insight" (*da tong* 大通).[20] Indeed, his miraculous virtuosity makes him seem "like unto the gods" (*rushen* 如神)[21] to mere onlookers lacking the same level of exquisite discernment. Practice and perspicacity go hand-in-hand with all the technical arts, for constant practice affords greater insights, whose rewards prompt still greater devotion to practice of the arts and so on.

Looking more closely at the specific methods involved, the relevance of order and the technical arts to governance are often obvious. Taxing land, devising redistributive mechanisms, and water control efforts are among the most basic functions of the early empires throughout the antique world, even in countries lacking the uniquely massive administrative infrastructure that was dedicated to good governance in early China.[22] Making war is another basic function, and while military texts fall outside the *Hanshu* bibliographic category of the "technical and computational arts," conceivably because there were far too many military classics to cram in under the heading, outside the bibliographical treatise it was usually assigned to the technical and the computational rubric, proving that bibliographic categories do not reflect sociopolitical theories, let alone realities on the ground.[23] It was meanwhile the duty and privilege of the capital regimes to issue the calendars and almanacs that fixed the proper times for rituals and for planting and harvesting, and to generate the sorts of tables, registers, and genealogical lists to which several of our authors refer.[24] The palace bureaus issued the maps and charts—literary, mathematical, and visual—for use by the civil and military officers.[25] But even to make a living (*zhi sheng* 治生) demanded special sorts of sociability, with alliances and patron-client relations signaled by elaborate ritual acts and sealed, as often as not, by contracts in writing.[26] Meanwhile, the primary sign of good governance, as the *Documents* classic made abundantly plain, was that legitimate regimes consult widely with different constituencies, both living and dead. Hence the necessity to divine and to consult omen experts regarding any irregular signs in the sky or on the earth as potential messages sent to warn. This brings the wide variety of methods for contacting the unseen spirit powers securely into the realm of the technical arts used in governing.

For that reason, the need to keep all three registers of *zhi* in play, from the most theoretical to the most mundane, is vital to full understanding of the chapters collected in this volume, particularly in relation to the healing arts, to philosophical approaches to life, and to manuscript production and repair.[27] Regarding the first, "Perfect bodies . . . required perfect politics," as Nathan Sivin noted.[28] The healing arts were designed to heal (*zhi*) a single body, members of a family or a larger community or even the body politic, and those who governed an area well, like healers, diviners, and purveyors of immortality recipes, assumed a powerful spiritual dimension that on occasion could rival, challenge, or supplant those of the imperial courts.[29] (We should not forget that good administrators like Li Bing 李冰, governor of Shu, were offered cult during their own lifetimes.) Classicists who laid claim to lineage transmissions from the antique sages could also parlay the sages' perfect insights into positions of power, although there were always doubters.[30] And in text learning, the same ruling metaphor of governing was applied once again: one had to "put it in order," or even "heal" or "cure" a manuscript (*zhi shu* 治書), when the strings for the bamboo slips rotted or the worms had nibbled away at the silk. Massive efforts directed toward remedying (also *zhi* 治) the flaws in hand-copied manuscripts led editors in the late Western Han palace libraries to try to devise more reliable methods for evaluating competing variants and editions, with the result that those same editors became ever more self-conscious authors, ever more intent upon imposing order on the transmission and circulation of texts.[31] There was a technique for every undertaking, apparently, even techniques for "ordering the soul" by finding a balance in the first, chaotic reactions (*zhi xin* 治心). And just when it seemed there was too much to know, another technical art, with its own special vocabulary and disciplined practices, came into being: the art of learning to embrace the Mystery 玄.[32]

An important point we editors seek to register here: during the two Han dynasties, those who trained in any number of professional capacities (including the classically trained) were never hired for "pure" scholarship.[33] There were pragmatic needs to be fulfilled urgently by men with the requisite expertise, even the entertainers and classicists.[34] Exponential growth in the numbers of students (or, more likely, clients of powerful people) at the Imperial Academy (Taixue 太學) did not signal, then, exponentially heightened interest in Five Classics learning per se.[35] The court classicists—and nearly all high officials borrowed some classical learning, whether or not they followed the teachings of Confucius[36]—styled themselves as men of action wielding the "great tool" that was the empire.[37] By the *Shiji* and *Hanshu* accounts, the leading classicists devised and carried out such initiatives as getting rid of poll taxes for children; reducing outlays for the imperial palaces and parks; and redistributing funds to the poor troubled by floods, famines, and bandits. At the same time, of course, they were to know when attractive policy proposals would prove counterproductive, or worse. They would have to "manage to unify it [policy] by reference to the inherent patterns" of the problem 統之以理.³⁸ And no one who had truly mastered an art would deny the real complexities at hand. Interestingly, Ban Gu 班固, chief compiler of the *Hanshu*—usually a proponent of classical learning and of the Ru (construed not as a discrete social group but plainly as an epithet for those claiming expertise in ritual practices and historical precedents), if ever there were one—names the classicists as but one of twelve sorts of technical experts needed to contribute to good governance (water-control engineers, legal experts, calendrical experts, military leaders, judges of men's fitness for office, etc.). For Ban, the classicists figure in this particular list primarily as "elegant" theorists and performers of the ritual proprieties,³⁹ with the "constant arts" 經術 simply the arts of governing.⁴⁰ This underscored the importance of using history as a mirror to discern which policies are workable and which are not.⁴¹ Xunzi, as usual, put it succinctly when he insisted that there were but three major arts to bringing people together in a common project (the use of charismatic example, the use of brute force, and the use of wealth), but only the first could keep them harnessed to a common purpose for very long.⁴²

To the men of high cultural learning in the early empires, there were no neat disciplinary lines dividing the poets from the healers, the administrators from the "harmonizers of yin and yang *qi*," the estate managers from the musicians. Though modern academics often find it hard to see beyond their respective fields, the ideals and practices of the governing elite in antiquity clearly envisioned worlds in ways we do not. The contributors to this volume, noting some of the manifest disadvantages of our current ways of slicing pies, urge students of early China, as best they can, to immerse themselves in the broad-ranging thinking of the polymaths whom they study so diligently.

II. On the *Shiji* and *Hanshu* Projects

Nearly all surveys of history writing in early China begin with the terse chronicle of middle Zhou period interstate and intrastate affairs called the *Annals* (*Chunqiu* 春秋, traditionally ascribed to Confucius), and proceed swiftly to the two first standard histories for the imperial period, the *Shiji* by Sima Qian 司馬遷 (originally *Taishi gong shu* 太史公書) and *Hanshu* by Ban Gu and other members of his family, noting that the *Shiji* and *Hanshu* share a hybrid Basic Annals-Biographies format. Surveys typically pay little attention to the other sections in these two standard histories, but we contend that their contents are vital to any deep understanding of the *Shiji* and *Hanshu* projects. These additional sections—three in *Shiji* and two in *Hanshu* (one fewer because the *Shiji*'s third section devoted to the history of the major pre-unification kingdoms is omitted from the *Hanshu*, as the unified history of a single centralized empire)—show specific contributions from technical fields interwoven into larger arguments about the techniques of good governance. But they do far more, in that they inform us about ways of thinking in early Chinese historiography. In this section, we focus on the structure of the two shared sections we are calling the "Tables" and "Treatises," because they are less familiar to non-specialist readers, and nearly all of the contributors to this volume refer to them constantly.

The *Shiji*'s five-part structure begins with twelve chronicles in the Basic Annals (*benji* 本紀) section, which together narrate a history spanning more than three millennia, the time between the legendary primeval emperors and Sima Qian's own ruler, Emperor Wu 武 of the Han. To compile this section (like the others), Sima Qian drew on whatever records existed in the palace archives, but supplemented them with local lore and hearsay. His second section comprised of ten Tables (*biao* 表) he compiled while correlating diverse names and events (more on this below); for example, the "Yearly Table of the Twelve Lords" covers the Chunqiu period treated by the *Annals* classic, from 841 to 477 BCE. Sima Qian's third section consists of eight Treatises (*shu* 書, or "monographs"), explored at greater length below.[43] Thirty chapters follow, called the Hereditary Houses (*shijia* 世家), which mainly supply the histories of the noble houses in pre-unification times (e.g., the "Hereditary House of the State of Zhao"), with two noteworthy exceptions: one Hereditary House focuses on Confucius, the "uncrowned king," whose disciples down through the ages are configured as his "descendants"; and one Hereditary House is allotted to Chen She, the rebel upstart who began the process of bringing down the short-lived empire of Qin.[44] (As noted above, the Hereditary Houses section does not appear in the *Hanshu*.) A fifth and final section, which receives by far the most attention today, is the Biographies (*liezhuan* 列傳) section, which collects memorable accounts of the lives of noteworthy individuals or groups, such as the biographies for the wandering knights.

Some historians have found the word *shu* 書 (here translated as "treatise") to be significant, thinking that Sima Qian intends to relate the contents of this section to the *Documents* classic (aka *Venerable Documents* or *Shangshu* 尚書). One should note, perhaps, that some of the contents (e.g., those related to the *feng* and *shan*) can be traced back to that classic, but the connections remain rather tenuous. Certainly Sima Qian had multiple writings (also *shu* 書) about ancient political and legal methods in mind when he created this section, and perhaps we can say that, like those hallowed writings, he emphasized ritual and music, and then punishments and war, in that order. He did not foreground his own expertise in calendrical calculations, which is interesting, because some legends had culture heroes inventing the calendar long before ritual and music.[45] Scholarly attention has focused on his last three treatises, in large part because of the fundamental importance of the *feng* and *shan* sacrifices to notions of legitimate rule and ever-present worries over water control and economic measures in budgetary matters.

Here are capsule synopses of the eight *Shiji* Treatises:

1. Treatise on Ritual (*Li shu* 禮書): A brief overview of the historical attitudes of Qin and Han rulers to classical ritual forms, followed by a discussion of the function of ritual in society along with illustrations of the symbolic aspects of particular rites. The treatise extensively quotes from and expands on the *Xunzi*'s "Discussion of Ritual" (*Li lun* 禮論) chapter.

2. Treatise on Music (*Yue shu* 樂書): Following a brief comparison of Qin and Han theories of music's cosmic resonances, the bulk of the treatise concerns the psychology of music and its effect on people's states of mind and heart, providing rich parallels to the *Xunzi* and other Han-period ritual texts. It then turns to a discussion, framed as a set of dialogues, concerned with music's social functions.

3. Treatise on Pitch Pipes (*Lü shu* 律書): The twelve notes of the pitch pipe are of interest to Sima Qian mainly insofar as a sophisticated understanding of their theoretical and practical potentials depends on the ability to explore *qi* theory (yin/yang) in relation to the eight directional "winds" (both human and heavenly) and the Five Constants (*wuchang* 五常) that order human relations.[46] That theoretical understanding, in turn, allows a practical benefit: a master can then deploy the pitch pipes to diagnose or predict a range of present and future

situations correctly. Aptly, then, this treatise begins with a set of anecdotes, arranged by chronology, designed to situate the military applications of the pitch pipes within the broader discussions pertaining to the social roles of war. Two brief technical sections follow: the first, composed of paragraphs that detail the eight types of resonances that correspond to the directional winds, and the second, some specifications for a pitch pipe designed to produce the twelve tones with their perfect proportions.

4. Treatise on the Calendar (*Li shu* 歷書): The first part of this treatise traces the development of the ancient calendars in relation to political legitimacy, ending with the "Grand Inception" (*Taichu*) reforms of 104 BCE, in which Sima Qian played a role.[47] The chapter ends with a technical section providing annual calendrical information for the years between 104 and 29 BCE, which clearly was a section "continued" long after the death of Sima Qian.

5. Treatise on the Celestial Offices (*Tianguan shu* 天官書): This lengthy treatise begins by enumerating and describing key asterisms and starry apparitions assigned to the five offices governing the sectors of the sky (the four directions plus the center). It moves on to the movement of the Five Planets and secondary celestial phenomena, often correlating the location of the "sign" with different events transforming human society. It finally turns to identifying past techniques related to such phenomena, establishing a strong link between good rule and favorable celestial omens.[48]

6. Treatise on the *Feng* and *Shan* Sacrificial Tours (*Fengshan shu* 封禪書): An elaborate historical survey of imperial sacrifices and tours from legendary times to the Qin and Han empires, this treatise focuses on the divergent precedents presented to the emperors as models for their own practices. The treatise pays particular attention to Emperor Wu's ritual reforms, ending with his performance of the *Feng* and *Shan* sacrifice on Mount Tai.[49]

7. Treatise on the Yellow River and Canals (*Hequ shu* 河渠書): This treatise is a historical survey from earliest times of the

main transportation and communication lines that allowed circulation between the major economic regions of the North China Plain, ending with Han Wudi's flurry of new roads and canals, designed to reduce serious flooding, better provision the capital at Chang'an, and irrigate more farming tracts.⁵⁰

8. Treatise on Fair Standards (*Pingzhun shu* 平準書): This treatise supplies a fine-grained historical overview of economic activities that starts and ends with the Western Han period. Topics include quality controls for currency, the two separate budgets for the operations of empire (with the first designated for the ritual obligations of the imperial house, and the other for general administrative purposes), innovations in taxes and expenditures, and the regulations for the domestic economy and for foreign trade.⁵¹

Naturally, the eight *Shiji* Treatises differ somewhat in structure and tone, because each is rooted in a distinct form of technical expertise related to statecraft. The first pair of chapters is devoted to discussions of *li* 禮 and *yue* 樂 (ritual and music), drawing on the *Odes* classic, as well as *Xunzi*, *Han Fei*, and Han-period ritual texts, to explain how ritual and music can affect personal cultivation in socially efficacious ways. The second pair of treatises treats *lü* 律 and *li* 曆 (pitch pipes and calendars), two closely related technical fields incorporating harmonics and astronomy.⁵² The fifth treatise represents an exhaustive verbal map of the night sky, linking celestial changes with terrestrial phenomena. The last three treatises are structurally similar, perhaps because each essentially is a chronological record of the particular responsibilities and prerogatives of empire: the imperial progresses designed to secure the local gods' support for the imperial person and his patrilineal line, vast construction projects to improve agricultural production, and measures taken for the control of coinage and the imperial finances. While criticism of Emperor Wu is sometimes muted, Sima Qian's disdain for his ruler's excesses and intolerance for corruption is clearest in these last three treatises, which focus more on Sima Qian's times (which may in itself say something about legitimate rule).

Despite the title change in the Treatises section,⁵³ Ban Gu's basic reliance on Sima Qian's model in composing the *Hanshu* 漢書 is readily apparent, not just in his decision to adapt the four relevant sections in the *Shiji* structure, but equally in his choices of what materials to assign to each section. Given the *Hanshu* focus on Western Han, the twelve Annals (*ji* 紀) begin with the Han founder, however, not with the legendary culture heroes of remote antiquity. The Bans' second section is also composed of ten Tables (*biao* 表) (see below), and the third, of their ten Treatises. Omitting the Hereditary Houses section, Ban Gu then ends the compilation with seventy Biographies (*lie zhuan* 列傳, literally "arrayed traditions"). The individual and collective biographies in this section include border nations, imperial consorts, and a lengthy, three-part biography of Wang Mang 王莽 (45 BCE–23 CE), clearly denying the "usurper" Wang the status of divinely-elected emperor.

Ban Gu introduced several innovations to the Treatise section. The most obvious change is that he expanded their number from eight to ten, even as he merged the first four *Shiji* treatises into two chapters, the first devoted to ritual and music and the second to the pitch pipes and calendar, and then gave primacy to the eternal pitch pipes and calendars by moving them before the transitory rites and music. The remaining four *Shiji* topics were basically preserved, even as they were given new titles, and four new topics were then added (nos. 3, 7, 8, and 10 in the list below). The ten treatises of the *Hanshu* are as follows:

1. Treatise on Pitch Pipes and the Calendar (*Lüli zhi* 律曆志): While this two-part chapter corresponds to the third and fourth *Shiji* treatises, it covers several areas of concern not found in the *Shiji* treatises. For example, the first part expands the numerological correlations provided by the *Shiji* discussion of the musical pitches, adding new sections on standardized measures, capacities, and weights. This first half concludes with a chronological survey of revisions to the calendar leading up to Liu Xin's 劉歆 (d. 23 CE) Triple Concordance (Santong lipu 三統曆譜), which embeds the annual calendar in several interlocking sets of macro-calendrical cycles. The second half begins with a list of explanations of key constants used in the Triple Concordance calculations, followed by material from Liu Xin's *Classic of the Ages* (*Shijing* 世經), tying historical events to the Triple Concordance system[54] and underscoring Ban Gu's dependence on Liu Xin's work.

2. Treatise on Ritual and Music (*Liyue zhi* 禮樂志): This chapter combines the topics of the first two *Shiji* treatises,

adapting segments of its prose and quotations from ritual texts, while developing a more didactic tone. Meanwhile, Ban Gu ties the benefits of ritual and music more closely to their performances at court, to their imperial sponsorship, and to their goals, education and acculturation. Both the first section on ritual and the second on music contain sets of chronologically arranged anecdotes and quotations. The latter includes the lyrics for two sets of songs used by Emperor Wu for making offerings at his ancestral temple.[55]

3. Treatise on Punishments and Law (*Xingfa zhi* 刑法志): Beginning with two chronological surveys devoted to general treatments of the role of punitive campaigns in good governance, and to the prevailing attitudes toward the laws, the treatise dwells on the mutilating punishments (how much is too much punishment, in relation to competing notions of justice and deterrence), providing a fascinating set of case studies and anecdotes about legal thinking during the Han. The treatise ends with a prose essay on punishment that makes liberal use of the *Analects* and especially the *Xunzi*, to condemn excessive punishment and excessive leniency equally.[56]

4. Treatise on Food and Wealth (*Shihuo zhi* 食貨志): This two-part chapter begins with an essay on general terms and principles in farming and the realm's interest in maximizing agricultural production, key issues because both social structure and imperial revenues depended on abundant annual harvests. Then, expanding on the *Shiji*'s Treatise on Fair Standards and the *Shiji* 129 "biography" of assets accumulating,[57] which offer brief surveys of relevant early Western Han events, the treatise continues with a chronological survey of successive courts' interventions in economic policies, including price controls, regulating coinage, and imposing taxes. The second part of the chapter treats trade and currency, reaching back into pre-imperial times before moving to the Qin and Han.[58]

5. Treatise on the Suburban Sacrifices (*Jiaosi zhi* 郊祀志): In 4 BCE, Wang Mang moved many of the imperial sacrifices

from their local cult sites to altars erected in the suburbs of the capital, disrupting the long-standing pattern of stately imperial progresses conducted all around the empire to worship the full range of the deities, including Heaven and Earth, the Five Emperors, and the most famous mountains and rivers. This treatise's very detailed survey of imperial sacrifices and their cult objects leads up to this change, even as the long treatment divides into two parts, at the point where Gongsun Qing 公孫卿 proposes sacrificial reforms. The *Shiji* "Treatise on the *Feng* and *Shan*" had described the excesses of Emperor Wu in light of the ideals ascribed to antiquity, and Ban Gu ends on a similar note, describing the excessive disruptions he attributes to Wang Mang, who revised the sacrificial liturgies when consolidating his power before proclaiming himself emperor.[59] (Tian Tian's chapter engages this treatise.)

6. Treatise on Celestial Patterns (*Tianwen zhi* 天文志): Completed by Ban Gu's sister Ban Zhao 班昭 (b. 49 CE) and Ma Xu 馬續 (fl. 111 CE), we think, this treatise begins with a list of celestial phenomena that cleaves closely to the structure and language of its *Shiji* counterpart, the Treatise on the Celestial Offices. The following section, a chronicle of celestial anomalies and their interpretations, is unique to the *Hanshu*. (Jesse Chapman's chapter in this volume focuses on this treatise.)

7. Treatise on the *Wuxing* (*Wuxing zhi* 五行志): Drawing primarily on the view that anomalies reveal disruptions of the cyclical alternations of the cosmic phases, and arranged with reference to the "Great Plan" chapter of the *Documents*, this lengthy, five-part chapter begins with a list of famous omen experts who posited schemes of natural cycles, only to move on to confront readers with the discrepant readings of baleful events by such eminent Han authorities as Dong Zhongshu 董仲舒, Jing Fang 京房, Liu Xiang 劉向, and Liu Xin 劉歆, when reading the floods and fires mentioned in the *Annals* classic or the Han irregularities.[60] (Nylan's chapter in this volume examines this treatise and explains why it is best not to call it "Treatise on the Five Phases.")[61]

8. Treatise on Geography (*Dili zhi* 地理志): Traditionally divided into two parts, the Treatise on Geography begins with an explicit quotation of the "Tribute of Yu" (Yu gong 禹貢) chapter of the *Documents* classic and recapitulates that chapter's descriptions of Yu's circuit of the Nine Provinces (*jiuzhou* 九州).[62] The second part lists the Qin and Han reconfigurations of the early spaces into the commanderies and counties (*jun xian* 郡縣) belonging to the early empires, enumerating the households and population of each commandery as well as their constitutive counties. It concludes with a survey, region by region, of the local cultures and customs, and their astronomical correlations, sometimes connecting these to particular historical or geographical features. (This is the subject with which Lee Chi-hsiang engages.)

9. Treatise on [Irrigation] Ditches and Canals (*Gouxu zhi* 溝洫志): Following the initial twelve segments that roughly match the counterpart passages of the *Shiji* Treatise on the Yellow River and Canals, the next seventeen segments cover discrete events from 111 BCE to 4 CE that were important to flood control and large-scale waterway improvements. (This treatise is central to Luke Habberstad's chapter.)

10. Treatise on the Classics and Writings (*Yiwen zhi* 藝文志): By its own internal count, this treatise preserves titles, length and format, and brief notes for 596 works totaling more than 13,000 *juan* (scrolls). This treatise, which clearly preserves elements of Liu Xiang's *Seven Summaries* and Liu Xin's *Separate Record*, two catalogues prepared for the palace libraries and archives,[63] arranges titles under six main headings: the Six Classics (*Liuyi* 六藝), Many Masters (*Zhuzi* 諸子), Poetry and *fu* (*Shifu* 詩賦), Military Writings (*Bingshu* 兵書), Calculations and Techniques (*Shushu* 數術), and Methods and Skills (*Fangji* 方技), under which headings there appear thirty-eight subdivisions.[64] (This treatise is the main focus of the essay by Mark Csikszentmihalyi and Zheng Yifan.)

While the section title and some of the topics of the *Shiji* treatises connect to the *Documents* classic, Ban Gu makes his connections to the same classic explicit in the *Hanshu*. The *Hanshu* Treatise on the Pitch Pipes and Calendar opens with a quotation from the Han-era "Canon of Yao" of the *Documents* classic; likewise, the Treatise on Geography with an explicit quotation of the "Tribute of Yu" chapter from the *Documents*. Two additional treatises (Treatise on Punishments and Laws and the Treatise on the *Wuxing*) continually refer to the *Documents*, and two more (Treatise on the Suburban Sacrifices and Treatise on Food and Wealth) begin by quoting the "Great Plan" chapter of the *Documents*. Even the title of the fourth *Hanshu* treatise echoes the first of the eight policy concerns (*ba zheng* 八政) in the "Plan." Thus, it becomes even clearer in the *Hanshu* that the treatises would situate themselves in what the compiler believed to be a set of hallowed traditions whereby the sages applied technical mastery to the diverse spheres of imperial administration.

What is curious, however, and deserving of a monograph in its own right are the subtle differences between Sima Qian and Ban Gu, even where their topics and treatments at first glance seem to coincide. To single out a significant discrepancy, the class and status distinctions between classicist and functionary play a prominent role in Ban Gu's treatise on ritual and music, whereas they are absent from Sima Qian's analysis. Both treatises discuss a number of proposed ritual reforms, in light of the many failed attempts to emend the imperial rites during each reign, so as to attain perfection and completeness (*bei* 備). Unlike Sima Qian, Ban Gu repeatedly blames this failure on "vulgar functionaries" (*su li* 俗吏) whose objections hamstrung the well-meaning classicists. For Wendi's reign, for instance, Ban Gu includes a memorial by Jia Yi, which states that "Now clearly, changing prevailing customs and seeing to it that the hearts and minds of those throughout the realm incline towards the correct Way is not the sort of thing that a vulgar functionary can accomplish."[65]

If the Treatises of *Shiji* and *Hanshu* are seldom read (and seldom translated), the Tables seem to have been almost entirely overlooked, except when readers need to verify a date for an enfeoffment or official appointment.[66] And yet Sima Qian's contemporaries and near-contemporaries singled out his Tables out for particular praise, and it is easy to see why.[67] Particularly with the Tables he devoted to the long pre-unification period, but even with the Tables for early Western Han, he had to wrestle with multiple competing calendars, as each kingdom had generated its own, and correlate historical records compiled by the rival courts, each touting the achievements of its own ruling house, sometimes to the point of falsifying facts.[68] Thus Sima Qian's achievement is comparable, if not greater than that faced by Caesar when he created his unified calendar.⁶⁹ While some of the earliest records we have for China are king lists and family registers, Sima Qian, so far as we know, had to invent the Tables format *de novo*, leaving Ban Gu with the huge problem of how to adapt the format to Ban's different project. As Sima Qian plainly states, he was searching for larger cyclical patterns and key nodal points of conjunction that had shaped historical evolution over the *longue durée*, trends not captured by a single list or even a set of lists.⁷⁰ Add to this that the main theater of action prior to unification in 221 BCE often shifted rapidly from one court or region to another, so that Sima Qian had to find a graphically complex visual layout to magnify and minimize affairs of the kingdoms by radically expanding or collapsing columns, depending on the number of state actors involved. Anyone who has wrestled with unwieldy Excel spreadsheets can begin to imagine the technical difficulties of devising the proper formats to encompass all of time and space in the known world for millennia.

The graph for Table (*biao* 表) as a verb means "to lay out clearly."⁷¹ The first observation that one can make when surveying the following list of Tables for *Shiji* and *Hanshu* is that time is presented as irreducibly sociopolitical, and not "natural." Second, while much of pre-unification "history" was culled from anecdotes whose ethical messages and didactic import often undercut the personal significance of the purported protagonist (so much so that the same story is ascribed to multiple agents, depending on their local utility in political persuasions), chronological Tables required specially devised dating formulae, probably under the impetus of the *Annals* classic, except with the unique, anomalous *Hanshu* Table of Figures. But more on that below.⁷²

It may be useful to examine the two sets of Tables given below.

Shiji Tables, from remote antiquity to Sima Qian's day

1. Table of the Three Dynasties, by era
2. Table of the Twelve Local Lords [*Chunqiu*], by year⁷³
3. Table of the Qin-Chu transition, by month
4. Table of Gaozu's Meritorious Offices, by year
5. Local Lords under Four Reigns, Huidi to Jingdi, by year
6. Local Lords under Wudi, by year

7. Princes under Wudi, by year

8. Generals, Chancellors, and Famous Officials in Western Han, by year

Hanshu Tables, devoted to Western Han only

1. Table of the Nobles and King, excluding those of the Liu clan [under Gaozu]

2. Table of the Local Lords

3. Table of the Imperial Princes

4. Table of Meritorious Officials for Three Reigns, Huidi, Dowager Empress Lü, Wendi

5. Table of Meritorious Officials for Six Reigns, Jingdi to Chengdi

6. Table of the Consort Clan Marquises

7. Table of High-ranking Officials

8. Table of Figures, Past and Present[74]

Often neglected as "mere sequence," the Tables underscore many of the underlying themes found in the two histories, sometimes to dramatic effect. Far from eliciting bored yawns, the Tables, along with their carefully crafted introductory and concluding paragraphs that occasionally, as paratexts, direct us to their proper interpretations, make for exciting reading, if only one makes the effort to plunge in. What astounds is this: that all the narrative strategies that make for memorable drama (juxtaposition; prefiguration and repetition; contrast and reversal; and integration of verbal portraits) are there, in spades, in the Tables, where stripped of excess verbiage, their impact can be all the stronger. Still more importantly, if the Ancients constructed their worlds in ways deeply unfamiliar to us, only acquaintance with the Tables affords moderns an unmediated view of the intense scrutiny Han scholars brought to the issues of timing, timely opportunities, and conjunctions of events (both predictable and irregular), as well as the "subtle words" the two first historians of unified empire saw fit to circle around. For that reason, the Tables direct our attention away from Sima Qian's castration (and our fascination with that story) and toward the problems that kept him writing.

Despite the differences in their writing styles, both Sima Qian and Ban Gu are deft masters of rhetoric, fully capable of wielding it to say or suggest the unsayable, yet the cunning arrangement of the Tables allows tricky points to become plain as day. Sima Qian never needs to bluntly state that in the early days Liu Bang, as commoner, hardly seemed the most promising of candidates to seize the throne; ranking him in the sixth place of nine contenders says it all. Likewise, Sima Qian brilliantly communicates the paucity of momentous deeds performed by Han Wudi in the latter half of his reign (r. 141–87 BCE), because there are nothing but blank spaces for the boxes allotted to "great affairs of state" from 112–91 BCE, and thereafter the main event is the rebellion by Wudi's own heir. Ban Gu, for his part, muddies the murderous record of Han Wudi (r. 141–87 BCE) toward his chancellors by tracking the postings to no fewer than fourteen high-ranking offices.

But what are we to make of the *Shiji*'s "Table of the Three Dynasties" versus Ban Gu's "Table of Figures, Past and Present"? Perhaps it is time to state the obvious: elaborate classification schemes, especially as they regard morality, are so culturally specific as to make them all sound mad to denizens of another time and place, in the style of Borges's *Encyclopedia*. With that firmly in mind, the first *Shiji* Table and the last *Hanshu* Table are worth another look. Curiously, Sima Qian tells us openly that the sources he used for this Table, which starts with the Yellow Emperor (aka Huangdi), supply plenty of "precise" names and dates, but they contradict one another. Elsewhere, in the Basic Annals section, he has already told us that no reliable information exists before Yao and Shun, who supposedly reigned a full half-millennium after the Yellow Emperor whose descent line is the main story followed in Sima Qian's first Table. "Is it possible the sources are lying?!"—Sima Qian innocently asks, a sure tipoff in case his readers are asleep. By this elaborate graphic genealogy, Sima Qian apparently wants to demonstrate that his entire cast of characters—Xiongnu or southeast "barbarian," as well as Han—have descended from a single ancestor and thus partake of equal humanity; that is how Han classicists took the Table, at any rate.[75] Moreover, the Table plainly shows us that for the nineteen generations between the Yellow Emperor and early Western Zhou, almost no reliable information exists.[76]

Regarding the last *Hanshu* Table (far more bizarre to us than any effort by Sima Qian), more than 2,000 characters "whose names can be known," thanks to the Classics, commentaries, and other records,[77] are sorted into nine ranks that most critics have deemed "moral," but that rather seem to reflect a mix of ethical and pragmatic considerations.[78] As the introductory paragraph boasts, so long as the figures had appeared in texts, oral and written, they were included in the Table, and, indeed, Ban Gu assesses even famous fictional characters alongside the fully historical. Not surprisingly, the highest rank is reserved for a few sage-rulers and to the "uncrowned king" himself, Confucius. To the lowest are consigned, equally predictably, the notorious miscreants whose infamy and stupidity have made them bywords for a lack of prudence. It is the middle ranks that somehow intrigue, perhaps perversely: by what criteria did Ban judge who was in rank 5? We can imagine Ban Gu (and possibly others) hoping to construct an edifice that would prove to be a "possession for all time," firmly in the hands of the Eastern Han court. As George Orwell noted ominously, "Who controls the past controls the future: who controls the present controls the past."

But other possibilities beckon, surely, given how unlikely it is that modern control mechanisms were in the sights of the early thinkers. Perhaps the Table was intended to convey something similar to Bernhard Karlgren's monumental "Legends and Cults in Ancient China" of a half-century ago. As readers may recall, the Swedish Sinologist grappled in that essay with the multiple genealogies provided for the earliest sages and worthies, evildoers and scoundrels, despite his belief that "very full and detailed accounts" may be nothing more than "caricatures" when compared with "scientifically established" histories, they being "systematized" beyond recognition or any semblance of the truth.[79] Still, given Ban Gu's profound love of good writing and his pride in himself as poet and eulogist,[80] it is equally likely that he meant to provide literary enthusiasts like himself (treasures at the court, in his view) with a useful crib to the immensely rich but hopelessly messy store of anecdotes at the learned writer's disposal by the first century CE. After all, historical reputations could rise and fall with astonishing speed, as the Ban family had seen, and it simply would not do to misplace your exempla. And so it behooved men writing serious pieces for the Han throne to know the relative standing of such disparate characters as Lord Chunshen, Guan Zhong, and the Second Emperor.

Historians like Liu Zhiji 劉知幾 (661–721) might quibble that the Table contained no "present" figures (i.e., no contemporaries of the Bans), but that was really beside the point.[81] To learned men with an interest in history and in fine writing, history was a "mirror," and all exemplary figures of the past, no matter how remote or near, were relevant to the present, because their characters, deeds, and outcomes served either to encourage good deeds or to demonstrate the futility or worse of certain initiatives. For these purposes, it was inconsequential if Huntun/Chaos from *Zhuangzi* was real or not. The question was, what lesson did his story convey? And that would explain why so many mythical, semilegendary, or fabulous characters inhabit the same Table, to our consternation. It would also explain a notable fact: after Ban Gu's Table, gradually of course, the best writers of the era no longer felt the need to cite an entire anecdote to prove a particular point. They could allude to the anecdotal figure, with the lightest of touches, to register a strong point of view. And this potential seems to have facilitated a trend made stylish with Yang Xiong's *Fayan*, where profound lessons could be encapsulated in a few graphs indicating personal names, a trend that leads straight to the "pure talk" conversations of the post-Han period.[82]

Increasing the plausibility of the last hypothesis regarding the Table of Figures, Past and Present is the classical-era thinkers' belief that the heart functioned as a storehouse of sensory perceptions, with hearing and seeing setting off complex reactions consisting of one part recognition and one part assessment. An apple perceived as red was simultaneously dubbed good, at least by the woman whose memory and experience predisposed her to like red apples. But judging the inherent properties of human beings is so much more complex than judging the right qualities to assign to familiar tastes or sights.[83] While there can be visceral responses to a given person, where, by habit or experience, one senses that a person with whom one has had no direct contact should be liked or disliked, sought out or shunned, the trappings of power or the appearance of affability can deceive most people most of the time. Time alone and a considerable expenditure of energy facilitate correct judgment. Close observation of a person's interactions and responses to disparate people, situations, and events will, sooner or later, reveal the true mettle of that person's character—or so the early thinkers insisted. Interestingly enough, by Eastern Han all the "people" rated in the Table of Figures, Past and Present had had their deeds debated at length, so their utility as rhetorical precedents referring to effective action in the real world was duly assured. In the words of Paul Veyne, this disparate collection of figures had thereby gained the authority of the Vulgate.[84]

Nearly all of our authors cite treatises in their chapters on technical matters. A few venture into the relatively uncharted waters that are the sixteen Tables in the *Shiji* and *Hanshu*. This book is meant to encourage further exploration, to nudge attention away from the chronicles in the Basic Annals and biographical narratives, in the hope of capturing something of the excitement that already has propelled many to engage in technical subjects in the Greek and Roman classics.[85]

III. Synopses of the Chapters in the Volume

The chapters of this book focus on a group of Treatises (geographic, astronomical, omenological, and sacrificial) and Tables compiled for the two Han dynasties (with many passages in those drawn from considerably earlier writings) before turning to the topic of medical practices and healing, as well as the tropes of classical learning in manuscript culture preserved in the *Hanshu* bibliographic treatise. Although this volume does not take up as a major topic the laws of the early empires, given the volume of good research in English on recent finds and received texts,[86] we might indulge, for the moment, in a bit of contextualization. The Song scholar Su Xun 蘇洵 (1009–1066) once idealized the technical literature of the early empires when he wrote that its laws were "few and simple . . . like the prescription manuals of the day, insofar as they explained the general outline, but left the adjustment of dosages to the healers."[87] No thinker of the early empires, so far as we know, would have characterized the laws of his day in this offhand manner,[88] just as no early thinker would have cast the medical classics he knew simply as "general outlines." On the contrary, during the early empires, there emerged a dizzying array of problems and procedures, prescriptions and theories, whose applicability to the specific issues at hand demanded the utmost care in deliberation. To be a worthy practitioner, then, meant long practice and a huge fund of erudition, a willingness to balance the use of past histories as mirror for the present with an alertness to the exigencies of the precise moment, to arrive at a judicious selection of competing remedies. This was true regardless of technical discipline, in land tenure measurements, mathematics, astronomy, omen reading, calendrical computations, and library cataloguing. And it was equally true at the court and in the provinces, in official and unofficial initiatives.

Here, then, is the list of chapters that make up this book:

1. Loewe, on land tenure

2. Habberstad, on water control

3. Lee, on Luoyang as the providential capital of a "restored" Eastern Han

4. Chapman, on reading starry signs in the heavens

5. Nylan, on omen theories and practices at court

6. Tian, on changes to the sacrificial schedule and pantheon

7. Chemla, on mathematical calculations and the "essentials"

8. Brown, on healing traditions, oral and written

9. Csikszentmihalyi and Zheng, on the *Hanshu* bibliographic treatise

We provide below abstracts for the chapters, for the convenience of readers:

Michael Loewe, as is typical of the master, rethinks the problem of land tenure from the ground up, quietly offering conclusions with major revisionist implications. As Loewe's chapter shows, several factors suggest a decline in the strength of the Han government from ca. 130 CE or even earlier, or roughly a century before the official "downfall" of Eastern Han. Official supervision over land holdings was weakening from 105 CE; alleviation of the plight of the vagrants and landless stopped after 132 CE. Meanwhile, large estates held in private hands and worked by hired laborers were growing in number; deeds for the purchase of land became more frequent from 125 CE; and imperial grants of orders of honor (*jue* 爵) with their allocations of land all but ended by 147 CE. There is even suspicion that official counts of the lands and the tax-paying population in 140 CE may have been falsified to some degree. By the registered taxpayers' count of 140, households in the north had grown larger in number, which likely means they took in individuals (clients, tenants, servants) who had been rendered landless. At the same time, in comparison with a 2 CE survey of taxpaying households, a higher proportion of taxpayers was registered for the south, but with markedly fewer members per household, likely the result of tax evasion and significant expansion into new territories (one or both) propelled by fears of ruin and hopes of gain. A number of officials and others voiced their criticisms of these very developments, some of which threatened the health and legitimacy of the body politic based in the North China plain.

Previously, the work of A. F. P. Hulsewé, Brian McKnight, and Nathan Sivin showed how delicate yet necessary to the body politic was the attainment of a fine balance between punishments and commutations, on the one hand, and grants, boons, and amnesties, on the other.[89] Such a balance provided a dual face for the emperor, as a virtual "life-giving" deity and as staunch defender of the laws, and, if need be, a fierce general. Michael Loewe's work now adds to this picture, suggesting that the sharp decline in the Eastern Han dynastic fortunes may have begun much earlier than we once thought, given the Eastern Han court's triple inabilities to withstand the rapid growth of large estates at its expense, to staff its large administration in the localities, and to dole out amnesties and orders of honor. All these phenomena would surely have decreased the impact of the imperial presence, in both its life-giving and life-threatening aspects. The court's inability to enforce its will much beyond its chief administrative seats goes a long way toward explaining, in turn, the otherwise inexplicably narrow focus of the standard history for Eastern Han, which seems to collapse the wide-angle lens of the *Shiji* and *Hanshu* while reducing the vast empire to a few key sites of fierce contestation. Runan and Yingchuan loom large in this narrative, needless to say, as the sites where the Proscribed Partisans began to gather force, but mainly the *Hou Hanshu* is engrossed by the actions taking place in the imperial court at several capitals, and what it styles (far too simplistically) as the fight-to-the-death of the three main contending interest groups for court power: the regular court officers, the *waiqi* consort clan members, and the eunuchs. (Meanwhile the *Hou Hanshu* highlights the prominence of the consort clans through its decision to assign the clans their own collective Basic Annals chapter.)

As this unduly bright light on capital and court has up to now impeded understanding of the crucial roles played by the old land-based magnates in governance of the localities, research on the land tenure system is long overdue.[90] Loewe clarifies the definitions of three terms whose meaning and significance has eluded moderns: *gongtian*, *mingtian*, and *sitian*, meaning "lands not yet assigned to households but liable to be disposed of by officials," "lands allocated to a named individual by a grant and possibly by purchase" (generally allocated to nobles and not magnates), and "private lands" contracts, which were supervised by local officials. In this way, Loewe raises the larger question: Are we justified in retrojecting back two thousand years our modern property institutions and values? His implied answer is "no."[91] Certainly, Loewe's conclusions challenge older theories that make the buying and selling of land by private "owners" after Shang Yang (ca. 350 BCE) central to any narratives purporting to analyze Chinese state and society, Marxian and not.[92] To some degree, they also undermine work by Charles Holcombe and others by suggesting that the formal end of Eastern Han may not have struck many members of the governing elite as momentous when it came in 220 CE, because for several generations by then the court had become increasingly irrelevant to provincial life.[93]

Luke Habberstad's chapter circles around the idea whose ramifications we moderns have yet to fully appreciate: that members of the governing elite in early China seldom divided the world up into "manmade" versus "natural"; ergo, water becomes "the blood and *qi* of the land, akin to what freely flows within the vital channels of the body." While water management has only lately reentered the academic discourse about early China (it being in ill repute because of Wittfogel's thesis alleging Oriental Despotism), Habberstad reminds his readers just how ubiquitous water managers were in the official ranks at the Han court and in the provinces, and just how frequently approaches to water management served as important markers of the quality of the ruler-minister relation. By design, it seems, Sima Qian and Ban Gu drew attention to the fault lines between different approaches to water control, and equally to the characters of the chief policy makers. And as the ruler-minister relations changed so substantially over the four centuries of Western and Eastern Han, it should hardly surprise us that the *Shiji* treatise provides a very different portrait of good governance from the *Hanshu* treatise. The *Shiji*, in an older, pre-unification spirit, emphasized close collaboration between subject and king as key to achieving the single best outcome for commoners. By contrast, the ruler envisioned by Ban Gu in Eastern Han commands a wealth of technical expertise at hand, thanks to his ready supply of elite advisers. Indeed, the sheer quantity of technical expertise accumulated and displayed by would-be policy-makers figures more obviously in the rhetoric as time goes on. While the ruler does not himself execute his plans, he alone is capable of making the wisest decisions when it comes to choosing among several alternative plans offered by rival officials avid for career advancement. At the same time, the Eastern Han histories seek always to avert blame from the emperor for disastrous decisions. For the foregoing reasons, the rhetoric of water control, like water itself, is unexpectedly subtle and shifting over time, which is why it pays to construct such comparisons with the utmost care.

Lee Chi-hsiang's chapter plunges the reader into one small section of the *Hanshu* geographic treatise devoted to Luoyang, the Eastern Han capital, as it relates to the wider world of classical learning in the two Han dynasties, for Lee demonstrates how much the mental world of Sima Qian's *Shiji*, with its particular presuppositions about the fabulous Western Zhou forms of governance, differs from the surmises that Ban Gu brought to his *Hanshu* compilation two centuries later. The chapter inevitably considers the importance of antique precedents to the self-image and legitimacy of the two Han courts, especially in the matter of selecting the site for their capital (a topic taken up in the *Baihu tong* 白虎通 compiled during Ban Gu's lifetime, if not by Ban himself). Professor Lee's contribution notes the stunning absence of corroborating archaeological evidence for a walled Western Zhou "secondary" capital near the site of present-day Luoyang. But Lee does not stop there. In a masterful survey of the surviving literature, he carefully traces the evolution of a possible factoid down through the ages. Lee's analysis shows us that the most impeccable authorities in the early empires were apt to engage in flights of fancy when they reconstructed events from a thousand years before, and that through a set of small tweaks to the record—linguistic, geographic, political, and emotional—they lent credence to somewhat outlandish suppositions, some of which continue to bedevil the best scholars of the present day.

Jesse Chapman's chapter opens with a frank acknowledgment that there are obvious commonalities between Sima Qian's (ca. 86 BCE) "Treatise on the Celestial Offices" (ca. 100 BCE) and Ban Zhao's (d. ca. 120 CE) "Treatise on Celestial Patterns" (ca. 100 CE). Both treatises use similar phrasing to describe the significance of a range of omens, including the movements of the stars and planets, and the formation of strange clouds, comets, halos, rainbows and *qi* manifestations. Chapman's chapter therefore aims to direct attention to the differences between the two treatises, which seem to indicate the shifting relationship between classical authority and technical discourse over the course of Western and Eastern Han.

Tracing a line of practitioners going back to remote high antiquity, the *Shiji* treatise frames the reading of celestial signs as a practice essential to good governance, while rarely invoking the Classics to support its claims. By contrast, Ban Zhao's "Celestial Patterns" liberally cites the Five Classics to support its vision of heavenly and historical processes. The "Celestial Patterns" treatise arranged its omen readings—including numerous celestial signs such as comets and eclipses—in a chronology like that of the *Annals* classic, with the result that nearly all its omen readings, implicitly or explicitly, seem to comment on events leading up to the downfall of the Western Han dynasty and the rise of the "usurper" Wang Mang. Ostensibly a mere copy or continuation of the *Shiji* "Celestial Offices," the *Hanshu* "Celestial Patterns" treatise offers a different vision of the politics of omen reading, wherein technical knowledge increasingly requires authoritative confirmation by the Five Classics, even as celestial signs reveal the moral and ritual failings of rulers and the high-ranking members of their courts.

Building on some of the issues raised in Jesse Chapman's chapter, Michael Nylan focuses on the puzzles and significance of the *Hanshu* omen treatise,[94] a work of immense length and complexity that has been the subject of previous secondary studies, all of them assuming that the treatise represents a comment on the Western Han rulers' successes and failures. Taking full advantage of the recent monumental contributions by Su Dechang (Taiwan) and by Chen Kanli (PRC),[95] Nylan seeks to complicate the usual reading of the omen treatise by placing it within a new context, that of Ban Gu's larger rhetorical project about Western Han, as that project re-imagines the authority properly accorded omen experts in the court's deliberations. More than Su or Chen, Nylan consistently queries the appropriateness of equating the term *wuxing* in the title with the Five Phases (or worse, Five Elements) on the grounds that the treatise plays on the multivalence of the term *wuxing*, which means Five Planets, Five Material Resources, and Five Modes of Conduct in the early sources, as well as Five Phases. She finds, as does Su, that no small effort was required on Ban Gu's part to merge the originally contradictory *wuxing* theories within a single omen treatise. Probing the insights of Su and Chen further, in light of recent Japanese scholarship, Nylan then analyzes the main contents of the treatise, as the treatise portrays relations between three major *Annals* traditions: the *Gongyang*, *Guliang*, and *Zuo* interpretive traditions. She asks readers to consider why Ban Gu chose to incorporate an omen treatise in his court history (*guo shi* 國史), because no such treatise appears in the *Shiji*. Finally, her chapter considers the grave objections to Ban's omen treatise registered by Liu Zhiji's authoritative *Compendium of Historiography* (*Shitong* 史通), since Liu's remarks colored its subsequent reception by scholar-officials.

Tian Tian's chapter makes its principal focus the sacrifices to Taiyi 太一 offered during Western Han. As she states, before 112 BCE, roughly midway through the reign of Han Wudi (r. 141–87 BCE), no sacrifice to Taiyi was ever offered by the ruling house or the central court, so far as we know, although Taiyi was already recognized as a deity in pre-unification times. (In this well-substantiated view, Tian Tian parts company with some other historians more prone to speculation and extrapolation.) In 112, Han Wudi took the unprecedented step of establishing a special altar for Taiyi called the "Taizhi" 泰畤 at Ganquan (aka Sweet Springs), his palatial estate located some distance away from the capital. With the construction of this altar, Wudi and his court in effect installed a brand-new supreme god as the object of worship in the court-sponsored sacrifices. During the last century of Western Han, the worship of Taiyi departed significantly from earlier practices under Wudi, necessitating a survey tracing the evolution of this cult based on the extant sources from the period, in particular the "Treatise on the *Feng* and *Shan* Sacrifices" in Sima Qian's *Shiji* and its counterpart chapter in Ban Gu's *Hanshu*, the "Treatise on the Suburban Sacrifices," whose approaches and assessments of the Taiyi cult differed sharply.

Karine Chemla's chapter begins with the nineteenth-century European construction of the "Chinese people" as a collection of "lacks," the most relevant to mathematics being their alleged innate or learned incapacity to formulate abstractions and hence to theorize. Chemla's earlier work noted that the early mathematicians tended to prize generality more than abstraction per se. That serves as necessary context for her main effort in this chapter, which shows that (a) abstraction was part of the mathematicians' repertoire of conceptual tools; and (b) there were specific types of instances when they were more prone to resort to abstraction during the two Han empires. Relying on two received mathematical classics, as well as recently excavated materials, Chemla gives us a four-part chapter premised on the basic distinction between problems and procedures in mathematics. Part 1 describes the difference between the observer's and actor's category, while Part II identifies instances of abstraction in the *Nine Chapters* classic. Part III turns to the earlier excavated *Writings on Mathematical Procedures* from Zhangjiashan, which shows evidence of abstraction (while an unprovenanced manuscript does not). Part IV circles back to the larger issue of abstraction to reassess its significance.

Chemla's chapter quickly establishes that the *Nine Chapters* classic was read, from early Tang, via Liu Hui's commentary (compiled 263) and Li Chunfeng's subcommentary (ca. 656). She then hones in on the precise language used in the mathematical classic and in Liu Hui's commentary, language that would have seemed very familiar to any student of the *Annals* classic or of Confucian lore, in juxtaposing the notions of *kongyan* 空言 (literally, "empty words") and *shi* 事 ("actual events or tasks"). By her analysis, both the mathematical classic and an early Western Han Zhangjiashan mathematical text artfully play with the same terms that ground the classical exegesis and Sima Qian's investigations into history. For, as Chemla shows, Liu and Li fully expected specific procedures in the *Nine Chapters* to be expressed using "abstract/empty expressions" (*kongyan*), seeing abstraction in the treatise, but seeing it in a way that they felt required further explication. Thus, in the exegetes' view (and in Chemla's as well), already by early Han times a specific sense of abstraction commanded extraordinary epistemological value within the technical field of mathematics. Additionally, resort to this classical expression *kongyan* in mathematics yields an important insight into the very processes of textual formation in antiquity: to wit, how commentators transferred concepts from one form of textual practice that earlier sources had reserved for Confucius in the *Annals* traditions and then applied them in a very different technical field, at once elevating the status of the mathematical classic and enriching its meaning in the process.

Miranda Brown's chapter moves us into the technical practices for healing as viewed initially from the biography of the Granary Master preserved in *Shiji* chapter 105. There the *Shiji* provides a compelling model of master-disciple transmission of medical learning in the early Western Han dynasty: the initiation of younger novices into long-standing "secret" traditions by aged masters. Indeed, so powerful was this *Shiji* biography that many historians of medicine have taken its model of master-disciple transmission to be the *main* (or even the *only*) way that the transmission and compilation of medical classics occurred in early China. This chapter therefore disputes the prevailing notion that initiation was the *only* route by which the healing arts circulated in early China. Through an analysis of the medical manuscripts discovered near a military colony at Wuwei 武威 (Gansu), in northwest China, dating to the first century CE, Brown shows how much explanatory matter these medical manuscripts actually conveyed, obviating for everyday purposes the need for formal transmission from a master. Brown next points out how fully integrated the medical arts were with the administrative arts, because both were required for the health of the body politic, as well as the health of individual persons. (Perhaps the earliest instance demonstrating this integration comes from the Zhangjiashan legal texts, whose statutes stipulate the minimum success rate for court healers and diviners.) Judging from the Wuwei manuscripts, the circulation of authoritative medical writings may well have gone through official channels, as well as through the more familiar (to us) context of master-disciple initiations. The ultimate aim of this chapter is thus to correct and complicate our standard picture of knowledge production and circulation, and ultimately canon formation.

The contribution by Mark Csikszentmihalyi and Zheng Yifan takes up the "Classics and Writings" catalogue in the *Hanshu*, which is based on two earlier catalogues of palace library holdings. To better contextualize the *Hanshu* bibliographic treatise, the chapter compares the treatise with two other texts that supply taxonomies of techniques for good governance: the *Zhuangzi*'s "All under Heaven" chapter and the *Shiji*'s "Essential Points of the

Six Types of Experts" section of Sima Qian's "Personal Narrative" (*Shiji* 130). Taken together, these three sources develop a multistage model of decline, designed to integrate the Classics (*jing* 經), presumed to date mainly to Western Zhou or early Eastern Zhou, the masterworks (*zhuzi* 諸子), rooted in Zhanguo writings, but advancing discrete methods of governing, and, finally, the late Zhanguo, Qin, and Han writings ascribed to those "one-sided" thinkers equipped with a mastery of "mere" skills (*ji* 技) and techniques (*shu* 術). These last hearken back to the Master, but have lost his comprehensive vision. This three-stage narrative of decline constitutes the framing device utilized in the *Hanshu* bibliographic treatise, which uses the Han trope of the "Way and its techniques" to support a persistent contrast between the sages' public-minded adaptation of particular techniques and the later, more self-serving techniques reflecting progressively shallower understandings of the methods of the Five Classics. Notably, the *Hanshu* narrative of decline and fragmentation simultaneously imposes a hierarchy that places the Classics and masterworks distinctly above the texts labeled as "technical arts," and integrates all three categories of manuscripts within a single overarching framework.

To demonstrate how this complex rhetoric works, the chapter examines a single striking feature of the *Hanshu* bibliographic treatise: the reappearance of "Calendars and Registers" (*lipu* 曆譜) and "Immortal Ascendance" (*shenxian* 神僊) techniques in two different sections of the catalogue. Applied to the realm of governing, this type of manuscript appears first in a section devoted to the masterworks, but when applied to the domestic and personal domains, comparable texts appear under subsequent catalogue headings in the treatise.

The contributors to this volume offer these preliminary reflections, fully aware that without the excavated and "found" sources that moderns command at the turn of the twenty-first century, no historians would have dared to push these topics in these particular ways. Moreover, a century ago, few scholars would have identified sacrifices, astronomy, or water works, for example, as integral components of their evolving picture of the early empires—one need look no further than the way Hans Bielenstein's deep dives into Ban Gu's history in the 1930s were dominated by concerns with military and political history. The editors and contributors only hope that their essays are judicious enough to serve the next generations of hard-working scholars. Thus, their fondest hope is that their essays will be duly corrected and revised as more and better resources become available in the future.[96]

Finally on a distinctly personal note, the editors wish to note the multigenerational nature of this volume. It was Ouyang Xiu 歐陽脩 (1007–1072) who painted this compelling portrait of expertise:

> The classicist, by definition, models himself on the Sage. The Way of the Sage is straight and simple. Therefore, when a classicist finds that the Way has become entangled, he extends it, in order to penetrate the pattern of the world, in order to exhaust the transformations of yin and yang, of heaven and earth, of humans and spirits, of all the myriad things and beings, as well as the great human relations, those between ruler and minister, father and son, between the auspicious and inauspicious rites, between life and death. . . . The learning of the classicist is doubtless wide-ranging and impressive, but he labors hard. That said, it is certain that he [like Kongzi, the Supreme Sage, himself] cannot master every particular.[97]

Ouyang's emphasis on both the hard work involved in mastering the past through continual inquiry into the present, and on the inevitability of imperfection, applies to the experts contributing to this volume, all of whom have styled themselves as "classicists," students of the past. In this volume we editors have gathered both the work of our teacher, Michael Loewe (though few who work in Han studies would fail to identify him as such), and a number of present and former students: Miranda Brown, Luke Habberstad, Jesse Chapman, and Zheng Yifan. Mindful that there is always too much to know, the editors hope and trust that we have nonetheless made a start, and that the next generations of scholars can fill in all the many particulars the editors have yet to comprehend.

—M. Nylan and M. Csikszentmihalyi

Notes

1. Nylan made her first foray outside the *Shiji* biographical section, noting that early critics deemed the treatises and tables of Sima Qian to be his most important achievement, in her essay titled "Sima Qian: A True Historian?" The letter to Ren An is the subject of a multi-authored book by Stephen Durrant, Wai-yee Li,

Michael Nylan, and Hans van Ess titled *The Letter to Ren An and Sima Qian's Legacy*. The introduction to that volume translates the letter and assesses the role of Sima Qian, the ascribed author.

2. G. E. R. Lloyd (personal communication, August 8, 2020) adds a note of warning: "Any skill or craft or art, not just one focused on the practical or the concrete, can be deemed a *techne*. It has a special linkage with rhetoric for instance, as also with *dialektike* and in lots of cases of skills where the Greeks refer to them with compounds ending in *-ike* we supply or understand *techne*, though in some cases it is *episteme*. . . . [T]he theoretical component varies from one instance to another, and one collocation to another, but it should certainly not be claimed that *techne* is always (just) practical." The editors would also draw readers' attention to Hannah Arendt's *Between Past and Future* (esp. pp. 104–109) where the nature of authority in relation to various metaphors (including that of the craftsman) is discussed.

3. See, e.g., Smith, "Sima Tan"; Queen, "Inventories of the Past"; Csikszentmihalyi and Nylan, "Constructing Lineages and Inventing Traditions."

4. We emphasize that there were two dynasties, as Western and Eastern Han ran on fundamentally different principles.

5. We have the 140 CE tax registration figures (not a population census, mind you!) for the Eastern Han court, but the differences between the size of households north and south should give us pause, as should the variations shown in the taxpaying populations under Western and Eastern Han. See Loewe's chapter in this volume; also Bielenstein, Maps for 2 CE and 140 CE, in his "The Census of China." Also relevant is the fact that Chengdi (r. 33–7 BCE) ordered a single map to be produced for his entire realm, and again, so far as we know, no such order went out from any Eastern Han court.

6. Several theories of Tianming were on offer in early China, and Ban Biao's version offers a major contrast to the Tianming theories articulated by Mengzi 孟子 or by the *Gongyang* commentary exegetes of late Western Han; see, for example, *Hanshu* 100.4208–12. For that reason, we prefer to leave it in the untranslated form, given that the reflexive translation of "Mandate of Heaven" obscures the range of theories about political legitimacy that the term connotes.

7. The Xiongnu, however, gave no little trouble to Wang Mang's Xin dynasty, 9–23 CE.

8. See the steles ascribed to Qin Shihuang for this "civilizing" language.

9. Slingerland's 2007 examination of the Zhanguo period metaphors for *wuwei* led him to render the term as "effortless action." Given the transportation and communication realities of the time (which were roughly the same for the two Han dynasties), the courts could not aim to induce perfect conformity of views among their subject populations, but rather aimed for sufficient acceptance of the courts' legitimacy to carry out the basic functions of government, such as infrastructure maintenance and enforcement of the laws.

10. We avoid the term "bureaucracy" whenever possible, to avoid the Weberian implication of "rationality." Though there were theorists who argued for such rationalization and routine testing of candidates and officials, the administration in the two Han dynasties depended heavily on hereditary status and learning, and it was deeply involved in such cosmic projects as "harmonizing yin and yang *qi*," unlike Weber's model bureaucracy.

11. A list of the fine scholars writing in Chinese would have to begin with Gu Jiegang 顧頡剛 and with Lü Simian 呂思勉, but many China scholars would nominate other favorites, living and dead.

12. Equally recommended is Raymond Williams's *Keywords*.

13. I.e., *zhi tian* 治田, *zhi shen* 治身. Arguably, "therapeutic governance" was the ideal, for the individual person and the body politic, if we mean "therapeutic" in the sense that Pierre Hadot uses it. NB: We have no means to ascertain whether certain types of technical treatises, say on medicine or agriculture, were in general circulation, either in private hands or in the manuscript markets mentioned in Wang Chong's *Lunheng*.

14. Ergo the identification of the arts of governing as reliant on the "great intention" or meaning (*da yi* 大意/大義), and the Classics as embodiments of the "main relations" 大倫. Such "constants" were thought to have wide applicability, and hence great epistemological value. Confucius himself, in compiling the *Annals* classic, was associated with the *weiyan dayi* 微言大義. In this connection, one might cite *Hanshu* 75.2172, which has the sage-ruler capable of tying all the techniques together, subordinating them to good governance: 臣聞之於師曰 . . . 聖人見道、然後知王治之象。故畫州土, 建君臣、立律曆、陳成敗。以視賢者, 名之曰經. . . . Also worth citing is Jia Yi's memorial, where he styles the Six Arts as "six techniques," whose greater meaning, once abstracted, justifies the name "Six Arts" (故內本六法, 外體六行, 以與《詩》、《書》、《易》、《春秋》、《禮》、《樂》六者之術, 以為大義, 謂之六藝). For further discussion, see Wang Baoxuan, *Jin guwen jingxue xinlun*, 54–55.

15. The graph *shu* 數 refers to the "regularities" that can be calculated, because they are regular, and therefore to mathematics, astronomy, omenology, and any skill reliant upon ratios and proportions (e.g., physiognomy). Curiously, the section titled "Shu shu" 術數 is then taken to describe *shu shu* 數術 experts (凡數術百九十家), with the characters reversed. While some of these techniques deal with the starry bodies, these belong to governance in the Han view, where mutual resonances prevail between heaven and earth.

16. *Literate Community in Early Imperial China*, Charles Sanft's new book on the northwest frontier materials (mainly excavated), speaks of this third sense of "technical," derived from manuscripts: "Many Han-era texts that archaeologists have recovered from sites of various types are technical, in the [limited] sense that they convey information about how to do things . . . and drew from a body of specialist knowledge and theory" (95). Accepting this narrow sense, Xu Gan's *Zhonglun*

(chapter 15) defines the technical arts as mostly having to do with physical prowess, in contrast to the arts of governing, that require more cultivation of the heart, the *xin* 心. One needs such people but not in the highest offices, because "What benefits does possession of these six capacities and skills bring to good order?" What one needs, evidently, is skill in the Great Way. See Makeham, *Zhong lun*, p. 215.

17. These have nothing particularly to do with "Daoist" or "proto-Daoist" arts, contra the common wisdom. The term *dao* 道, to take one example, is identified with Kongzi, as A. C. Graham's *Disputers of the Tao* showed.

Dao and *shu* are synonyms in many early writings, as ways of doing things to accomplish a particular end, which can be noninterference or *wuwei*. In a number of Han texts, they are related to each other by part-whole metaphors, as examined in Csikszentmihalyi, "Chia I's 'Techniques of the Tao.'" Ban Gu appears to use them interchangeably. The *Xunzi* provides one of the clearest definitions of the *dao*, disputing those who would make loftier claims for it (in English, capitalizing it): "It is the *means by which people make their paths*, and what the noble man has as Path" (人之所以道也、君之子所道也). For the *Xunzi*, "Dao" refers to neither the Way of Heaven nor the Way of Earth, in other words; it refers rather to the unitary Way. While Dao may be used in a plural way, as in the *Fayan's* 法言 contrast between "other Daos" 它道 and the "correct way" 正道 ascribed to Yao, Shun, and King Wen, discerning the best methods to use in a given situation is almost exclusively a matter of *shu*, as in the *Shiji's* contrasts between the techniques of the Classicists or Ru 儒術 and those of the Yellow Emperor and Laozi 黃帝老子之術, or even the "techniques of guiding and pulling *qi*" derived from Master Red Pine [Chisongzi 赤松子] and Wang Ziqiao 松喬導引之術.

18. See, e.g., *Hanshu* 36.1947, 72.3081, 78.3273.

19. Some texts do distinguish "techniques" from the One Way, and where they do, they tend to say that One Way is preferable to many techniques, which otherwise may work against one another. See, for example, Xun Yue's formula in *Hanji* contrasting the disordered techniques with the culminating Way of the sage: 放百家之紛亂、一聖人之至道、則虛誕之術絕、而道德有所定矣. The idea is simply that sociopolitical unity is always preferable to competing voices, an Eastern Han view.

20. The "San jiao" (Three Teachings) chapter of Ban Gu's *Baihu tong* says that no single technique can induce good governance; cf. Xun Yue, *Hanji*, which says that those who command the expertise (*shu*) to make wise and efficacious policies (立策決勝之術) must take into account a constellation of factors.

21. Some women are sages, but the rhetoric of the time mainly envisions men in office. However, most members of the governing elite (pace Puett) did not think it likely that they would become gods. As the *Lüshi chunqiu* explains it, the sages' much vaunted "foresight" (*xianzhi* 先知) in no way differs from the foresight accessible to other people, in that *all* knowledge requires a careful assessment of signs and symptoms. For that reason, "The sage cannot run ahead of the crowd," and is only viewed as "god-like" because most people do not take the trouble to

examine the signs. Put another way, the sage is not of a different ontological order; instead, he differs in having the command of "techniques" and "regularities" (*shu* 數)," which he has worked to acquire and which, in theory, can be shared.

With modern-day constructions imposing disciplinary boundaries on the past (often in the name of anachronistic sectarian disputes), we tend to overlook the antique view: that an open-ended attitude to learning, which ends in virtuosity (as opposed to regulated behavior), is celebrated everywhere, even in *Analects* 19/22: 賢者識其大者、不賢者識其小者。莫不有文武之道焉。夫子焉不學？而亦何常師之有? Here, the idea is that the master learns from anyone who is superior to him in any form of undertaking.

22. See Mann, *Sources of Social Power*; Crone, *Pre-industrial Societies*.

23. Obviously, the editors here decry the convention rendering "Daojia" as "Daoists" and "Rujia" as "Ruists" or "Confucians," but a failure to notice the gap between text and reality plagues many fields, not just Early China studies. On this, see de Pee, *Losing the Way*, chapter 1.

24. Feeney, *Caesar's Calendar*; Cullen, *Foundations of Celestial Reckoning*.

25. Frequently, a literary map supplies the evidence needed to reconstruct seating charts, for example, and ritual hierarchies at court.

26. Michael Nylan and Trenton Wilson, in a 2021 essay titled "A Brief History of Daring," give their translation of an elaborate *Yili* ritual designed to facilitate peer-to-peer exchanges.

27. NB: *Shuowen jiezi* 8A.3b associates the "classicists" (Ru) with *shu shi* 術士 ("masters of techniques"), meaning experts in governing. *Hou Hanshu* 54.1780 is but one text showing that, at each level, governing is a technique (or a combination of techniques). *Hanshu* 30.1723 calls the *Documents*, the "queen" of the documents relating to political life, "the means to induce broad listening [in court], a technical manual for understanding" 書以廣聽, 知之術也.

28. Sivin, "Text and Experience," 190. This observation is confirmed in Wang Chong's *Lunheng* in several passages, especially the three chapters titled "Plague Demons" (Yi gui 疫鬼), "On Exorcism" (Jiechu 解除), and "Appraising Demons" (Ding gui 訂鬼), where Wang argues, "If [rulers] practice the virtues of Yao and Shun, the world will enjoy perfect peace, and the myriad calamities will vanish. And although you do not expel the pestilences, the plague demons will not go forth" 行堯、舜之德, 天下太平, 百災消滅, 雖不逐疫, 疫鬼不往.

29. This "body politic" relationship is developed with specific reference to *shu* in Wang Chong's "Defining Worthies" chapter (Dingxian 定賢), starting with the basic statement of the metaphor: "Sages and worthies all have their technical arts to govern the age. Those who have these technical arts succeed, while those who do not use them fail. This may be compared with a physician healing disease. Those who have preparations may even heal serious diseases, those who do not use them may not even improve minor complaints . . ." 夫聖賢之治世也有術, 得其術則功成, 失其術則事廢。譬猶醫之治病也, 有方, 篤劇猶治; 無方, 甚微不愈。That these *shu*

are in some way the same is asserted as early as the Qin *Lüshi chunqiu*: "Healing the body and governing the realm [both *zhi*] use technical arts (*shu*) sharing an inherent pattern" 治身與治國，一理之術 ("Shenfen" 審分).

30. Communication (personal) to Nylan from Stacey Van Vleet (June 2020). For the doubters, see *Shiji* 128.3224.

31. This is not to say that some techniques are not to be used; see, e.g., Xun Yue, *Hanji*: 加以好神仙之術.

32. Nylan thinks here of *Zhuangzi* and Yang Xiong's Mystery Learning, which tempered the positivism of the *Analects* with the awareness of "unknowing" found in the *Zhuangzi*.

33. The "pure Classicists" (*chun Ru* 醇儒) are "exceptional people"; "purity" here cannot mean "absence of admixture," because the leading classicists all used *yin/yang* theories (e.g., Dong Zhongshu and Liu Xiang) and, later, Five Phases theory (e.g., Liu Xiang, Jing Fang, and Liu Xin), and some figured as *fangshi* (as did many at Wudi's court). It is important to register this point, as, oddly, some China-boosters have suggested that today's *zhishi fenzi* are no different from Han Ru or Song *shi dafu*.

34. Of course, it is Yang Xiong who likened men like himself (classicists and literary figures) to the lowliest entertainers at the court.

35. Even as late as 132 CE, the reforms instituted for the selection of candidates for office mandated that they demonstrate a "knowledge of the Classic(s?)" *or* the ability to draft documents. See Bielenstein, *Bureaucracy*, 136. Michael Nylan and Nicholas Constantino, in a forthcoming paper on the rites controversies during Zhangdi's reign, show that good scholars have long known that the stated figures must be wildly inflated anyway.

36. *Hanshu* 89.3623, commenting on three major figures, defines the Ru (classicists) as those who wield the "classical/constant arts" (*jingshu*) to administer, so their ruler finds them useful: 儒者，通於世務，明習文法，以經術潤飾吏事，天子器之.

37. *Da Dai Liji* analogizes the realm (aka Tianxia) to a "big tool" 大器也.

38. "Xin yu," *juan* 4 ("Wuwei" 無為); comparable examples abound in Han texts of all persuasions, including *Huainan zi*, "Taizu xun" chapter 泰族訓 (故有道以統之，法雖少、足以化矣; 无道以行之，法雖眾、足以亂矣); also *Hanshi waizhuan*, 8.20.

39. *Hanshu* 58.2634, and this was important enough a statement to be imported into *Shiji* 112.2964.

40. See, e.g., *Hanshu* 75.3155.

41. See Xun Yue's *xu* (then postface, not preface) to his *Hanji* 漢紀, where he argues that regardless of a person's situation, in court or out, everything useful exists in good histories, and that has been the case since pre-unification days: 其揆一也.....凡《漢紀》有法式焉。有監戒焉。有廢亂焉。有持平焉。有兵略焉。有政化焉。... 有常道焉。有權變焉。有策謀焉。有詭說焉。有術藝焉。有文章焉. . . . 可以興、可以治。可以動、可以靜。可以言、可以行.

42. *Xunzi*, juan 16 ("Yi bing"): 凡兼人者有三術: 有以德兼人者, 有以力兼人者, 有以富兼人者。 . . . 以德兼人者王, 以力兼人者弱, 以富兼人者貧。古今一也。 Hutton (*Xunzi*, 161–62) translates this as: "Overall, there are three methods for capturing a people. There is capturing a people by means of virtue. There is capturing a people by means of force. There is capturing a people by means of wealth . . . he who captures a people by means of virtue will become a true king, he who captures a people by means of force will become weak, and he who captures a people by means of wealth will become poor. Both ancient times and the present age are one and the same in this."

43. The term "monographs" is sometimes used to distinguish the *Shiji* treatises from the *Hanshu*, which uses *zhi* instead, but the two counterparts are virtually identical in purpose and approach; presumably, *Hanshu* emended the title of the Treatises so as not to have *shu* as both a section and manuscript title.

44. On the Qin, the most recent view is that provided by Nylan, "Han Views," 79–80; contrast Yuri Pines, "To Die for the Sanctity."

45. Somewhat surprisingly, the treatises on ritual and music have drawn comparatively little attention, and most certainly less than the Basic Annals or Biographies sections. Two commendable treatments of the treatises are Mansvelt-Beck, *Treatises of Later Han*, and the recent volume of essays in *Monographs in Tang Official Historiography* [hereafter *Monographs in Tang*]. Beck's volume includes studies of the "Forerunners" for each of the *Suishu* treatises with precedents in the *Shiji* and *Hanshu*. While the *Suishu* monographs are not always precise counterparts to the *Shiji* and *Hanshu* treatises, they do provide interesting contrasts and comparisons. (As relevant, they are mentioned again below.)

46. Noa Hegesh's essay on Han musical theory (forthcoming in *Isis*) is a superb piece of technical writing on this topic.

47. The old Zhuanxu Calendar was no longer serviceable as the appearances of the new moons lagged by half a day. This undercut the throne's claims to legitimate rule, but Wudi's pursuit of such reforms also had to do with his pursuit of immortality. Cullen, "Numbers, Numeracy, and the Cosmos"; Haab-Schanke, "Crisis and Reform."

48. David Pankenier has translated this treatise in *Astrology and Cosmology*, 458–511.

49. Burton Watson includes this treatise in *Records of the Grand Historian*, 3–52. He may have based his work on Edouard Chavannes's classic, *Le T'ai chan*.

50. Watson translated this treatise in *Records of the Grand Historian*, 53–60.

51. Watson translated this treatise as "The Treatise on the Balanced Standard" in *Records of the Grand Historian*, 61–88.

52. Formal inconsistencies between the first two pairs of chapters and the other treatises, as well as the presence of material that clearly comes from later times, such as in the "Treatise on the Calendar," have led scholars to question whether it

is possible that the last three chapters represent the original formal structure for all eight treatises, and to speculate that the content of the first four chapters were lost and reconstructed at a later date.

53. As noted above, from *shu* to *zhi* 志.

54. Christopher Cullen has translated most of this treatise in *Foundations of Celestial Reckoning*. The first part of the treatise is found in chapter 5 and elsewhere, and the second part in chapter 3, while Liu Xin's *Shi jing* is also the subject of Cullen's "Birthday of the Old Man." More recently, Daniel Patrick Morgan and Howard L. Goodman ("Numbers with Histories") summarized the treatise (56–59) and addressed the question of why the two topics were placed in the same chapter.

55. These are the "Interior Palace Songs to Secure the Age" ("Anshi fangzhong ge" 安世房中歌) and "Nineteen Sections of Songs for the Suburban Sacrifice" ("Jiaosi ge shijiu zhang" 郊祀歌十九章). The text also mentions songs commemorating imperial tours. See *Hanshu* 22.1046.

56. Uchida Tomoo translated this treatise into Japanese in *Kanjo keihōshi*. Three years earlier, Hulsewé translated it into English in *Remnants of Han Law*, superseding the much older Italian translation by Alfonso Andreozzi, *Le leggi penali degli antichi cinesi*.

57. See Michael Nylan, "Assets Accumulating."

58. Nancy Lee Swann translated this treatise in *Food and Money*. As Yang Lien-sheng noted, Ban Gu's attitude is much more critical than that of Sima Qian when it comes to commerce and the accumulation of wealth. See Yang's "Notes on Dr. Swann's *Food and Money*," esp. 526.

59. Kano Naosada and Nishiwaki Tsuneki translated this treatise into Japanese in *Kanjo kōshishi*. Marianne Bujard translated it into French as number 187 of the Monographie series published by the École française d'Extrême-Orient: *Le Sacrifice au ciel dans la Chine Ancienne*. A comparison of the *Shiji* and *Hanshu* treatises on imperial sacrifice that focuses on the specific functions the sacrificial system performed to promote political unification is Yang Hua, "Qin Han diguo de shenquan tongyi."

60. Tomiya Itaru and Yoshikawa Tadao translated this treatise into Japanese in *Kanjo gogyōshi*. Michael Nylan's "'Treatise on the Wuxing'" is essentially a comparison between the *Hanshu* treatise and the *Suishu* version.

61. Notably, this treatise has drawn strong criticism. In his history of histories, the *Shitong* 史通 (Comprehensive Analysis of Historiography), the mid-Tang historian Liu Zhiji devoted one entire chapter of a total of twenty *juan* (scrolls, chapters) to criticizing Ban Gu for including this treatise in his *History*, on the grounds, essentially, that Ban's writing was too speculative and too inconclusive to be of any practical use to the courts he served.

62. This imaginative reconstruction of a Xia-era geography is augmented by details attributed to the archaic Zhou office of the "Bao Zhang shi" 保章氏, and this treatise also shares passages with the ritual *Zhou li* 周禮.

63. For the difference between "libraries" and "archives," see Nylan, "On Libraries."

64. Li Ling's *Lantai wanjuan: du Hanshu* is essentially a densely annotated scholarly translation of this treatise into modern Chinese.

65. *Hanshu* 22.1030. For Wudi's reign, Ban Gu includes a memorial of Dong Zhongshu that laments, "Nowadays, we have abandoned the virtuous teachings of the former kings, and only employ functionaries who cling to laws to govern the people" 獨用執法之吏治民. For Xuandi's reign, Ban Gu includes a remonstrance of Wang Ji 王吉 that states, "Today, using vulgar functionaries to lead the people is not the ritual standard which can carry across multiple eras." See *Hanshu* 22.1032, 22.1033.

Admittedly, these two ritual texts differ in multiple other ways, with Sima Qian's treatise including more discussion of both ritual theory and ceremonial protocols. (Here perhaps Ban Gu thought it unnecessary to repeat what Sima Qian's treatises had already said.) Both texts include a history of efforts to develop and reform the rites. For early Western Han, Sima Qian tells of 1) Shusun Tong's emendations to the rites; 2) a failed conference to reform the rites during Wendi's reign; 3) Chao Cuo's recommendation to reduce the authority and independence of the local lords during Jingdi's reign (a topic to which Sima Qian repeatedly returns in many chapters of the *Shiji*, as he sees Jingdi's reign as a major watershed); and 4) a wide range of ritual reforms under Wudi. Ban Gu continues this story of "incomplete" reforms.

66. Only three in-depth studies of the Tables have been done, so far as the editors know. Wang Liqi 王利器 and Wang Zhenmin 王貞珉 wrote on the "Table of Figures, Past and Present," in the *Hanshu* (a Table critiqued earlier by Liu Zhiji); that by Wu-shu Hui, which focuses on the Tables in the *Shiji*; and Nylan, "Mapping Time," which compares the *Shiji* and *Hanshu* Tables. Multiple short studies focus on minor features of the Tables.

67. See Nylan, "Mapping Time," for these people.

68. See Petersen, "Which Books," for the changing meaning of *baijia* 百家 (Hundred Experts) in relation to *zhuzi* 諸子 (the Masters), two terms that may have merged as early as the first-century of Eastern Han. Zheng Qiao singled these Tables out for praise, thinking they represented (as they doubtless did) the hardest-won achievement of the *Shiji*. See Nylan, "Mapping Time," 1n1.

69. Compare Feeney, *Caesar's Calendar*. *Hanshu* 62.2723n6 acknowledges the problems.

70. *Shiji* 14.511.

71. *Shiji* 13.487n1.

72. For the dating formulae and its significance in the *Chunqiu/Annals* classic, see van Auken, *Commentarial Transformation*. *Shiji* 130.3304 treats the *Shiji* as the continuation of the *Annals* tradition.

73. *Shiji* 14.509 gives the single mention of Zuo Qiuming in the *Shiji* proper, where *Zuo*'s history is but one of several authoritative chronicles devised in various states (the only one for Lu).

74. Zhang Handong 張漢東, "Hanshu Gujin renbiao shuzheng ding wu," 28–29, on Wang Liqi 王利器 and Wang Zhenmin 王貞珉 (Qi Lu shushe, 1988).

75. This point was disputed by Master Chu (Chu Da, the *Odes* master), who insists that the *Odes* speak of the divine progenitors of the culture heroes (*Shiji* 13.504–6).

76. This *Shiji* 13.500 states explicitly, if the empty boxes signifying units of time and space did not already speak volumes.

77. *Hanshu* 20.861. Many other sources are mined for data, judging from the extant text of the Table.

78. That would explain why the First Emperor of Qin is given a much higher rating than the moralizers would assign him, while the hapless Second Emperor is harshly condemned.

79. Karlgren, "Legends and Cults," 199. Karlgren thought a watershed could be established after unification, c. 200 BCE. Excavated texts would push systematization back much earlier (as should have been evident from the *Lüshi chunqiu*, compiled c. 238 BCE).

80. How much Fan Ye also saw him this way is shown by *Hou Hanshu* 60.

81. *Shitong*, 3:7/5, 16:7/19.

82. van Els and Queen, *Between History and Philosophy*; in particular his essay notes the changing citation style for anecdotes.

83. On this, see Nylan, *The Chinese Pleasure Book*, introduction and chapter 1.

84. For this, see Veyne, *Did the Greeks Believe*, esp. 8.

85. Three favorites of Nylan's are Kaster, *Guardians of Language*, on grammarians and pedagogy; Lehoux, *What Did the Romans Know?*; and Anderson, *Realness of Things Past*, on the misapplication of modern abstractions to ancient realities. G. E. R. Lloyd has supplied us with unfailingly intelligent studies of technical aspects of the classical Mediterranean. Scholarly work on prose prefaces in the technical literature is a "hot topic" in Classics, for example, as evidenced in Santini and Scivoletto, *Prefazioni, prologhi, proemi de opere tecnico-scientifiche latine*.

86. Hulsewé, of course, was responsible for producing two translations still of immense use: *Remnants of Han Law* and *Remnants of Ch'in Law*; and today, there is also Lau and Staack, *Legal Practice*, which is itself exemplary; but there is less-good work as well, including Liu, *Origins of Chinese Law*, an influential work that badly misconstrues the basis for Han-era legal thinking, sedulously avoiding talk of both the *Documents* and the *Xunzi*, while resorting instead to various works that were lionized later, but not during the two Han dynasties (e.g., the *Mencius* and the *Zhongyong*).

87. Su Xun, *Jiayou ji jianzhu*, juan 5, 114–15.

88. Attempts in late Eastern Han by Ying Shao and company to devise a "code" from the disparate and contradictory laws and precedents did not succeed, judging from the extant sources. Here our ideas differ from those of Yates and Barbieri-Low, who presume a "code" underlies the court cases found in the Zhangjiashan cache. Here we use the standard definition of "code": a type of legislation that purports to *exhaustively cover a complete system of laws* or a particular area of law as it existed at the time the code was enacted, *by a process of codification*.

89. McKnight, *The Quality of Mercy*; Hulsewé, *Remnants of Han Law*; Sivin, "Cosmos, Society, and State"; compare Lloyd and Sivin, *The Way and the Word*.

90. This may explain the immense influence now in China of Watanabe Yoshihiro 渡邊義浩, some of whose writings have been translated into Chinese.

91. Loewe suggests that while Sui and Tang landholdings may look somewhat similar to those of the two Han empires, rather different land tenure systems were in place, precluding direct comparisons.

92. Archaeological analyses still tend to reflect Marxist ideas. In the 1930s, the followers of the "Shihuo" 食货 movement also saw land tenure and food distribution as two of the key elements shaping the early empires. Tao Xisheng 陶希聖 (d. 1988) in China and Katō Shigeshi 加藤繁 (d. 1946) in Japan were two of the "Shihuo" leaders, who insisted that more material culture finds (not Marxian theories) were needed for further research. For additional details, see Masubuchi Tatsuo, *Chūgoku kodai no shakai to kokka*. These "Shihuo" scholars were in continual conversation with Dong Zuobin, Guo Moruo, Lü Zhenyu, and Hou Wailu, not to mention Yang Lien-sheng.

93. In this connection, Loewe's view undercuts the central thesis of Charles Holcombe, *In the Shadow*, and many others, arguing that the formal abdication of the last Eastern Han emperor in 220 CE may not have changed the lives of people outside the capital all that much.

94. As this is a follow-up study to her recent work on the *Suishu* omen treatise, the chapter ideally will be read in conjunction with that analysis.

95. See Su's *Hanshu Wuxing zhi yanjiu* and Chen's *Ruxue, shushu, yu zhengzhi*. NB: "Book" is a bad translation for *shu*; it refers to the codex (something that did not exist in manuscript culture in China); the translation of Five Elements has been out of fashion for decades in Chinese medicine, with Five Phases the proper term. Aside from the English title for Su's work, these two new studies represent major contributions to Chinese scholarship, in terms of their depth and breadth. Nylan's approach differs from theirs, however.

96. Michael Nylan and Mark Csikszentmihalyi thank Michael Loewe, Luke Habberstad, and Trenton Wilson in particular for their initial comments on this introduction, from which they have profited.

97. Ouyang Xiu, *Ouyang Xiu quanji*, juan 45, 594.

Asian-language Bibliography

Ban Gu 班固. *Baihu tong shuzheng* 白虎通疏證. Compiled by Ban Gu 班固 32–92. Edited by Chen Li 陳立 (1809–1869). Beijing: Zhonghua shuju, 1994.

Ban Gu et al. *Hanshu* 漢書. Compiled by Ban Gu 班固 (32–92), Ban Zhao 班昭 (45–ca. 116) et al. 12 vols. All refs. to the punctuated edition. Edited by Yan Shigu 顏師古. Beijing: Zhonghua shuju, 1962.

Chen Kanli 陳侃理. *Ruxue, shushu yu zhengzhi: zaiyi de zhengzhi wenhua shi* 儒學、數術與政治：災異的政治文化史. Beijing: Beijing daxue chubanshe, 2015.

Hanshu: see Ban Gu et al.

Kano Naosada 狩野直禎, and Nishiwaki Tsuneki 西脇常記. *Kanjo kōshishi* 漢書郊祀志. Tōkyō: Heibonsha, 1987.

Li Ling 李零. *Lantai wanjuan: du Hanshu Yiwen zhi* 蘭臺萬卷: 讀漢書藝文志. Beijing: Sanlian, 2011.

Masubuchi Tatsuo 増淵龍夫. *Chūgoku kodai no shakai to kokka* 中国古代の社会と国家. Tōkyō: Iwanami Shoten, 1996.

Ouyang Xiu 歐陽修. *Ouyang Xiu quanji* 歐陽修全集. Beijing: Zhonghua shuju, 2001 rpt. of 1191/1196 rpt.

Shiji 史記: see Sima Qian.

Shitong: All references are to *Shitong tongshi* 史通通釋, annot. Pu Qilong 浦起龍 (1679–1762). Taipei: Yiwen yinshuguan, 1978.

Sima Qian. *Shiji* 史記. Compiled by Sima Qian 司馬遷 et al. 12 vols. All refs. to the punctuated edition. Beijing: Zhonghua shuju, 1972.

Su Dechang 蘇德昌, *Hanshu Wuxing zhi yanjiu* 漢書五行志研究 [English title: A study on "The Treatise of the Five Elements in the *Book* [sic] *of Han*"]. Taipei: Guoli Taida chuban zhongxing, 2013.

Su Xun 蘇洵. *Jiayou ji jianzhu* 嘉祐集箋注. Shanghai: Shanghai guji chubanshe, 1993 rpt. of original c. 1066.

Tomiya Itaru 冨谷至, and Yoshikawa Tadao 吉川忠夫. *Kanjo gogyō shi* 漢書五行志. Tōkyō: Heibonsha, 1986.

Uchida Tomoo 内田智雄. *Kanjo keihōshi* 漢書刑法志. Kyōto: Harvard-Yenching Dōshisha, Tōhō Bunka Kōza Iinkai, 1958.

Wang Baoxuan 王葆玹. *Jing guwen jingxue xin lun* 今古文經學新論. Beijing: Zhongguo shehui kexue chubanshe, 1997; rpt. 2004.

Wang Chong 王充. *Lunheng jijie* 論衡集解. Beijing: Guji chubanshe, 1957.

Wang Liqi 王利器, and Wang Zhenmin 王貞珉. *Hanshu Gujin renbiao shuzheng* 漢書古今人表疏證. Jinan: Qi Lu shushe, 1988.

Yang Hua 楊華. "Qin Han diguo de shenquan tongyi: chutu jianbo yu 'Fengshan shu' 'Jiaosi zhi' de duibi kaocha" 秦漢帝國的神權統一: 出土簡帛與《封禪書》《郊祀志》的對比考察. *Lishi yanjiu* 5 (2011): 4–26.

Zhang Handong 张漢東. "Hanshu Gujin renbiao shuzheng ding wu" 漢書古今人表疏證訂誤. In *Guji zhengli yanjiu xuekan* 古籍整理研究學刊 3 (2000): 28–29.

Western-language Bibliography

Anderson, Greg. *The Realness of Things Past: Ancient Greece and Ontological History*. Oxford: Oxford University Press, 2018.

Andreozzi, Alfonso. *Le leggi penali degli antichi cinesi; discorso proemiale sul diritto e sui limiti del punire e traduzioni originali dal cinese, dell' avvocato Alfornso Andreozzi*. Firenze: G. Civelli, 1878.

Arendt, Hannah. *Between Past and Present.* New York: Viking Press, 1961.
Bielenstein, Hans. *The Bureaucracy of Han Times.* Cambridge: Cambridge University Press, 1980.
———. "The Census of China during the Period 2–742 A.D." *Bulletin of the Museum of Far Eastern Antiquities* 19 (1947): 125–63.
Bujard, Marianne. *Le Sacrifice au ciel dans la Chine ancienne: Théorie et pratique sous les Han occidentaux.* Monographies 187. Paris: De Boccard, 2000.
Chavannes, Edouard. *Le T'ai chan: essai de monographie d'un culte chinois; appendice "Le dieu du sol dans la Chine antique."* Paris: E. Leroux, 1910; rpt. Farnborough: Gregg, 1969.
Chemla, Karine, Damien Chaussende, and Daniel Patrick Morgan. *Monographs in Tang Official Historiography: Perspectives from the Technical Treatises of the History of Sui (Sui shu).* Cham: Springer, 2019.
Constantino, Nicholas, and Michael Nylan. "On the Rites and Rites Controversies in Mid-Eastern Han." In a publication edited by Anne Cheng and Stephane Feuillas for the Collège de France. Forthcoming.
Crone, Patricia. *Pre-industrial Societies.* Oxford: Blackwell, 1989.
Csikszentmihalyi, Mark. "Chia I's 'Techniques of the Tao' and the Han Confucian Appropriation of Technical Discourse." *Asia Major* series III, vol. 10, nos. 1–2 (1997): 49–67.
Csikszentmihalyi, Mark, and Michael Nylan. "Constructing Lineages and Inventing Traditions in the *Shiji*." *T'oung Pao* 89 (2003): 59–99.
Cullen, Christopher. "Numbers, Numeracy and the Cosmos." In *China's Early Empires: A Re-appraisal*, edited by Michael Nylan and Michael Loewe, 321–38. Cambridge: Cambridge University Press, 2010.
———. "The Birthday of the Old Man of Jiang County and Other Puzzles: Work in Progress on Liu Xin's *Canon of the Ages*." *Asia Major* 14 (2004): 27–70.
———. *The Foundations of Celestial Reckoning: Three Ancient Chinese Astronomical Systems.* New York: Routledge, 2017.
de Pee, Christian. *Losing the Way.* Amsterdam University Press, forthcoming.
Durrant, Stephen, Wai-yee Li, Michael Nylan, and Hans van Ess, eds. *The Letter to Ren An and Sima Qian's Legacy.* Seattle: University of Washington Press, 2016.
Feeney, Denis. *Caesar's Calendar: Ancient Time and the Beginnings of History.* Berkeley: University of California Press, 2007.
Graham, A. C. *Disputers of the Tao: Philosophical Argumentation in Ancient China.* Chicago and La Salle, IL: Open Court, 1989.
Haab-Schanke, Dorothee. "Crisis and Reform of the Calendar, in *Shiji* 26." *Oriens Extremus* 45 (2005/06): 35–47.
Hegesh, Noa. "Mind the Gap: Acoustical Answers to Cosmological Concerns in the First Century B.C.E. China" (forthcoming in *Isis* journal).
Holcombe, Charles. *In the Shadow of the Han: Literati Thought and Society at the Beginning of the Southern Dynasties.* Honolulu: University of Hawai'i Press, 1994.

Hulsewé, Anthony François Paulus. *Remnants of Ch'in Law: An Annotated Translation of the Ch'in Legal and Administrative Rules of the 3rd century B.C. Discovered in Yün-meng Prefecture, Hu-pei Province, in 1975.* Leiden: Brill, 1985.

———. *Remnants of Han Law, Volume 1.* Leiden: Brill, 1955.

Hutton, Eric. *Xunzi: The Complete Text.* Princeton: Princeton University Press, 2014.

Karlgren, Bernhard. "Legends and Cults in Ancient China." *Bulletin of the Museum of Far Eastern Antiquities* 18 (1946): 199–365.

Kaster, Robert A. *Guardians of Language: The Grammarian and Society in Late Antiquity.* Berkeley: University of California Press, 1988.

Lau, Ulrich, and Thiess Staack. *Legal Practice in the Formative Stages of the Chinese Empire: An Annotated Translation of the Exemplary Qin Criminal Cases from the Yuelu Academy Collection.* Leiden: Brill, 2016.

Lehoux, Daryn. *What Did the Romans Know? An Inquiry into Science and Worldmaking.* Chicago: University of Chicago Press, 2012.

Liu, Yongping. *Origins of Chinese Law: Penal and Administrative Law in Its Early Development.* Oxford: Oxford University Press, 1998.

Lloyd, Geoffrey, and Nathan Sivin. *The Way and the Word.* New Haven: Yale University Press, 2003.

Makeham, John. *Zhong lun* 中論. References are to the translation titled *Balanced Discourses* by John Makeham. New Haven: Yale University Press, 2002.

Mann, Michael. *The Sources of Social Power.* 4 vols. Cambridge: Cambridge University Press, 1986.

Mansvelt-Beck, B. J. *The Treatises of Later Han: Their Author, Sources, Contents, and Place in Chinese Historiography.* Leiden: Brill, 1990.

McKnight, Brian E. *The Quality of Mercy: Amnesties and Traditional Chinese Justice.* Honolulu: University of Hawai'i Press, 1981.

Morgan, Daniel Patrick, and Howard L. Goodman. "Numbers with Histories: Li Chunfeng on Harmonics and Astronomy." In *Monographs in Tang Official Historiography: Perspectives from the Technical Treatises of the History of Sui (Sui shu),* edited by Karine Chemla, Damien Chaussende, and Daniel Patrick Morgan, 51–87. Cham: Springer, 2019.

Nylan, Michael. "Assets Accumulating: Sima Qian's Perspective on Moneymaking, Virtue, and History." In *Views from Within, Views from Beyond: Approaches to the* Shiji *as an Early Work of Historiography,* edited by Hans van Ess, Olga Lomova, and Dorothee Schaab-Hanke, 131–69. Wiesbaden: Harrassowitz, 2015.

———. "Beliefs about Seeing: Optics and Moral Technologies in Early China." *Asia Major* 21, no. 1 (2008): 89–132.

———. "Han Views of the Qin Legacy and the Late Western Han 'Classical Turn.'" *Bulletin of the Museum of Far Eastern Antiquities* 79–80 (December 2013/actually 2020), 51–98.

———. "Mapping Time in the *Shiji* and *Hanshu* Tables." *East Asian Science, Technology, and Medicine*, Special Issue on Numerical Tables and Tabular Layouts in Chinese Scholarly Documents: Part I: On the Work to Produce Tables and the Meaning of their Format, edited by Karine Chemla (2016): 1–65.

———. "On Libraries and Manuscript Culture in Western Han Chang'an and Alexandria." In *Ancient Greece and China Compared*, edited by G. E. R. Lloyd and Jingyi Jenny Zhao, 373–408. Cambridge: Cambridge University Press, 2018.

———. "Sima Qian: A True Historian?" *Early China* 23/24 (1998–99): 203–46.

———. "The 'Treatise on the Wuxing' (Wuxing zhi)." In *Monographs in Tang Official Historiography: Perspectives from the Technical Treatises of the History of Sui (Sui shu)*, edited by Karine Chemla, Damien Chaussende, and Daniel Patrick Morgan, 181–235. Cham: Springer, 2019.

Nylan, Michael, and Trenton Wilson. "A Brief History of Daring," accepted for a volume comparing the antique Greek and Chinese emotions, edited by David Konstan and Huang Yang (to be published in Chinese and English).

Pankenier, David W. *Astrology and Cosmology in Early China: Conforming Earth to Heaven*. Cambridge: Cambridge University Press, 2013.

Petersen, Jens Østergaard. "Which Books Did the First Emperor of Qin Burn?—On the Meaning of *Pai chia* in Early Chinese Sources." *Monumenta Serica* 43 (1995): 1–52.

Pines, Yuri. "To Die for the Sanctity of the Name: Name (*ming* 名) as Prime Mover of Political Action in Early China." In *Keywords in Chinese Culture*, edited by Wai-yee Li and Yuri Pines, 169–215. Hong Kong: Chinese University of Hong Kong Press, 2020.

Puett, Michael. *The Ambivalence of Creation: Debates Concerning Innovation and Artifice in Early China*. Stanford, CA: Stanford University Press, 2001.

Queen, Sarah A. "Inventories of the Past: Rethinking the 'School' Affiliation of the *Huainanzi*." *Asia Major*. 3rd series. 14, no. 1 (2001): 51–72.

Sanft, Charles. *Literate Community in Early Imperial China: The Northwestern Frontier in Han Times*. Albany: State University of New York Press, 2019.

Santini, Carlo, and Nino Scivoletto, eds. *Prefazioni, prologhi, proemi de opere tecnico-scientifiche latine*. 2 vols. Rome: Herder, 1992.

Sivin, Nathan. "State, Cosmos, and Body in the Last Three Centuries B.C." *Harvard Journal of Asiatic Studies* 55, no. 1 (June 1995): 5–37.

———. "Text and Experience in Classical Chinese Medicine." In *Knowledge and the Scholarly Medical Traditions*, edited by Don Bates, 177–204. Cambridge: Cambridge University Press, 1995.

Slingerland, Edward. *Effortless Action: Wu-wei As Conceptual Metaphor and Spiritual Ideal in Early China*. New York: Oxford University Press, 2007.

Smith, Kidder. "Sima Tan and the Invention of Daoism, 'Legalism,' et cetera." *Journal of Asian Studies* 62, no. 1 (2003): 129–56.

Swann, Nancy Lee. *Food and Money in Ancient China: The Earliest Economic History of China to A.D. 25 (Han Shu 24, With Related Texts, Han Shu 91 and Shih-chi 129)*. Princeton: Princeton University Press, 1950.

van Auken, Newell Ann. *The Commentarial Transformation of the* Spring and Autumn. Albany: State University of New York Press, 2016.

van Els, Paul, and Sarah A. Queen, eds. *Between History and Philosophy: Anecdotes in Early China*. Albany: State University of New York Press, 2017.

Veyne, Paul. *Did the Greeks Believe in Their Myths? An Essay on the Constitutive Imagination*. Chicago: University of Chicago Press, 1988.

Watson, Burton. *Records of the Grand Historian: Han Dynasty*. Vol. 2. Hong Kong and New York: Renditions-Columbia University Press, 1993.

Williams, Raymond. *Keywords: A Vocabulary of Culture and Society*. New York: Oxford University Press, 2015.

Wu Shu-Hui. "On *Shiji* 史記 22, Table Ten: A Year-by-Year Table of Generals, Chancellors, and Prominent Officials Since the Founding of the Han 漢 Dynasty," *Journal of Chinese Studies* 59 (2014): 121–64.

Yang Lien-sheng 楊聯陞 (1914–90). "Notes on Dr. Swann's *Food and Money in Ancient China*." *Harvard Journal of Asiatic Studies* 13, nos. 3/4 (1950): 524–57.

Yates, Robin D. S., and Anthony J. Barbieri-Low. *Law, State, and Society in Early Imperial China: A Study with Critical Edition and Translation of the Legal Texts from Zhangjiashan Tomb no. 247*. 2 vols. Leiden: Brill, 2015.

1

Land Tenure and the Decline of Imperial Government in Eastern Han

MICHAEL LOEWE 魯惟一

"It is on the work of the farmers that the world depends." So ran the opening words of decrees that were proclaimed in 178 BCE and subsequently. They conveyed a principle on which lay the decisions of the imperial government and the duties of officials and which determined the work of by far the larger part of the population, as well as the distribution of its produce. There were also individuals who would find the means to intervene for their own benefit in the complex economy devoted to agriculture.[1] Nonetheless, in both the Roman Republic and Empire and in China's early empires, large areas of the land lay under the control of a few highly privileged individuals, be they senators or high-ranking officials, merchants or members of the imperial family. In principle, and in all probability in practice, the disposal of land lay in the hands of officials, and it was perhaps only from late in Eastern Han that it became more readily available for acquisition by purchase. Readers may be left with the impression that such transactions were a regular and permissible procedure throughout Han times. Thus this work sets out to pose a number of research questions relating to land tenure in the cases of Qin, Western Han, and Eastern Han, from 221 BCE to 220 CE: how these large areas were worked and to whose benefit; by what means and according to what conditions either a large landholder or an individual farmer occupied an estate or worked a holding; what were the material circumstances and social status of those who undertook the hard labor of the ploughman or harvester; or in what way those stalwarts were able to secure a livelihood by doing so.

In an article published in 1936 in Chinese and in 1956 in an English version, Yang Lien-sheng drew attention to the social and economic conditions that prevailed in Eastern Han times, roughly the first two centuries CE. Yang wrote of land concentrated under the management of large landlords and worked by the dependent small peasant.[2] A little later, in his survey of the economic conditions of Western Han, Nishijima Sadao took the view that "the growth of large-scale land ownership was now clearly beyond control" and that "By the late Han period, the existence of great landholdings had become accepted as a matter of course, and the state made no further attempts to restrict it."[3] As far as may be known, no serious attempt has been made to counter these conclusions, and hence the consideration here of the evidence we have to answer the basic questions. In doing so, I am grateful to scholars such as Professors Balazs, Bielenstein, and de Crespigny, whose research has clarified the dynastic and political developments of the time together with the intellectual reactions of some of those who witnessed them.

The regulations of Western Han provided for officials of the commanderies and counties to control the occupation of the land and to assess the extent of its produce. Toward the end of that dynasty, what may be called 'private tenure' was developing, to become more extensively practiced by perhaps 100 CE. The principle adopted early in Western Han, that the tenure of land depended on allocation by officials, was being overtaken in Eastern Han in the face of private enterprise and perhaps by the growth of acquisition of land by purchase.

We may ask why the steps taken to allocate land that had been practiced since the foundation of the Western Han virtually ceased in 147 CE; why in the counts that we have for the years 105 to 145/146 the number of individuals decreased, while that of households rose; how it came about that, as between 1–2 and 140 CE,[4] the proportion of persons and households registered north and south of the Yangzi River varied so widely; and why from 132 CE the attention paid to the vagrant elements of the population seems to have declined or even ceased. We may also note a further difference: the ordinances and statutes controlled the allocation of land and its inheritance. But the evidence concerning the sale of land between individuals that we possess includes no reference to the part that responsible officials played in such transactions.[5]

There is a marked decrease, from 57,671,400 to 49,150,220, in the numbers of those whom the officials registered in the two extant population registers of 1–2 and 140 CE. Reports of natural catastrophes that might have accounted for so great a change have yet to be found, and it is possible that, so far from showing a decline in the number of the inhabitants of the empire, the registers evince the inability of local officials of the commanderies or counties to complete accurate returns for the lands under their jurisdiction. Such inability, to be tentatively dated from 130 CE, was due to encroachments of non-Han peoples in the north and the independence of large-scale landholders in the center and the east. Meanwhile individual initiative with the cooperation of local officials was leading to a more effective control and exploitation of land in the south.

Basic differences underlie the ways in which land was worked in Han territory and elsewhere. In the pages that follow, the terms "landholder," "occupier," and "tenure" carry no implication that the statutes and ordinances, or any accepted but unwritten principle of government during Han times, had the power of validating ownership of the land or restricting its usage.

Yet the conditions in which land was occupied, worked, and owned and the control that a governing authority exercised over such matters must surely rank as a prime factor in determining the political, social, and economic conditions in which a community thrived or failed. To the dismay of historians of China's early empires, essential as such matters are to understanding how a society was regulated, all too frequently we possess no definitive, let alone comprehensive, statement of how the work of the fields was organized; or of how and why a farmer was entitled to occupy and work parts of the land; or of how and why a ruling authority controlled these undertakings and provided security for those who were engaged in them.

For the Sui and Tang dynasties (589–618 and 618–907), we possess far more definite information than for their predecessors, as may be seen in the records that we have for the systematic ways in which imperial government devised and imposed its controls. For the earlier empires we have some statements of fact and some statistics, but we are left to reconstruct a picture of how the fundamental work of agriculture operated from information that is fragmentary. In examining the evidence, historians search in vain for a definition of any concept of land ownership, other than a claim or boast that one short-lived and discredited emperor once put forward.[6] Provisions of government included the annual compilation of registers of the land and those who were settled there. These documents were designed to provide a living for the population, to organize and control the use of the land and its produce, to increase the revenue, and to discourage vagrancy. They conveyed acknowledgement of a farmer's settlement on the land; they did not convey his possession of it.

Administering large areas as the kings did, the statutes did not provide them with the use of lands for their own cultivation and enrichment; and while housing was made available on the estates of the nobles, these were not accompanied by land holdings for their own use. We may, nonetheless, take note of several ways in which the land was being occupied and worked. One was due to the positive steps that the imperial government took to allocate land and thus encourage a greater production of staple goods and of revenue; another drew from the private initiative of certain prominent persons who were able to occupy land by various means, at times including purchase; and some persons might hold land thanks to grants that an emperor made to a favorite. While some large areas of land apparently lay within the hands of a few rich individuals, the Han governments operated a deliberate means of making plots available for the less privileged individuals or families to work. They might take such steps to mitigate the effects of a natural disaster that left large numbers of the inhabitants landless, homeless, and with no means of earning a livelihood.

We may note how the laws of the land laid down the conditions and provided the means of taking such measures, and we benefit from occasional positive records of imperial orders to implement such instructions. We may also note the growth of large estates in private hands and the court's failure to limit landholdings or alleviate suffering as such estates expanded in number and strength. A major lack in the extant sources concerns the conditions in which tenants worked the land and in what ways a landowner depended on wage earners to till the fields and gather the harvest. We do note that positive measures to allocate land to families mostly fell into abeyance from 147 CE and were not reintroduced in the dynasties that followed Han. It was left to the emperors of Sui and Tang, acting on some earlier precedents, to institute their own systematic method of distributing the land to the best advantage for both the people and the government.

We may bear in mind the labor and hardships that were involved in the preliminary stages of reclaiming land from marsh or forest for tillage, in shaping the fields both to receive sufficient water and withstand the ravages of nature, such as flooding. This work of land reclamation was in many cases essential, but it must surely have been nothing other than dire distress, compulsion, or a miserly payment that could induce anyone to undertake it. For often a year or longer, laborers could not hope to garner a harvest on which they could live, so we may wonder who instigated and paid for these pioneering steps. We are occasionally told how, in these early stages, those who were engaged in these tasks were provided with equipment and tools, or the seed with which to sow their first crop, or perhaps an immediate provision for the sustenance of their families.[7] Our histories do not reveal precisely how a large landholder benefited from leasing out land to tenants, or the dues that such tenants would be obliged to pay.[8]

Thanks to the survival of fragments of archives combined with references in the standard histories, it may be possible to examine some of these problems for the Han empire, but we can reach no definitive conclusions for the four centuries of Qin and Han rule. The Statutes and Ordinances of 186 BCE, found in fragments at Zhangjiashan 張家山, include regulations for the distribution of land and the ways in which it could be inherited. Records in the *Hanshu* and *Hou Hanshu* inform us of the times when such regulations were put into effect and leave us to infer when they were not enforced; they also tell us of how some elements of the population stood to benefit. We may attempt to put such information within a wider context, that of the increasing expansion of the Han government's interests in the south, as well as the initiative taken by members of the population to evade the hardships of a farmer's life and to seek a more lucrative means of subsistence.

In considering these questions, we shall fasten on the allocations of land that Han officials could order, the possible extension of officials' administration in the south, and certain changes that are apparent in the counts of the farming population. Developments and changes took place within the context of the decreasing force of imperial rule and its disciplines, and a tendency for the easier life of commerce and industry to attract workers away from the land.

I. The Produce of the Land

We may consider some of the theoretical statements about the produce of the land, perhaps framed for rhetorical purposes rather than as precise calculations; the records of instances in which the imperial government decreed measures that affected the disposition of lands; the statutory provisions whereby such measures were implemented; as well as various accounts of the extent of the land's produce as allocated for the support of a household. Some of the evidence that we shall quote includes figures, and we must ask how reliable these are likely to be. Where these appear in round numbers, we may suspect that they derive from estimates or calculations or even that they were generated for purposes of persuasion or perhaps from mere guesswork. Where they are given down to the last digit, as they are in some entries from both the archives and the chronicles, we may presume that they result from reports that officials rendered to their superiors.[9] So while such precise figures may have been subject to some degree of manipulation or textual corruption, they are more likely to represent the results of an actual count that was made and, hence, may be more credible. We must, however, note that these figures were preserved as a record of the duties that an official had undertaken; they were not put forward as material on which policies should be based.

A statement that is recorded as coming from Li Kui 李悝 (ca. 400 BCE) contains two sets of figures. What we cannot know is whether or not the compiler of the *Hanshu* was anachronistically applying figures of his own time to an earlier age.

(a) 100 *li* 里 of land amounts to 90,000 *qing* 頃 that have been demarcated. Given that one-third of the area is occupied by mountains, scrub, and settlements, there remain 6,000,000 *mu* 畝 of arable land. Energetic cultivation will produce 3 *dou* 斗 from 1 *mu* (approximately 3 x 2 liters from 0.1 English acre), with a corresponding decrease if the cultivation is not energetic.[10] As a result, the average yield of unhusked millet of the *li* is 1,800,000 *shi* 石.[11]

(b) Given one able-bodied man supporting a total of five persons, cultivating 100 *mu* of arable land and harvesting, 1½ *shi* (approximately 30 liters) per 1 *mu*, totaling 150 *shi*.[12] Deduct one-tenth for tax, leaving 135 *shi*; allow for the monthly need of food for five persons at 1½ *shi*, totaling 90 *shi* for five persons annually. This leaves a surplus of 45 *shi*, i.e., 1,350 cash (*qian* 錢), with one *shi* valued at 30 cash (*qian*). Deduct 300 cash for religious dues. There remain 1,050 cash.

The second account states, on this basis, that most farming families are in a constant state of want. Its figures compare with those seen in a statement attributed to Chao Cuo 鼂錯 ca. 178 BCE, where Chao concludes that a household of five members, of whom at least two were liable for statutory service, could perhaps harvest 100 *shi* if they could work 100 *mu*.[13] We note also that, much later, Zhongchang Tong 仲長統 (180–220) referred to an ideal yield of 3 *hu* (approximately 60 liters) per *mu*.[14]

Theoretical, imaginary, or ideal as these figures may be, they may be compared with some that derive from practical experience. Fragmentary records survive of the grain that was distributed as a ration to servicemen and their families who were stationed on the garrison lines of the northwest, at dates from 100 BCE onward. Monthly allowances (stipulated in *shi*) were set on the following scale:[15]

officers and other ranks	3.3
guardsmen	3.0
male child, adult female	2.16
male infant	1.6
female infant	1.16

Records from the same source give figures for the monthly distribution of grain to households of five members. These vary between 4.8 and 6.0 *shi*, i.e., annually between 57.6 and 72 *shi*.[16] If these figures were to be applied to the statement that is ascribed to Chao Cuo, the family of five members that he had in mind would each year harvest produce that would be sufficient for twenty months for feeding (if not for additional expenses). That would presumably have been subject to deduction of the types mentioned by Li Kui.

II. The "Orders of Honor" (*jue* 爵)

Attention follows below to the landholdings of exceptionally large sizes that rich merchants and others may have been able to acquire or those that were available to an emperor's favorites or the relatives of his consorts. We should first turn to the circumstances and conditions in which the majority of the population were able, and in some cases entitled, to work the much smaller plots on which they depended for their livelihoods.

They worked such plots subject to the supervision or even control of officials of the commanderies and counties who were implementing the statutes and ordinances of the empire. In general, these officials did not make arbitrary decisions by so doing. The standard histories record instances when a decree emphasized the basic importance of working the land, from the time of Wendi (r. 157–141 BCE) onward.[17] In addition, on repeated occasions, an emperor issued orders for the grant of bounties of various types. These included the conferment of *jue* 爵, i.e., "orders of honor," or sometimes "orders of aristocratic rank," which featured from 202 BCE until 147 CE, and then sporadically (in 168 and 215). Such privileges were granted either to one adult in each household, or to broader populations, or to certain specified groups in elements of the population, and, in exceptional cases, to a named individual only.

While historians have long drawn attention to these measures and their significance, the references made thereto in the Zhangjiashan Statutes and Ordinances of 186 BCE have transformed our understanding of the privileges or gifts that the orders of honor carried with them.[18] The system of the *jue*, which had been introduced before the imperial era in the kingdom of Qin, may be summarized as follows.

1) A series of twenty orders or titles of honor ran from that of *gong shi* 公士 at the lowest level to that of *chehou* 徹侯, or *liehou* 列侯, at the top.[19] Every few years an imperial order conferred the gift of one or perhaps more orders upon the heads of households or others, who were specified as *nanzi* 男子 (adult males), or described simply as *ren* 人 (human beings) or *min* 民 (members of the populations).[20]

2) Some forty-five grants are recorded for Western Han between 202 BCE and 4 CE.[21] One grant was made at Wang Mang's accession in 9 CE, but no others were issued during the Xin dynasty (9–23 CE).[22] Some thirty were made in Eastern Han between 27 and 147 CE. Thereafter two grants are recorded, for 168 and 215.[23] As against the number of one, two, three, and occasionally four or five grants given in a ten-year period, grants were made frequently during two periods, 67–42 BCE (fourteen grants in twenty-five years) and 53–79 CE (eleven grants in twenty-six years).[24]

3) A rare, and possibly unique, reference to the grant of *jue* nearly two centuries after Eastern Han, during Eastern Jin (317–419), dated for 405 CE, when two orders were conferred upon the *bai guan* 百官 or *bai xing* 百姓.[25] There are apparently cases after 405 CE, but no known cases for the Sui and Tang dynasties when other systems of land occupation operated.[26]

4) On each occasion when a grant was made, the grant could include a promotion for those who had already received

one, so that the positions of some rose to higher positions on the scale of the twenty orders. Such increases did not go beyond the eighth-highest order, that of *gongcheng* 公乘. In special cases, a decree could order the gift of one of the higher orders, such as that of the highest but one, known as *guanneihou* 關內侯 "Noble of the Interior."

5) An allocation of land and housing accompanied each grant of one of these orders, with the allocation lying between 1½ *qing* 頃 of land and 1½ units of housing for those of *gongshi* rank, and 20 *qing* and 20 residences for those of *gongcheng* rank.[27]

The full allocation was as follows:[28]

Order	Number of *qing*	Number of residences
chehou 徹侯, *liehou* 列侯	[105]	105
guanneihou 關內侯	95	95
da shu zhang 大庶長	90	90
si ju shu zhang 駟車庶長	88	88
da shang zao 大上造	86	86
shao shang zao 少上造	84	84
you geng 右更	82	82
zhong geng 中更	80	80
zuo geng 左更	78	78
you shu zhang 右庶長	76	76
zuo shu zhang 左庶長	74	74
wu da fu 五大夫	25	25
gong cheng 公乘	20	20
gong da fu 公大夫	9	9
guan da fu 官大夫	7	7
da fu 大夫	5	5
bu geng 不更	4	4
zan niao 簪裊	3	3
shang zao 上造	2	2
gong shi 公士	1½	1½
gong zu 公卒	1	1

There are perhaps signs that these allocations were not always carried out in full. Bu Shi 卜式 was twice rewarded for his services, once with the order of *zuo shuchang* and then with that of *guanneihou*. On each occasion he received ten *qing* of land.[29]

Right at the top of the scale the Nobles were thus allocated 105 residences, but there is no statement that they also received a measure of land.[30] The *guanneihou*, "Nobles of the Interior," received 95 *qing* of land and 95 residences. Such estates would have required the labor of and been able to provide a living for 95 households. We are not told in what ways the Nobles and *guanneihou* benefited from such large allocations. The ninety-five households were doubtless liable to pay tax. Whether or not the householders paid any additional dues to the Nobles is not stated. Nor can we know for certain whether it was the officials of the counties who collected tax directly from those households or whether they drew it, in bulk, from the Nobles and *guanneihou*. Possibly we discern here the beginnings of a system of tenantry whereby the farmers paid a rent to the virtual occupier of the land that they worked. We lack information as to whether or not wage earners were employed and paid to work by the day or the month or in accordance with the extent of the land.

Our histories of Western Han record forty-five occasions, between 202 BCE and 4 CE, when a general bestowal of *jue* was ordered,[31] and in all but eleven cases these were granted to *min* 民 or *ren* 人 without further specification. In some cases they were accompanied by grants of *jue* to officials.[32] In at least nine cases they were granted to heirs.[33] In fifteen cases they were accompanied by a gift of meat and strong drink to household members, and, notably, seven of these took place between 65 and 52 BCE, during the reign of Xuandi (r. 74–48).[34] These years followed the downfall of the once-powerful Huo 霍 consort family in 66 and Xuandi's decision to take a personal part in ruling the empire. Significantly, it was precisely at this time, from 65 to 49, that the *nianhao* were adopted to signify the well-being of the reign and the blessings that it had received, such as (*yuan kang* 元康 65–62), the appearance of auspicious birds with their spiritual messages (*shen jue* 神爵 61–58), the five phoenixes (*wu feng* 五鳳 57–54), and the fall of sweet dew from the heavens (*gan lu* 甘露 53–50). Presumably, the gifts of meat and strong drink, accompanied by the propagation of happy messages, were intended to display the fine state of the empire that the emperor wished to claim as his own contribution to the times.

While a general grant of *jue* brought a means of livelihood and a relief from want to some, it may also have involved dispossessing others from their lands and rendering them destitute. We may perhaps observe some signs that officials at the capital city of Eastern Han were not entirely oblivious of the hardships that many families suffered, and that they expressed some sense of responsibility for relieving such distress. On all but five of the thirty occasions of a general grant of *jue* in Eastern Han, between 27 and 147 CE, these were accompanied by a gift of grain to those who were rendered homeless and lacking a means of livelihood, such as widows and orphans.[35] We may reflect that such relief measures may not necessarily have arisen from purely philanthropic or charitable motives. Instead, they may well have been intended as a gesture that would attract the loyalties of the benefited and dissuade those who had been dispossessed from taking to crime, as with the case of those who were described as vagrants *liu min* 流民, for whom see below (section V).

This scheme may be seen as a means of settling the population, the better to have them earn their livelihoods, and of increasing both the extent of the land that was under production and the amount of tax that could be collected in kind. The lowest allocation of land and housing presumably would have been deemed sufficient for one registered household (*hu* 戶), usually of four or five members. As the allocation was increased with successive grants, it would not have been possible for that one household to work all the land, especially when there was no certainty that the newly allotted holdings would adjoin those that were already held. The allocation of twenty *qing* and twenty residences that accompanied the grant of the order of *gongcheng* was presumably intended for twenty households.

We may suppose that the intention that lay behind this system was to enrich the governing authority by expanding and controlling the cultivation of the land and the production of staples and that it was not primarily inspired by altruism. Essentially it was an instrument of purposeful government, during the Qin and Western Han, designed to draw as much wealth as could be extracted from the land so as to increase the material resources that would be at the disposal of the officials. The grant of a second or subsequent order of honor might well provide a means of subsistence for a farmer's second or third son, yet it imposed a duty on the recipient to ensure that the new allocation was being cultivated; such an obligation might well involve problems.

Wang Mang all but ignored the institution of granting *jue*, posing as he did as a restorer of the ways of the preunification Zhou (ca. 1050–256 BCE); perhaps he was biased against the *jue*, which had been introduced in the kingdom of Qin before 221 BCE. From 147 CE onward, grants of *jue* and the allocations of land that went with them were mostly stopped, although we hear, from that time onward, of proposals—apparently unheeded—from Wang Fu 王符 (ca. 90–165) and others to restore the institutions and practice of a more highly disciplined way of government.³⁶

The system of *jue* also may have acted as a means of maintaining occupation of plots of land within one and the same family. In ten instances in Western and six in Eastern Han, orders of honor were conferred on persons who were described as *hou zhe* 後者, a term that was used in various ways but that basically means "heirs."

(1) We read the phrase *dang wei fu hou zhe* 當為父後者, which we may understand as "qualified to be the heir."³⁷ Yan Shigu 顏師古 (581–645) explains it as persons who are the heirs, even though a son has not yet been born to the acknowledged wife (嫡 *di*). He Zhuo 何焯 (1661–1722) takes precisely the opposite view, with the term designating just such persons born to the acknowledged wife. More specifically, we also see *chang zi dang wei fu hou zhe* 長子當為父後者 ("eldest son qualified to be the father's heir").³⁸

(2) In at least six instances, the expression *wei fu hou zhe* is used.³⁹

(3) In two cases, we read *liehou si zi* 列侯嗣子 "heirs of the nobles."⁴⁰

With the possible exception of (3) above, it is not certain whether one or more than one son would receive the honor and its allocation of land. For (3), insofar as the son of the acknowledged consort would be due to inherit the nobility, it would make sense if the grant of an honor were to be made to a number, or even all, of his brothers and half-brothers, so as to provide them with a living. We do not know whether or not such steps were taken, as primogeniture may have been operative in some or all cases.

As noted above, with the two exceptions of 168 and 215, the practice of decreeing grants of *jue* was discontinued after 147. It has yet to be determined whether or to what extent this change was due to dynastic or political antagonisms or to the increase of large landholdings that lay in the hands of the privileged few. We may perhaps ponder whether by 147 it had become difficult, or even impossible, for officials of Eastern Han to fulfill the purpose of the *jue* by allocating land to males who had just reached the age of maturity (*nanzi*), simply because the independent landholders were able to prevent such use of farmland that they in fact held under control.

We may thus see the system of *jue* as a means of retaining supervision over the farmers' use of the land within the authority of the central government, and we shall take note of the circumstances in which such intentions were overridden. But we may first turn attention to some of the difficulties in operating the scheme. Changes in the allocation of the land would become necessary at the death of the farmer to whom a plot of land had been allocated. If his son stood to inherit this, or if the land was reallocated to a new holder, such a fact had to be acknowledged in the official registers. In addition, it might well be desirable, or even necessary, to provide a means of living for the widow and other surviving members of the original household. Possibly, such a death might leave land without cultivation for a time. In such a case, reallocation to another household would have involved it in the task of reclamation without the benefit of the land's yield. In other cases, the grant of forest land or marsh may have accompanied that of one of the orders of honor, and the holder would face the same type of difficulty.

The bookkeeping that officials would need to complete, by the approved date in the eighth month, might well prove irksome. For whether a household took on cultivated land from a predecessor or would face the task of reclamation, such matters would have to be entered on the registers of the land and their inhabitants that were compiled every year. In addition, the frequent decrees that bestowed an order of *jue* would have involved a full examination of landholdings, at least in theory. Such administrative work would have placed a severe burden on the shoulders of the local officials, whether they were appointed at the level of the county (*xian* 縣), district (*xiang* 鄉), or village (*li* 里). In practical terms, we may consider the circumstances in which the relatives of a farmer were placed at his death and the reallocation of his holding. Personal charity may have helped some of them to live. In an alternative scenario, a hard-working or disgruntled official may well have harbored little inclination to add to his burden by listening to the cries of anguish and providing some relief. With no home in which to live and no means of livelihood, the surviving relatives may well have been reduced to becoming vagrants (*liu min* 流民) or criminals. And here, on some occasions at least, the system of the *jue* itself might serve to alleviate just such instances of distress, as well as crises caused by nature's disruption of the farmers' work. For 57 and 132 CE, our texts record the gifts of *jue* to unregistered persons or vagrants who were willing to submit themselves to registration and its consequential duties, with twelve such grants being made in the intervening years.[41]

III. The Large Estates

Considerable evidence, couched sometimes in general terms rather than with reference to particular situations, shows that alongside the allocation of plots of land by means of the system of *jue*, large estates existed under the occupation or control of persons whom our historians describe in the terms of *hao jia* 豪家 or *hao zu* 豪族, i.e., magnates, or that of *bing jian* 并兼, concurrent tenure of a number of plots. Our sources do not inform us whether the term *bing jian* necessarily implied plots that were separated from each other. If they were spatially connected, they could easily have formed a single enclave that would be subject to supervision and control by the landholder more easily than if they were separated.[42] By Eastern Han times it certainly seems likely that some of the large estates were indeed formed by one single expanse of territory, such that the landholder might well be able to act with an independent and comprehensive control. But while the term *bing jian* may be understood in this way, it applies to the actual, rather than legal, conditions in which the land was held. We cannot say for certain how far occupation of large areas carried with it any idea of a right of ownership or of any restrictions that the statutes and ordinances might impose in legal terms.[43]

In the hands of a single, powerful magnate, tenure of large tracts of land might well have caused hardship to the poorer neighbors, and we shall hear of protests that were registered against it. Such complaints, like calls for limiting the size of the estates, seem to have been voiced more openly in Western than in Eastern Han times. Apparently, in Western Han the growth of the large estates was indeed arousing some criticism. By Eastern Han times it had become accepted as a normal or unavoidable, if in some eyes deplorable, condition of the countryside.

As early as ca. 100 BCE, Dong Zhongshu 董仲舒, who held no high office, had protested against the imbalance in the landholdings of the rich and the poor.[44] At the accession of Aidi in 7 BCE, Shi Dan 師丹, the Imperial Counsellor (*yushi dafu* 御史大夫), raised the same question, blaming the imbalances on the practice of *bing jian*. There followed a proposal made by Kong Guang 孔光, the Chancellor (*chengxiang* 丞相), and He Wu 何武 the Imperial Counsellor (*da sikong* 大司空), to restrict the extent of land named for an individual (*ming tian* 名田) to thirty *qing*. Some of the leading families of the day, such as the Ding 丁 and the Fu 傅 clans, who were headed by an emperor's consort, stood open to suffer from such a restriction and prevented its acceptance.[45] Earlier, during Xuandi's reign (74–48 BCE), we hear a hint of the oppressive behavior that some of these landlords were wont to exhibit. "Better suffer the treatment of the senior officials than that of those magnates," some said.[46]

Early in Eastern Han, General Ma Yuan 馬援 (14 BCE–49 CE), a close ally of the Eastern Han founder, apparently enjoyed huge profits from large landholdings. He is said to have been a large-scale stock breeder of cattle, horse, and sheep and to have harvested grain by the ten thousand *hu* 斛.[47] As sons of Ma Yuan and half-brothers of Mingdi's Empress Ma 馬, Ma Fang 馬防 (d. 101) and Ma Guang 馬光 (d. 91) both commanded considerable wealth and bought fine, fertile land close to Luoyang, where they erected their mansions and presumably grew wealthy from the produce.[48]

We may also consider how these figures compare with those of some of the large landholdings that some few and highly privileged members of the empire enjoyed, but we do so with some caution, as the sizes given for such landholdings can in no way be checked.[49] We read of the 300 *qing* held by Fan Zhong 樊重 (ca. 20 BCE to ca. 20 CE); the 400 held by the family of Zhang Yu 張禹 (d. 5 BCE); the 700 held by the forebears of Yin Shi 陰識 (d. 59 CE); the 800 held by Liu Kang 劉康, son of Guangwudi; and the 400 held by Zheng Tai 鄭太 (152–192). In theory, these major estates would have been capable of producing grains between 30,000 and 70,000 *shi*, respectively; those of average size, between 6,500 and 7,000 *shi*.[50] Professor Ebrey estimates that the average holdings of 144 CE reached between 65 and 70 *mu*; a different estimate gives a figure that is twenty times greater, of 1.4 *qing*.[51]

It may well be asked by what means it was possible to gain occupation or control of these large estates. Occasionally, perhaps in response to a particular need or emergency, it was possible to acquire one of the orders of honor by supplying grain for official relief in times of hardship or by delivering it for the supply of troops at the borders. Alternatively, in 158 BCE, permission was granted to sell orders of honor.[52] It may well be asked whether transactions of these types carried with them occupation of the land that had accompanied the original grant of the *jue*, and, if land was transferred in that way, whether this was one means of building up a large estate.

The grant of high-ranking orders of honor during Han times may well have resulted in bringing large areas of land under the control of a single holder and thus setting a precedent for what came to be known as *bing jian*. We cannot know how far large estates actually existed during the short-lived Qin Empire, but we read of the allegation made by Du Du 杜篤 (d. 78 CE) that some of the rich men who had been accumulating estates had been actively oppressive and causing disorder during Qin times.[53] References to such practices in Western Han accuse the *bing jian* men of profiteering.[54] Around 127 BCE Zhufu Yan 主父偃, Counsellor of the Palace (*zhong dafu* 中大夫), proposed that those with multiple estates (*bing jian*) should be moved to the newly established settlement of Maoling 茂陵, with a view to reducing their pernicious activities and influence.[55] At a time of popular distress ca. 120 BCE, consideration was given to lessening the power and influence of such persons.[56] Around 10 CE, Wang Mang ordered steps to be taken to suppress the *bing jian*,[57] and Huan Tan 桓譚 (43 BCE–28 CE) submitted a memorial to Guangwudi advising him to take such measures.[58] We nonetheless hear again of their arbitrary activities during Hedi's reign (88–105).[59] Writing with the forceful views of an independent mind toward the end of the dynasty, Zhongchang Tong expressed his views clearly, complaining of the wealth that the magnates flaunted, with their mansions scattered around everywhere and their fields linked together throughout the land. Not so much like the leaders of a household or of a group of small residents (*wu* 伍),[60] they could call upon the services of a thousand houses and famous estates, so that their own glories and pleasures exceeded by far those of the entitled lords of the land (*feng jun* 封君), with their power and influence no less than that of the governors and the magistrates.[61]

Some of the large landholdings may well have come into being thanks to sale and purchase, but we cannot be sure how such transactions took place, what was the extent of land that changed hands in this way, or how common a practice this had become. In all probability, the sale of land became possible or even commonplace in Zhanguo (475–222 BCE) before a unified empire. We hear Dong Zhongshu (ca. 198–ca. 107) ascribing permission for such transactions to Shang Yang 商鞅, prime minister of Qin (d. 338).[62] We may recall, as noted above, that purchase of one of the orders of honor may have brought with it its allocation of land.[63]

We have a further reference to the practice for late Western Han.[64] Appointed Imperial Counsellor in 44 BCE, Gong Yu 貢禹 put forward a cogent account of the economic distress of his day. He alleged that, to avoid payment of the poll tax in coin, parents were putting their children to death at birth. At the same time, far too much attention and energy were being spent on mining metals and on manufacturing coins. The way to wealth lay in trade; many farmers, unable to find the tax with which to meet the demands that they faced, were deserting their fields altogether. Even if the peasantry were to be given land, they would be likely to sell it cheap and take to commerce. Presumably, it was the rich merchants who were accumulating their land holdings by such purchases. Conceivably, with the growth of lands that were held and managed in private hands, land allocation by means of the orders of honor may have been becoming a less meaningful and effective means to increase the produce of the fields; in some areas, in all likelihood, it had become less easy, or even impossible, to find land that lay outside the hands of large estates and was, in fact, available for such allocations.

The question arises of how those persons who had acquired a control of large landed estates contrived to administer them, maintaining law and order over a large number of inhabitants, and organizing the work of the farmers in such a way that they themselves would benefit.

IV. *Gong tian*, *Ming tian*, and *Si tian*

When our early historians used the terms *gong tian* 公田, *ming tian* 名田, *si tian* 私田, and *zhan tian* 占田, they were writing in the full expectation that their readers would be familiar with their meanings and implications; they thus felt no need to explain what these terms signified. Without such a basic knowledge, not to mention a definitive statement on which moderns can call, historians today are obliged to search our sources for hints that will lead to a clarification.

We find a main distinction between *gong tian* and *ming tian*. The *gong tian* were lands, whether arable, scrub, marsh or forest, that had not yet been assigned for particular use that lay within the three metropolitan divisions, the commanderies, and kingdoms. These *gong tian* were distinct from lands that had been specially treated by allocation, gift, by way of reward, or as a loan to named individuals (*ming tian*).[65] Since *gong tian* lay at the disposal of the emperor and his officials, occupation of such lands might in effect be a first step toward reclamation. *Ming tian* were lands that had been made over to the care, occupation, and tillage of specially named persons. The term *ming tian* may also have applied to land acquired through purchase, such as the lands of General Ma Yuan and in the examples adduced below. The meaning appears clearly in one of the deeds of land sale that is cited below in the terms *suo ming you* 所名有 ("named as the owner").[66]

In Western Han and at least at the early years of Wang Mang's reign, from 9 CE, *gong tian* signified arable land that was subject to disposal by officials. Some of this was within the parks and surrounds of the imperial tombs and their settlements and was, thus, in the keeping of the Superintendent of Ceremonial (*Taichang* 太常).[67] Possibly, *gong tian* further included lands that had been open to activities of people such as the seminomadic Qiang 羌 and had been taken over by Han military forces.[68] *Gong tian* could be made over to another person by way of a gift, for example by Wudi himself, on the discovery of his long-lost half-sister;[69] to Su Wu 蘇武 on his return to Chang'an from detention by the Xiongnu (81 BCE);[70] or to Cheng Zhong 成重, a potential rebel who had submitted to the Han in 2 CE.[71] *Gong tian* could also be lent out to impoverished families, together with the seed and food that they would need to survive and make a livelihood and farm (as in a case from 67 BCE).[72] (We are not certain how those who actually worked these lands were taxed.)[73] When Zhao Guo 趙過 ordered cultivation of *gong tian* within the Three Metropolitan areas (ca. 85 BCE), he may have been using such lands to try out the experimental methods of agriculture that he was initiating.[74] A report says that in 29 BCE Wen Shun 溫順, Commissioner for the Lesser Treasury (*shaofu* 少府), was charged with buying *gong tian* and imprisoned. This incident may suggest that at times official supervision of these lands was irregular or lax.[75]

In 9 CE, Wang Mang, newly enthroned as emperor, claimed that at an earlier stage he had ordered *gong tian* to be distributed and its work organized according to the ancient and revered (but partly fictionalized) *jing tian* 井田 "system."[76] References to *gong tian* are less frequent for Eastern Han than for Western Han. Some *gong tian* was granted to the poor in 66 CE,[77] and parts were loaned temporarily to those who had been afflicted by a natural disaster in 107.[78] Following a cattle plague in 84 CE, an imperial decree provided for landless persons who were willing to be allowed to move to fertile lands, with the gift of *gong tian*; they were to work there as hirelings, with the loan of seed and food.[79] Zhi Shou 郅壽 (d. 91?), deputy Director of the Secretariat (*Shangshu puye* 尚書僕射), faced a charge that he had made a formal request to buy some *gong tian*.[80]

The term *ming tian* occurs twice in the *Shiji* and five times in the *Hanshu*, but it does not appear in the *Hou Hanshu*. It seems to signify lands that were held in a particular category, being registered in the name of an individual. The term was not applicable to lands held by the magnates (*hao zu*) holding great estates. Merchants were debarred from holding *ming tian*.[81] In 7 BCE, some suggested that the *ming tian* that lay in the hands of the king, nobles, and princesses should be limited to 30 *qing*.[82]

Used perhaps in the same way as *ming tian*, the term *si tian*, which is seen but rarely, first appears in a passage from the *Odes* classic with a pronounced distinction from *gong tian*.[83] In a spirit of public service, the poet calls for rain to fall first on the *gong tian* and only then on the *si tian*. This passage was cited to support arguments advanced by the ritual experts Xiao Wangzhi 蕭望之, around 60 BCE, and by Zuo Xiong 左雄, around 126 CE.[84] Gu Yong 谷永 (d. 9 BCE), a fearless critic of Chengdi, upbraided that emperor for setting up *si tian* among the common people; he also deplored the purchase of *si tian* by some of the kings.[85] Several steps were taken in 115 CE to repair some water channels, with a view to improve irrigation for the *gong tian*, *si tian*, and *min tian* 民田.[86] Land held by Liu Kang 劉康 in 83, whose site remains unidentified, was described as *si tian*.[87]

V. The Vagrants

We have noted the apparent discontinuation of regular grants of orders of honor, with their allocations of land, from 147 CE. We may also observe a further change that may date a little earlier. It concerned both the disposition of land and the solution of one of China's long-lasting problems, that of the vagrants.

These were the *liu min* who had been dispossessed of their farms and livelihoods, thanks to violent acts of nature, such as a flood, or the incursions of non-Han peoples, riding at will into the Han countryside. They constituted a problem that had afflicted the dynasties all too frequently and that an imperial government could ignore at its peril, because groups of homeless, starving families in flight from such conditions could only be expected to turn to a life of brigandage and violence.

Less attention is reported in this connection for Western than for Eastern Han, with some increased activity from 48 BCE.[88] Vagrants who had deserted or been driven from their official habitations now lacked legal recognition of their existence. On occasion, they were allowed to register themselves afresh, evidently obtaining by this the possibility of earning a living by working on the land. In Western Han, they were allowed entry into the Metropolitan Area from the east, through the passes. Some scholars have suggested that they may have been given a livelihood in the lands attached to the settlements built around the imperial tombs.[89]

Positive action took place far more regularly and frequently during the first hundred years of Eastern Han. Orders provided a grant of *jue* and its allocation of land for those who were willing to accept registration together with its obligations.⁹⁰ Such measures were taken on fifteen occasions between 57 and 132; they are not reported subsequently.⁹¹ We also read of relief measures for these displaced persons in 100, 102, 103, and 108.⁹²

Varying results followed attempts to rehabilitate vagrants. As governor of Danyang, a commandery on the periphery whose standard of life and cultural values were not as advanced, Li Zhong 李忠 is credited with registering 50,000 vagrants shortly after 30 CE.⁹³ Again, in the reign of Guangwudi, we hear of vagrants for whom agricultural settlements were established in the distant regions of Guiyang commandery.⁹⁴ Thereafter, we read of such successes twice, around 155 and 190.⁹⁵

We may note the almost complete absence of mention of efforts that Han officials took to solve the problem of the vagrants after 132. We recall that grants of *jue* were nearly halted after 147 and ponder whether such changes were due to a lack of firm government or to the weakness of provincial officials, or both. Some contemporary accounts would have us believe that they resulted from the actual retention of large landed estates in private hands. Such landholders may themselves have faced a problem of a type that was utterly at variance from that of the relief of the *liu min*, but that may have contributed to a partial solution of the problem. The landholders needed to find the labor with which to make their lands productive. Possibly they could do so by providing hospitality and a livelihood for some of the vagrants left to their own devices by local officials, who offered no measures of relief. In such circumstances, whatever the demands that the landholders may have made on the vagrants, these would not have been subject to the orders and dispositions of the officials.

Such a suggestion must remain a matter of speculation. We cannot necessarily expect the court-compiled histories to tell of successful, even praiseworthy arrangements set up by rich individuals, which ignored or defied the statutory procedures whereby officials were to regulate the tenure of land. We may also wonder whether some of the vagrants may have found relief by joining some of the popular movements, whose leaders promised a life of plenty or individual salvation after death. There are hints that the leader of the Yellow Turbans sought to attract them to join their communities.⁹⁶

VI. Inheritance and the Terms of a Will

As shown above, the Zhangjiashan Statutes and Ordinances of 186 BCE provided for near relatives to inherit the orders of honor held by a head of household, and we can only presume that those rules carried with them the use of the land that had been allocated to him. To this we may add the slight evidence on this subject seen in the rare, perhaps unique, example of a will, dated on the day *xinchou* 辛丑 of the ninth month, whose first day was *renchen* 壬辰, of the fifth year of Yuanshi 元始, corresponding with 31 October 5 CE.⁹⁷ The will sets out to provide for the livelihood of some of the dying testator's relatives, by conveying to them the use of plots of land with a standing crop (*dao tian* 稻田), land stocked with mulberry trees (*sang tian* 桑田), and land under water (*po tian* 波田).⁹⁸ Neither the position nor the extent of such plots of land are stated. They were to be made over for a period of a few months, with the provision that those who received them would not sell them. We may only suppose that this provision rested on the charitable intentions of the testator, who was in no position to make more than temporary arrangements.

The implications of the term *yu* 予 that has been rendered above as "convey" are by no means clear, but, as the will stipulates that the land was to be returned after a few months, it cannot be regarded as involving ownership, whether permanent or not.⁹⁹ Rather than signify a gift, in the present instance *yu* would seem to carry with it a temporary provision for maintenance of a livelihood.

The will states a request that is addressed to Du 都, an elder (*san lao* 三老) of the county (*xian* 縣) and its subordinate district (*xiang* 鄉), to Zuo 左, a salaried clerk (*you zhi* 有秩) of the district, and to Tian 田 and Tan 譚, two leaders of the village (*li shi tian* 里[師]田), to draw up the will in writing.¹⁰⁰ The witnesses named at the end of the document include the leader of the village, a member or members of the security group (*wu* 伍), and named relatives.¹⁰¹

The use of land in this way was to be brought about by the natural leaders of the countryside. The will was drawn up by three low-ranking clerks in response to the request of the testator, who, in all probability, was illiterate and needed their help. They were in no way performing duties that fell to them as junior members of the establishment of officials, who held a specified responsibility for administration. The will should be seen as representing the practical arrangements that were made on a personal level.

VII. Deeds of Purchase (*quan* 券)

We have paid attention to the regular allocation of land by officials, the tenure of holdings by the kings and nobles within the estates that they administered, the distribution of what are termed *gong tian*, and the attachment of holdings to named individuals. We have also noticed the occurrence of a few references to the purchase of land, whether by person to person or from officials' holdings, and we have considered the provisions made for inheritance and the slight evidence that we have for the force of a will. We have, however, carefully avoided use of terms such as "landed property," "ownership," or "possession," in view of the differing implications that such terms carry. Unable as we may well be to formulate any definite conclusion on the question of whether or in what way land was deemed to be property in Han times, we must take account of several factors that affect this question: the presence of graves on the land; the provisions for the inheritance of a farmer's orders of honor; and the existence of written deeds that assert an act of purchase, presumably with consequent ownership.[102]

We may first consider what may well have occurred on the death of the farmer who had been cultivating the plot that had been allocated to him, although it will entail a degree of speculation. The immediate concern of the household would be to provide burial in a manner that would satisfy two demands: first, a wish to comply with the normative ideal of devotion to a parent (*xiao* 孝) through the maintenance of rituals that they were taught; and second, the need for speedy encoffinment and burial in a hot climate. The farmer would doubtless be buried on or near the land on which he had worked; at the same time, with no other means of sustenance, his family would probably need to continue to earn their living by cultivating the same land. But they had lost the presence of the person to whom the land had been allocated.

The Zhangjiashan Statutes of 186 BCE recognize the existence of this very problem. They provide that the heir of a deceased holder of the *jue* himself receives such honors at that person's death, to some degree.[103] At the top of the scale, the heir retains the same position on the scale, as Noble (*chehou*) or Noble of the Interior (*guanneihou*). Lower down on the ranks, the heir receives an honor that was one or two degrees lower than that of the deceased person, for example, the son of a man qualified as *gong cheng* (no. 8 in the list of orders) receives the order of *guan dafu* (no. 6).[104] We can only presume that the allocation of land to the heir was adjusted accordingly. The Statutes also provide that the son of a holder of *jue*, who dies while carrying out duties imposed by officials, succeeds to his *jue*. And if the deceased has held no such order, the heir receives that of *gong shi* (rank no. 1). The legal heir is designated, in order, as the son, daughter, father, mother, or collateral, and, finally, wife.[105] Elsewhere, we read that a widow becomes the successor or heir of the household (*hu hou* 戶後), with its fields and residence.[106]

We have no means of estimating how effectively or universally these legal provisions were implemented, or how far such procedures varied throughout Western and Eastern Han. In evidence, we have a few written documents that may likewise provide for a continued occupation of land, as seen below. With a few exceptions, these do not concern land of a size that would provide a livelihood for a household; mainly, they concerned small plots, isolated for burial.[107]

In this way there were drawn up "deeds" (*quan* 券) that asserted the purchase of the land in question and with it, its occupation. Such deeds recorded the names of the purchaser and vendor and the situation of the land to be bought in relation to neighboring plots or roads. They named the price paid for the plot and recorded the fact that the sum had been duly paid; they declared that the produce of the land would belong to the purchaser and that the bodies or skeletons of any persons already buried in the plot would become his servile dependents. The deed named the witnesses and closed with the order that strong drink had been provided for their refreshment. Hong Liangji 洪亮吉 (1746–1809) wrote of such deeds being prepared ahead for a person's burial, inevitably with inflated values for the land and a call on heaven, earth, sun, and moon to bear witness. "Altogether laughable," he wrote, "but the practice had started in Han and Jin times."[108]

The wording and formulae seen in these deeds are also seen in documents that derive from other transactions, as may be seen in a number of records from Juyan 居延.[109] Some of these refer to articles besides land, such as items of clothing.[110] One example from Juyan, where the year is not recorded and the date is given as the seventh month, provides for payment to be completed by the twelfth month.[111] At least one other strip concerns the purchase of land.[112] This was for an area of 35 *mu*, for a price of 900 cash (*qian*). The record on the strip adds that, when measured, the land was found to be short, and the money was returned.

The deeds that are under consideration here were at times included within the grave, either as inscriptions on its building or as separate documents, engraved or written on stone, lead, brick, paper, or even jade. The deeds concerned the land on which the tomb occupant had been buried. Exceptionally, one concerned the purchase of land that would provide a farmer's living, either by sowing crops or keeping livestock.[113] The "tomb deeds" (sometimes called "tomb contracts") were couched in what seems to be legal terminology, often drawing on the same formulae.[114] They included dates, but may have done so retrospectively, and in a number of cases these are expressed inaccurately and cannot be valid. They may include references to mythological or fictional beings. The situation of the plot in question may be described in precise terms, and the deed may mention the existence of markers in the form of stones set up to indicate its boundaries. The figures of cash given for the purchase are in round numbers. Such documents would serve to identify the land and its "ownership," if that expression may be used. We may well imagine that a household might well be tempted to have a deed such as the following drawn up. The rendering that follows is tentative.

> 8th day (*yihai* 乙亥) of month 5, whose first day was *bingzi* 丙子, in the year 1 (*renshen* 壬申), of Huanglong 黃龍 [49 BCE]; one plot of extensive land in the quarter known as Qingluannian 青欒年, [acquired] by sale by Zhuge Jing 諸葛敬 from Ma Jiqing 馬吉慶, adult male, of Nanyang 南陽, named as the owner. Price 21,000 cash; cash delivered this day. The land adjoins [that of] He Fang 賀方 to the east and Shen Dayi 沈大義 to the south; it extends up to the great road in the west and adjoins land of Zheng Jiang sheng 鄭江生 in the north. Herbal produce and game to belong to Zhuge Jing; corpses that might be in the plot are to become the dependents, male or female, at Zhuge Jing's express service. Large stones will mark the boundaries east, west, south, and north. One-half measure of strong drink is for the attendants Ding Yang 丁陽 and Guo Ping 郭平, each being cognisant of the agreed terms of the deed.[115]

Irregularities in the dating and perhaps some other features suggest that this document was not drawn up in 49 BCE.

Surviving deeds during Han times carry the dates of 68, 49 BCE, and 56, 76, 81, 125, 169, 171, 176, 178, 179, 182, 184, 188, 198; perhaps the one case that is best known is from tomb no. 2 at Wangdu 望都 (modern Hebei province). Post-Han examples are known at least for 245, 274, 284, 439, and 442, and at least twenty-seven additional examples are dated between 837 and 1614. Some of the inscriptions include errors in the form of identifying a date in inaccurate terms or in a way that is at variance with the calendar. In some cases, there are comparable difficulties over place names that are mentioned. Such anomalies have aroused the question of whether some of these deeds should be regarded as authentic or as forgeries.[116]

These documents are anything but records of transactions that were sanctioned officially in accordance with legal provisions. However, they may be taken as evidence of actual practices that concerned the disposition of land. Rather than categorize the documents as "forgeries," we should see the documents as fabrications that were drawn up to sustain what had been or actually was taking place. So, while they may well include inaccuracies, they demonstrate ways in which some land passed from person to person in return for a payment of cash. There would be little reason to create such "tomb deeds" designed to validate certain practices of buying land if, in real life, similar or the same practices did not hold. Probably some of these deeds were fabricated long after the dates that they cite, and such examples may be regarded as "make-believe" assertions or prayers rather than as officially attested documents. Their "make-believe" status does not preclude their use as evidence attesting that exchanges of this type were actually taking place on a personal or household level rather than as being contrived by officials and recognized with their approval.

These deeds are described as *quan* 券, a term possibly signifying a document that had been engraved and then split into two parts, with each party holding one, until the time when the two parts, fitted together, would prove authenticity.[117] The suggestion is fascinating, as such a practice is immediately comparable with that of medieval starrs and charters of Europe, but no positive evidence has been cited to validate that documents were split in this way and for this purpose in China's early empires. Certainly, the deed that was dated 274 (see below) twice mentions breaking the deed, yet it appears to have remained intact.[118]

The surviving examples of these deeds are all on durable substances. None have survived on wood, silk, or paper, and, while such examples may yet be found, the choice of stone or metals for inscribing the texts may suggest a wish to ensure their more permanent survival. We may, however, note that such an intention did not prevent the general practice of inscribing official texts, such as the statutes and ordinances, on wood or perhaps silk. There are marked differences in the layout and style of writing between these "tomb deeds" and official documents, as the latter were inscribed on carefully measured strips of wood with full attention to reproducing a polished or formal type of calligraphy. The standard of script that is adopted in these documents is markedly inferior to that of official reports found at Dunhuang 敦煌 and Juyan 居延. In official writings, considerable care accompanied the expression of dates to obviate mistakes, but such precautions did not prevail in the composition of the tomb deeds, one of which includes a *nianhao* that had been replaced some months earlier.[119] No officials are named as taking part in the sale of these plots of land; reference to the Han statutes and ordinances, such as appear fairly regularly in official statements, are seen but once in the tomb deeds, in a document dated to 245 CE.

The deeds conceivably resulted from actual arrangements made by individuals on a strictly private level, possibly in preparation for a death. Some of them may be retrospective and perhaps in that sense spurious. It need not surprise us that there are no records of officials who took a part in these transactions, because the Han statutes and ordinances probably did not countenance such arrangements and may have forbidden them. A named official of senior rank, that of Governor of a commandery, is indeed mentioned in one deed, apparently conniving in the purchase of land for his own satisfaction. His name does not appear elsewhere in contemporary writings, and we may wonder whether he had retired from an official career in honorable circumstances or been obliged to leave under suspicion and disgrace.[120]

To summarize, we suggest that the inaccuracies that may be seen in these deeds, so far from proving that they are fabrications, may well have arisen in the course of originating these documents. These were often made retrospectively to assert retention of a site on which a relative or an ancestor had been buried, rather than as legal statements of a right to possess landed property. One may postulate that, in some cases, the purpose of making these inscriptions was to identify one site as belonging to a family's ancestor, particularly when a large number of graves were situated close together.[121]

VIII. Officials' Salaries

We may consider such figures as we have noted above in connection with the large estates in light of the scale for official salaries in the empire. In whatever way salaries were paid, whether in grain or cash or partly in both, the notation of their grades is in terms of measures of grain.[122] These applied to the total number of officials of the Han empire, close to 130,000 in 5 BCE, of whom some 30,000 staffed the central offices in Chang'an and 100,000 served in the commanderies and kingdoms.[123] The official posts were graded in a series of some twelve ranks, which were notated from "full 2,000 *shi*" at the top to "equivalent to 100 *shi*" at the end of the scale, with a few exceptions at each end.[124] Without full information at our disposal, we may give examples of a few correspondences:

Land Tenure and the Decline of Imperial Government | 75

Official post	Grade (*shi*)	*qing* needed to supply income	Reference
Commissioner for ceremonial (*Taichang* 太常)	*zhong* 中 2,000	20+	*Hou Hanshu* (tr.) 25.3571
Governor of a commandery (*Taishou* 太守)	2,000	20	*Hanshu* 19A.742
Magistrate, large county (*Ling* 令)	1,000	10	*Hanshu* 19A.742, *Hou Hanshu* (tr.) 28.3622
Senior Director, Archives (*Taishi ling* 太史令)	600	6	*Hou Hanshu* (tr.) 25.3572
Magistrate's staff	400–200	4, 2	*Hanshu* 19A.742, *Hou Hanshu* (tr.) 28.3622

It is not possible to calculate the total cost of the salaries paid to civil servants of the early empires, nor can we suggest an average figure of an official's income that can be applied with any certainty. Were the figure of 200 *shi*—right at the lower end of the scale—to be taken as an example, and were this to be interpreted as denoting the actual figures of the produce of the land, the cost would have been no more than a fraction of what was available. The total cost of officials' salaries would have been 26 million *shi*. If Chao Cuo's figures were to be taken as being applicable, this would correspond to the produce from 260,000 *qing* of land worked by 260,000 households. A figure that the *Han shu* gives for the total extent of land that was available for cultivation or reclamation, presumably for the end of Western Han, is 32,290,947 *qing*.[125]

IX. Size and Spread of the Population

For a few select years in Eastern Han, we possess figures from various sources for the numbers of households and individuals whose existence was recorded, and in some cases for the extent of the arable land. It is questionable how far these figures for households and individuals derived from attempts to count the whole population of the empire. They should be seen principally as the results of registers that were drawn up with a view to administering two forms of taxation, the one levied on the produce of the arable land and the other as a poll tax on those who worked it. There were, however, a number of categories of persons who may not have been included in such counts, such as the kings and nobles and their entourages, officials, merchants, craftsmen and laborers in the cities, conscripts, slaves, deserters, and vagrants. Some of these persons may have been included in the figures that we have for a mere ten of the counties of Western Han, some part of whom were engaged in activities other than farming.[126]

There are also reasons why the figures that we are given may not allow definite or general conclusions as to how the occupation of the land grew or declined. They survive for a few years, spaced at random, rather than at regular intervals of time; as a result, they cannot show a pattern or sequence of developments, nor are the examples that survive complete. There is one further weakness. For 1–2 and 140 CE, we have figures for each of the major administrative divisions of the empire, the commanderies and kingdoms. In many cases the titles of such units remained unchanged. However, direct association or comparison may not always be possible, as the extent and areas of those units were by no means necessarily identical for the two occasions when those figures applied.[127]

Hans Bielenstein observed a considerable change in the spread of the registered population, as is apparent from the counts of 1–2 and 140. The proportion of those registered north and south of the Yangzi River was that of 7.6 to 2.4 for the earlier count, and that of 5.4 to 4.6 for the later one.[128] With the caution that is due to the weaknesses that have just been mentioned, we may amplify this observation by noting a significant and pervasive difference, with some exceptions, in the ratio between individuals and households. Thus:[129]

(a) In the northern belt of the empire, running from Dunhuang to Liaoxi, the number of individuals who were registered and counted was less, while the households included more members. This area included Dunhuang, Jiuchuan, Zhangye, Wuyuan, Yanmen, Yunzhong, Dingxiang, Daijun, Shanggu, Youbeiping, Liaoxi, Liaodong, with exceptions (more individuals, fewer members of the household) for Wuwei, Yuyang, Xuantu, and Lelang. There was a major decrease in the number of individuals for Shuofang, where the ratio of household to individual remained at 3.9. Major differences in the household/individual ratio are seen:

	1–2 CE	140
Yanmen	1:4.0	1:7.8
Youbeiping	1:4.8	1:5.8

(b) The same anomaly applied to the central area south of (a), running from Jincheng to Hongnong and Shangdang, and from Changshan and Henei to Donglai and Beihai, and to Xiapi and Runan. This area included, to the north, Anding, Beidi, Shangjun, Xihe, and Taiyuan; to the west and center, Longxi, Youfu feng, Zuoping yi, Jingzhao yin, Hongnong, Lujiang, Hanzhong Hedong, Shangdang, Guanghan, with exceptions of the same type for Jiujiang, Wudu; to the east, Changshan, Zhuojun, Bohai, Julu, Qinghe, Lean, Weijun, Qiguo, Taishan, Langye, Henei, Dongjun, Jiyin, Shanyang, Donghai, Henan, Chenliu, Yingchuan, Zhongshan, Huaiyang, Dongping, Luguo, Chuguo, Peiguo, Runan, Xiapi; with exceptions of the type already mentioned for Anping, Pingyuan, Beihai, Donglai, Guangyang.

(c) To the south of the Yangzi the reverse conditions applied, with more individuals being registered but the households including fewer members. This area included Nanyang, Jiangxia, Shujun, Nanjun, Wujun, Danyang, Kuaiji, Yuesui, Zangke, Lingling, Wuling, Bajun, Changsha, Guiyang, Yuzhang, with exceptions (more individuals and more members of the household) for Yizhou, Lujiang, Hejian.

(d) In the extreme south (Nanhai Cangwu, Hepu, and Jiuzhen), there was a marked increase in the numbers of individuals and households, with a decline in the number of individuals in the households. Rinan was exceptional, with fewer individuals in the households of 1–2 CE (4.4) as compared with 5.5 per household in those of 140. (Figures for Jiaozhi and Yulin are not complete.)

In general, the proportionate numbers of individuals in the households did not vary greatly. Major changes occurred in the following cases (subject to the caveat that there may have been changes in the extent of the areas concerned). In the central area and east, and two commanderies in the northern belt, there were large increases, from a minimum of 2.7 to a maximum of 7.8:

	1–2 CE	140		1–2 CE	140
Jingzhao yin	3.4	5.3	Yanmen	4.0	7.8
You Fufeng	3.8	5.3	Daijun	4.9	6.2
Hedong	4.0	6.1	Zhongshan	4.1	6.7
Taiyuan	4.0	6.4	Xindu	4.6	7.1
Changshan	4.7	6.4	Hejian	4.1	6.7
Qinghe	4.3	6.1	Guangyang	3.4	6.2
Zhuojun	4.0	6.2	Liang	2.7	5.1
Qi	3.5	7.6			

In Xihe, and otherwise in the south, by contrast, there were huge decreases, from a maximum of 7.2 to a minimum of 3.6:

Henan	6.2	4.8	Yizhou	7.0	3.8
Yuzhang	5.2	3.9	Xihe	5.1	3.6
Guiyang	5.5	3.7	Huaiyang	7.2	4.8
Yuesui	6.6	4.7			

We may suggest some possible reasons for these differences. For the north, the decrease in the number of inhabitants counted and the large size of the households resulted from the infiltration or acceptance of large numbers of non-Han persons to live and move in the areas. Apparently, Eastern Han officials were not registering members of a mobile people. In addition, with such seminomadic peoples active in the areas, there was less opportunity for the second or third sons of a Han farmer to receive an allocation of land and settle there, and, therefore, greater reason, *faute de mieux*, for them to remain with the households where they had been raised.

For the intermediate areas further south and in the east, officials were becoming less active and efficient at times when individuals were able to occupy and work the land, perhaps independently of a government's control. If one may speculate, the disruption that was caused by the huge change in the Yellow River's course in 11 CE may have affected the growth of private estates in the east. This could well have tempted particularly enterprising and courageous individuals to occupy lands that had been deserted so as to reclaim them for farming.

For the third area, where the numbers of registered inhabitants had increased while the numbers of the households' members had decreased, there had been greater opportunities for promoting agriculture by reclaiming land, perhaps with the encouragement of local officials. Conceivably there were greater opportunities for men and women of a comparatively young age, perhaps with no more than one child, to take the initial steps of earning a livelihood on a newly acquired farm. Possibly the figures reveal that a broad migration had been taking place. We however have no means of determining how far an extension to the south was due to the deliberate policies, the leadership and supervision of Han officials, or the enterprise of individuals who possessed the initiative to move south and establish estates over which they could hold control. The sources provide one hint, perhaps. Around 150 CE, Cui Shi 崔寔 recommended the removal of some population from densely populated areas, but we are never told whether any positive steps followed his recommendation.[130] Possibly such a move brought migrants into lands that were not so closely under the control of officials as those elsewhere, making an actual occupation of lands and establishment of estates easier in such conditions.

A further significant feature is apparent in the figures that we are given for the years in which all the emperors up to Zhidi died (146 CE); as they are absent for such subsequent occasions, it appears that they were compiled before the death of Huandi in 168 CE. They may derive from the work of one Fu Wuji 伏無忌, who was one of those men who were commissioned to prepare historical records around 151. He suggested that on these occasions the figures for individuals, households, and arable land had been inflated.[131]

Date	Individuals (kou 口)	Household (hu 戶)	Ratio of hu/person	Extent of arable (qing 頃, mu 畝, bu 步)	qing per household
57	21,007,820	4,279,634	1:4.9		
75	34,125,021	5,860,573	1:5.8		
88	43,356,367	7,456,784	1:5.8		
105	53,256,229	9,237,112	1:5.7	7,320,170, 80, 40	1.2
125	48,690,789	9,647,838	1:5.0	6,942,892, 13, 85	1.3
144	49,730,550	9,946,919	1:4.9	6,896,271, 56, 194	1.4
145	49,524,183	9,937,680	1:4.9	6,957,676, 20, 108	1.4
146	47,566,772	9,348,227	1:5.0	6,930,123, 38, –	1.3

To these we may add the following figures, as given in the *Tong dian*:[132]

57	21,007,820	4,270,654	1:4.9
157	56,486,856	10,677,960	1:5.2

The foregoing figures may be compared with the figures for the end of Western Han (1–2 CE) and those given in the *Xu Han zhi* 23.3533) for 140 CE:[133]

1–2	57,671,400	12,366,470	1:4.6
140	49,150,220	9,698,630	1:5.0

They may also be compared with figures that have been calculated for total number of inhabitants of the Roman Empire:[134]

14 CE   45.5 million
164 CE   61.4 million

As given, these figures show a dramatic rise in the numbers of individuals and households registered between 57 and 75, 75 and 88, and 88 and 105, followed by a decline especially between 105 and 125, continuing until 146. The differences in the figures that we have for Han, which amounted to eight millions or one sixth, cannot necessarily be taken to be due to purposeful migrations. This change, moreover, is in contrast with the numbers of households, which rise steadily from 57 to 145. We may suggest, but cannot verify, that this anomaly arose from the unrestricted intentions of those men who occupied the large estates.

We may speculate that it would have been in the interests of such men to discourage or even prevent officials from observing the actual conditions that prevailed in their lands. In addition, it could well have been in the interests of the officials to spare themselves the tiresome work of carrying out such inspections, provided that they escaped a charge of misreporting. We have noted above the expressed preference for working with officials rather than with the landholders, thus suggesting that it was the latter who were more demanding. We may ask whether, in such circumstances, it suited both the officials who were drawing up the reports and the landholders for the latter to provide the figures needed for the registers and the tax dues. In such circumstances, the landholder might well be tempted to reduce the number of individuals attached to his household and thus the liability for paying poll tax *suan* 算 in cash out of his pocket. On the other hand, a landholder might happily increase the number of households, providing the land tax *zu* 租 in produce, as he could extract this tax from the other farmers to a large extent when necessary.

There may perhaps be another explanation for the anomaly whereby the numbers of registered individuals declined while those of the households increased. From the time of the Qin kingdom (in the sixth century BCE), there existed a provision that a household should not include more than one able-bodied male, with the intention that any others would be settled elsewhere as productive farmers. The extent to which such a provision was operative will surely have varied greatly over time. Possibly the figures suggest that by the time they were drawn up, the central government in Eastern Han was no longer in a position to exercise such a control, with the result that second or third sons remained as members of the household.

X. Conclusion

The evidence considered here for the early empires concerns one vital aspect of imperial government, social security, and public life that affected nearly all of the population and continually plagued successive administrations, judging from the contents of the *Shiji*, *Hanshu*, and *Hanji* 漢紀, with this last a compilation by Xun Yue 荀悅 (148–209) commissioned by the throne whose occupant wanted a convenient digest of the main historical lessons registered in the *Shiji* and *Hanshu*. Again, the impression we receive tallies with other sources: farming and land allocation are matters of utmost importance to the realm's well-being. In Western Han, repeated calls for restrictions on landholding were made, and Wang Mang's dynasty finally acted on those measures. Such measures had dual aims, to "enrich the state" and to protect small landholders, because large landholders all too often resisted paying their fair share of taxes. Nonetheless, the Eastern Han courts never thoroughly implemented them, to ruinous effect. To educate the throne, Xun Yue at the end of Eastern Han repeatedly returned to certain themes, showing, for example, how very low rates of taxation also empowered local magnates and weakened the throne by putting more money into the pockets of magnates who then snatched up the property of indigent farmers. Still worse in Xun's view, the Han courts had never acted to prevent powerful people from acquiring "combined" (*jian* 兼) bases of power in officeholding and landholding. The regulations in antiquity had been simple, he argued: they forbade officeholders and nobles to compete for profit with the ordinary people, because the distinction between *gong* versus *si* (what is in the public interest versus what is not) is the premise on which good governance in every era rests.[135] Xun was equally troubled about migrants, viewing them as a sign of administrative collapse.[136] In general, "whenever selfish acts outnumber those in the public interest, the rules and regulations are limited in their applications, and policies and orders falter, and this by definition endangers the ruler."[137] Yet lawlessness in any form has the power to quickly corrupt society.[138] The throne must act "to strengthen the trunk and weaken the branches,"[139] because the aggregation of land in the hands of the few threatens the good of all.

That Xun Yue had reason to be concerned is suggested by a summary of the findings presented in this chapter: grants of orders of honor with their allocations of land, as practiced from the early days of Western Han, were nearly stopped by 147 CE; relief measures designed to help the impoverished ceased a little earlier; an increase in the numbers of members of households in the north and east of the empire was matched by a decrease in the south; great estates are likely to have grown up from these times rather than from later, as has usually been assumed. Land could change hands by sale and purchase at the individual level, but without the direct supervision or control of officials; and while the extent to which land transfers took place cannot be measured, it is unlikely to have been large, in the author's view.

The concept of land as property that could be owned by individuals applied in no more than minor ways. In general terms, we may suggest that it was in exceptional instances and occasions rather than as general practice that personal ownership of land existed and was acknowledged during Western and Eastern Han. Some lands may have been separated so as to serve a ritual purpose, rather than to promote economic growth, whether for an individual or for the empire. Eventually, large estates were formed under the control of rich men, along with establishments that housed religious communities, Daoist and then Buddhist (not clearly distinguished in this period). It has yet to be shown in what sense and in what conditions those who occupied such estates may be said to have "owned" them, in the modern "rights-bearing" sense.

In the absence of an effective governing body and its officials, people could occupy land without hindrance and take their own steps to prevent encroachment or takeover. Such conditions may well have arisen to some extent in the latter decades of Western Han, then pertained in a still greater degree during the second century of Eastern Han. How far or for how long occupation in this way would survive one generation may well be in question, nor would an imperial government necessarily have recognized that such informal "possession" existed on a legal basis.

Starting under Shang Yang, ca. 350 BCE, a royal and later an imperial government took effective steps to control and promote the occupation of land and to increase its produce. The scholarly consensus holds that, by the very last decades of Eastern Han, the central government had lost sufficient strength and will with which to control the empire effectively. Then the way lay open for the rise of separatist movements, with effective power falling into the hands of individual leaders. This decentralizing process was well underway from 147 CE, somewhat earlier than has been generally recognized. This is seen in the way in which the government's control over landholdings grew weaker, while greater opportunities existed for the private occupation of the land.

Land Tenure and the Decline of Imperial Government | 83

The acquisition of land privately, through purchase, had likewise been introduced in preimperial times, under Shang Yang, if the legends are correct. Its practice increased during Western Han, drawing criticism first from Dong Zhongshu and later from Shi Dan. By the middle of the second century CE, private rather than official administration of the land had become a regular feature in the central and eastern parts of the empire, and it is against such a background that Cui Shi, Wang Fu, and Zhongchang Tong called for a reintroduction of the rigorous yet fair governing methods imposed by Qin. While private initiative may have been advancing economic growth in the south, officials serving in that area may have been consolidating a structure of administration whereby an independent kingdom of Wu could be created.

The foregoing account leaves an impression of a dynastic and political weakness affecting the government of the Eastern Han empire from the middle of the second century, an impression that is strengthened by a number of statements, allegations, or protests that senior officials put forward at the time. At the accession of Shundi (r. 126–44), for example, some senior officials are stated to have been lazy. Zuo Xiong, Director of the Secretariat (*Shangshu ling* 尚書令), wrote of how, long, long ago, the people hated the officials as if they were vermin. Here he was resorting to a well-known rhetorical device, that of assigning the evils, scandals, and disgraces of one's own time to the remote past, whose rulers he could castigate with both vigor and impunity.[140] Zuo Xiong was pointing to the need to improve the quality of officials and the efficiency of their administration. He was protesting against the freedom that allowed officials to "fulfill" their duties in a perfunctory fashion, ignoring some of the illicit activities that were taking place. He blamed officials of the counties and districts, some of whom were enjoying highly privileged positions, to the detriment of the population in their charge. The early historian, by contrast, tends to place the blame squarely on the eunuchs for thwarting any attempts to improve the situation. Magistrates of the counties were being changed each month. Newcomers were welcomed to such posts, old-timers were turned away, and some official posts were being left vacant.

A few years later (from 132 CE), Lang Yi 郎顗, who held no official post, was calling for positive action to improve the quality of the government.[141] He drew a contrast between the officials' laxness when taking steps to care for those in need, and the energy they showed for riding around Luoyang; too many people were being arrested for minor infractions, and the prisons were full, he said.

A further view of the times may be seen in the reaction of Zhou Ju 周舉 to a call for advice at the time of a great drought in 134 CE.¹⁴² Possibly serving in the Secretariat, Zhou argued that greater attention should be paid to the actual situation of the day rather than to its embellishments. The conduct of public affairs should be reformed, and deceptive practices excluded. Palace women who were no longer of service should be removed, the prisons should be put in order, and the outlays for palace banquets should be reduced.

It was during the turmoil that accompanied and followed the downfall of Western Han that the tombs of some of its emperors were despoiled. By contrast and perhaps exceptionally, Shundi's tomb was desecrated shortly after his death.¹⁴³ Such a deed may have been performed simply by robbers intent on the acquisition of loot, as the text implies; alternatively, it may be seen as a determined gesture of protest against the rule imposed during Shundi's reign. While the immediate causes of the desecration are unknown, the act tells of the readiness by some to take little notice of the respect due to the emperor of the reigning dynasty or to take heed of its ability to determine and enforce the punishment that was due.

Liang Shang 梁商, father of Shundi's Empress Liang Na 梁妠, was appointed Regent in 135 CE and succeeded by his son Liang Ji 梁冀 in 141. Liang Ji was eliminated in 159. Our sources paint a picture of the Liang family and particularly Liang Ji exercising power in a usurpatious manner, which authors may have felt a need to express loyalty to the ruling house. One may perhaps wonder whether a more generous view might be taken of Liang Ji's intentions and activities; that of attempting to reassert the power of the central government at a time when its own institutions had weakened and were failing to administer the empire.

Appendix I: The Extent of the Land

The figures that we are given for the extent of the land depend ultimately on the returns submitted by local and provincial officials, and there are known cases when these were not to be trusted.¹⁴⁴ Some reserve is therefore necessary in drawing conclusions from the figures that the *Hanshu* provides for the extent of the land for 1–2 CE, which run as follows (in *qing*), as per *Hanshu* 28B.1640; *Hanshu bu zhu* 28B(2).49a.

Marked area (*tifeng tian* 提封田):	145,136,405
Minus, as non-arable (habitation, roads, mountains, rivers, forest, scrub)*	102,528,889
Potentially arable*	32,290,947
Under the plough (*ding ken* 定墾)	8,270,536

*The text reads *ke ken bu ke ken* 可墾不可墾, with some editions omitting *bu ke ken*; *Hanshu bu zhu* 28B(2).49a.

These figures leave 2,046,033 *qing* unaccounted. They may be considered along with those given in the same passage for the registered population, which, as has been noted above, are not correct.

Appendix II: Examples of Deeds for the Sale of Land

Information is not available consistently for the following select deeds of sale, many of which are treated in Niida, *Chūgoku hōsei shi kenkyū* and Zhang, *Zhongguo lidai qiyue huibian kaoshi*. For general studies of the deeds, see, for example, Li, "Ye tan 'di quan' di jianbie," and Wu, "Handai mai di quan kao." Eight examples of these deeds are included in the library holdings of calligraphy that belong to the Shodō hakubutsukan (Museum of Calligraphy), in Tokyo. Editorial note from Nylan: It is interesting how often the people buying land in the deeds of sale listed below are low-ranking commoners. Perhaps their patrons and administrative superiors had them make "paper purchases," in order to avoid the appearance of official malfeasance.

Known Examples

68 BCE, inscription on stone, describing the purchase of land by Yang Liang 揚/楊量/量, described as Bazhou min 巴州民. Bazhou existed as an administrative unit at the end of Jin and in the kingdoms of Southern Qi (479), Liang (502–56), and Northern Qi (550–77). *Wenwu* 1964.12, p. 61, citing *Gu shi bao shou lu* 古石抱守錄 (not available to the present writer). On the conundrum presented by the dates and location, see Zhu Jiang 1964; Lu Yuesen 2001.

49 BCE, of lead, describing the purchase of land by Zhuge Jing 諸葛敬; dating is incorrect and identification of the year by the sexagenary "stems and branches" (*ganzhi*) system is odd. Huanglong is taken to refer to the reign period (*nianhao*) of Western Han; it might perhaps refer to that of the kingdom of Wu (229–31). Liu, *Xiao jiao jing ge jinwen* 13.76b; Fang, "Cong Xu Sheng mai di quan lun Handai 'di quan' di jianbie," 52 (op. cit.).

56 CE, of lead, describing the purchase of land by Xu Sheng 徐勝, described as *da nu* 大奴, of the office of the Governor of Guangyang, from Gao Jicheng 高紀成, a *nanzi* 男子, at the price of 25,000 cash (*qian*); dating inaccurate. Lu, "Handai Xu Sheng mai di qianquan jianjie," 60; Fang, "Cong Xu Sheng mai di quan lun Handai 'di quan' di jianbie," 52. Now held in the Shandong Provincial Museum.

76 CE, an inscribed stone, describing the purchase of a grave site by six brothers for 30,000 cash; no names or details given; dated as *Jianchu yuan nian* 建初元年. Found in a graveyard in Kuaiji. The text is headed *Da ji* 大吉. *Jinshi xubian* 金石續編 1.2b; Lu, "Handai Xu Sheng mai di qian quan jianjie," 60.

81 CE, inscription on jade, describing the purchase of burial ground for 102,000 cash (*qian*), measured precisely in paces (*bu* 步); dating correct. For the text, see Zhang, *Zhongguo lidai qiyue huibian kaoshi*, 45; *Wenwu* 1964.12, 61, Wu, "Handai mai di quan kao," 23. Rubbing of the jade in the National Library, in Beijing.

125 CE, on lead, describing the purchase of land bought for the removal and reburial of Li De 李德, Governor of Dongjun, the deed to be valid for fifty-six years; date of the deed incomplete and inaccurate. Li De is not known for Eastern Han. *Wenwu* 1964.12, 61.

169 CE, on lead, describing the purchase of land bought by Wang Weiqing 王未卿, described as *nanzi* 男子, from another *nanzi*, both of Henan; dating is correct. *Zhensong tang gu yi wen* 15.26b; Lu, "Handai Xu Sheng mai di qian quan jianjie," 60; Wu, "Handai mai di quan kao," 23.

171 CE, a deed for the purchase of land by Sun Cheng 孫成, described as a *da nu* attached to a government office, from Zhang Boshi 張伯始, a *nanzi* of Luoyang, for 15,000 cash (*qian*). Stored by the Ding 丁 family of Huangxian 黃縣, dating is correct. Chen, *Liang Han jingji shiliao luncong*, 279; Wu, "Handai mai di quan kao," 23.

176 CE, Liu Yuantai 劉元臺; inscription in seven columns on a seven-sided hollow rod of pottery, found in a brick-chambered tomb, Yangzhou; dating correct. Jiang Hua, "Yangzhou Ganquanshan chutu Dong Han Liu Yuantai mai di zhuanquan," 57.

178 CE, on lead, describing a purchase of six *mu* of land by Cao Zhongcheng 曹仲成 from a *nanzi*, to be a burial plot; dating correct. Now held in the Shodō hakubutsukan, Tokyo. Wu, "Handai mai di quan kao," 23.

179 CE, on inscribed lead, declaring a purchase of ten *mu* of land. The inscription, one of the longest yet known whose dating formula is correct, is addressed to Wang Dang 王當 and his relatives, in a different form from other deeds, such as the one dated 49 BCE. From a brick-chambered tomb east of Luoyang, Jiang. "Yangzhou Ganquanshan chutu Dong Han Liu Yuantai mai di zhuan quan," 55.

182 CE, inscription brushed on the brick wall of the multichambered brick tomb no. 2 from Wangdu 望都, Hebei. Some characters are illegible; comparable to the inscription of 179. *Wangdu er hao Han mu* 望都二號漢墓, 13–14, 20 (Fig. 16).

184 CE, on lead, describing the purchase of five *mu* of land by Fan Lijia 樊利家, *nanzi*, from another unknown *nanzi*, at 15,000 cash (*qian*), at a rate of 950 coins for 1,000. Once said to be in the possession of the Wengs 翁 of Jinling 金陵. Luo Zhenyu pointed out the inaccuracy of the text, some of which is unintelligible. Its dating is inaccurate, as it refers to the ninth month of Guanghe 光和 7, despite a change in the reign period (*nianhao* 年號) to Zhongping 中平. See *Zhensongtang jigu yiwen* 15.27b–29a; Lu, "Handai Xu Sheng mai di qian quan jianjie," 60.

188 CE, on lead, describing the purchase of one *mu* of land for 3,000 cash (*qian*) by Fang Taozhi 房桃枝, of Luoyang, described as *Da nü* 大女, from a woman described by the same term. By the contract, bodies on the land are to become slaves or people of servile status, and 950 coins are to be accepted in the exchange as 1,000. By the deed, the site adjoins that of Fan Hanchang 樊漢昌 to the north; Fan Hanchang also is named as one of the witnesses. Dating is incorrect. *Zhensongtang jigu yiwen* 15.29b, 30a; *Xiao jiao jing ge jin wen* 13.76a; Lu, "Handai Xu Sheng mai di qian quan jianjie," 60.

198 CE, on lead, describing the purchase of land by Cui Fang 崔坊, described as *ji zhu* 祭主 (not an official title), no price being specified. The inscription calls on Sovereign Heaven (Huang tian 皇天) and Ruler of Earth (Hou tu 后土) as witnesses; *Xiao jiao jing ge jin wen* 13.76b; Lu, "Handai Xu Sheng mai di qian quan jianjie," 60.

229 CE, see under 49 BCE above.

245 CE, inscription on lead-tin, describing the purchase of a large area of land (4 *qing*, 50 *mu*) by Xiao Zheng 肖整, in a brick tomb in Nanling 南陵 county, Anhui; dating correct. "Anhui Nanling xian Maqiao Dong Wu mu," 978, 1020.

274 CE, on inscribed tin, describing the purchase of burial site by a person or persons unknown for 500,000 cash (*qian*), from Dangtu 當涂 county, Anhui. The inscription twice mentions "breaking the deed." Dangtu xian wenwu guanli suo, 92.

284 CE, inscribed pottery tile, describing the purchase of land by Yang Shao 楊紹, described as *da nan* 大男, for 4,000,000 cash; no sexagenary *ganzhi* dates are specified; text ends with the telling words "a private contract between people, to be enforced as statutes and ordinances" (*min you si yue ru lü ling* 民有私約如律令). This inscription is cited by Xu Wei 徐謂 (1521–93) and Tan Qian 談遷 (1594–1658); Hong Liangji (1746–1809) comments that such deeds were made prior to death. See note 108 below. Li Shougang, "Ye tan 'di quan' di jianbie," 79. Note the stupendous sum mentioned here.

Further sketchy references for the post-Han will be found as follows:

439 CE, on brick, texts of three deeds (transcription not given), in Hubei. *Wenwu* 2005.10, 42–44.
442 CE, two deeds found in a village, in Guangdong. *Kaogu* 1989.6, 566–67.
837–1614 CE, 27 items. *Kaogu* 1987, 3, 223–31, 219.
1243 CE, see *Wenwu* 1977.7, 13.
1345 CE, see *Kaogu* 1989.6, 540.

See also *Kaogu* 1965.10, 529 for two deeds, one not later than 227 and one dated to 262; and *Kaogu* 1965.11, 573 for deeds of the thirteenth century.

Notes

1. See also *Guoyu*, "Zhou yu," 1:06, with the director of farming put first among the officers. Meanwhile, the king supposedly "devoted his attention entirely to agriculture." Editors' note: Bu Xianqun 卜宪群, "鄉論與秩序: 先秦至漢魏鄉里與國家關係的歷史考察," *Zhongguo shehui kexue* 中国社会科学 2018:12, esp. pp. 192–94, shows the changes in local administration, beginning with Huandi's reign, related to the decline of capital power and revenues, in particular the concentration of local decision-making powers in the hands of the magnates, which prompted Cao Cao to "reform" the administrative structure (196–97).

2. Yang, "Great Families of Eastern Han," 113–14. I am glad to record my thanks to Professor de Crespigny and Professor Nylan for their help over many matters and their comments on what follows below.

3. Nishijima, "Economic and Social History," 558.

Land Tenure and the Decline of Imperial Government | 89

4. Figures for the population that are given in *Hanshu* 28A are stated there to be for the year *Yuanshi* 1, i.e., 1–2 CE (*Hanshu* 28A.1543), and these dates are followed here; Zhou Zhenhe, *Xi Han zhengqu dili*, shows that they were based on counts made during Chengdi's reign (33–7 BCE).

5. The only extant reference that we have for an employee of the government taking part is to that of a *you zhi* 有秩, in answer to a personal request made by a testator; see note 100 below. While the *you zhi* evidently counted as possessing a "rank," they did not command a named salary.

6. See *Hanshu* 24A.1144 and Hsu, *Han Agriculture*, 165, for Wang Mang's assertion that all land belonged to the monarch.

7. For a few instances in which official orders were made to provide these necessities, see the discussion below.

8. Secondary sources often claim that Han tenants paid roughly half the crop to their landlord, but the evidence is thin. For consideration of some of these problems in Rome, see Harris, *Rome's Imperial Economy*, 44–46, 274–75.

9. See, e.g., the counts of the registered households and individuals retained in the *Hanshu* chapter 28 and *Xu Han zhi* chapters 19–23, for 1–2 and 140 CE; and the figures for arable land, orchards, and the distribution and receipt of grain in documents found at Yinwan, and dated ca. 10 BCE. See *Yinwan Han mu jiandu*, pl. 13, trans., 77–78; Loewe, *The Men Who Governed Han China*, 60.

10. For a correction of the textual reading *san sheng* 三升 to *san dou* 三斗, see *Hanshu bu zhu* 24A.7b, notes. The *qing*, which included 100 *mu*, measured 11.329 English acres; the *sheng* 199 cc and the *dou* 1.99 liters.

11. The *shi* (later the *hu*) measured 19.968 liters. *Hanshu* 24A.1124; Swann, *Food and Money*, 136–42; *Hanshu bu zhu* 24A.6b.

12. Bray, *Science and Civilisation in China* 6, part II, p. 448, gives a general equivalent of 1.5 *shi* to 6–700 kg/ha.

13. *Hanshu* 24A.1132; Swann, *Food and Money*, 162.

14. See Zhongchang Tong's "Sunyi pian" 損益篇, in *Hou Hanshu* 49.1656, Balazs, *Chinese Civilization and Bureaucracy*, 223. For the use of *hu* as a measure of capacity in place of *shi*, see Loewe, *Problems of Han Administration*, 254–59.

15. Loewe, *Records of Han Administration* II, 69.

16. Loewe, *Records of Han Administration* II, document MD 9, nos. 2, 5, and MD 10, nos. 3, 5, 8.

17. E.g., *Hanshu* 4.116 (for 178 BCE) and *Hanshu* 5.152 (for 141 BCE).

18. See *Zhangjiashan Han mu zhujian*, strips nos. 310–13 and 314–16; Barbieri-Low and Yates, *Law, State, and Society*, 790–93.

19. References in the histories do not include mention of the order of *gong zu* 公卒, which ranked below that of *gong shi*, the lowest in the series of twenty. *Gong zu* appears as part of the series in the documents from Zhangjiashan.

20. Men are designated as *nanzi* frequently in the *Hanshu* and *Hou Hanshu*, but the term is not seen in the *Shiji*. It is explained (1) as *hu nei zhi chang* 戶內之長 (citation from the *Qian shu yin yi* 前書音議, in *Hou Hanshu* 2.96, note 4; for

Qian shu yin yi, see Loewe, "Orders of Aristocratic Rank," 114–5n6) and (2) as those who were without an official title or an order of honor (Li Xian's comment in *Hou Hanshu* 50.1672, note 2).

21. *Hanshu* 12B.54, and 12.357.
22. *Hanshu* 99B.4114.
23. *Hou Hanshu* 1A.33, 7.289, and 8.328. For a complete list of occasions when a decree conferred the gift of *jue*, see Loewe, "Orders of Aristocratic Rank," 165–71.
24. The first of these occasions occurred during the reign of Xuandi (r. 74–48) at a time when the government was taking steps to advertise its benefits and success (see note 73 below).
25. The precise application of these terms is by no means clear, as it varies from text to text.
26. *Jinshu* 10.258; for the readings of *guan* or *xing*, see *Jinshu* 10.272, n. 22.
27. See *Hanshu* 1B.54, for what may be an early reference to the accompaniment of land and housing with some of the *jue*.
28. What is perhaps the earliest reference to a grant of *jue* in the *Hanshu* (1B.54) mentions both arable land and housing. No entry for the extent of land for the *chehou* is specified in the fragments that we have from the statutes.
29. Bu Shi was appointed Imperial Counsellor (*Yushi dafu*) in 111 BCE. *Shiji* 30.1439, *Hanshu* 24B.1167, 1173, *Hanshu bu zhu* 24B.18b; figures for the gold that accompanied the second gift vary as between the *Shiji* and the *Hanshu*.
30. Kings, nobles, and princesses may have received private plots of land that were designated *ming tian*; for a proposed limitation of such plots, see n. 79 below.
31. See, e.g., *Hanshu* 1A.33 (205 BCE), 12.357 (4 CE).
32. See, e.g., *Hanshu* 2.85 (195 BCE), 8.259 (62 BCE), 11.334 (7 BCE).
33. See, e.g., *Hanshu* 4.111 (179 BCE), 8.249 (67 BCE), 9.298 (33 BCE).
34. See, e.g., *Hanshu* 8.254 (65 BCE) and 8.269 (52 BCE). *Jiu* 酒, usually translated as "wine," may well refer to an alcoholic drink (more like sake) made from grain as well as from grapes, which had reached China from the West ca. 100 BCE.
35. See, e.g., *Hou Hanshu* 1B.80 (53 CE), 7.289 (147 CE).
36. For Wang Fu, Cui Shi, and Zhongchang Tong, see Balazs, *Chinese Civilization and Bureaucracy*, 187–225; Loewe, *Problems of Han Administration*, 290.
37. *Hanshu* 4.111, *Hanshu bu zhu* 4.6b (179 BCE) *Hanshu* 10.328 (9 BCE). For the establishment of heirs, see *Zhangjiashan Han mu zhujian*, strips nos. 367–91.
38. *Hou Hanshu* 1A.33 (27 CE).
39. *Hanshu* 8.265 (57 BCE); *Hanshu* 9.298 (33 BCE); *Hou Hanshu* 5.212 (109 CE); 6.252 (126); 6.256 (129) and 7.289 (147).
40. *Hanshu* 8.265 (57 BCE), *Hanshu* 9.298 (33 BCE).
41. *Hou Hanshu* 2.96 and 6.259.
42. See *Hou Hanshu* 26.913 for comments on the *hao hua* 豪猾.
43. For these questions, see section V.
44. *Hanshu* 24A.1137; Swann, *Food and Money*, 180–81.

45. *Hanshu* 24A.1142; Swann, *Food and Money*, 200–4. See also *Hou Hanshu* 28A.958 for Huan Tan's reference to a ban on merchants from holding office, as a means of preventing the accumulation of estates.

46. *Hanshu* 90.3668.

47. *Hou Hanshu* 24.828. For the use of the term *hu* rather than *shi* 石 as a unit of capacity, from Eastern Han times, see Loewe, *Problems of Han Administration*, 254–59.

48. *Hou Hanshu* 24.857.

49. There may be instances in which the figure of 100 *qing* is given in general or even rhetorical terms, rather than as an exact measurement; e.g., stringent plans to take over private sources of wealth ca. 115 BCE included the confiscation of land, which was measured by the hundred *qing* in large counties and at a hundred or so in the small counties (*Shiji* 30.1435, *Hanshu* 24B.1170); 100 *qing* is named in a case of wrongful takeover of land ca. 20 BCE (*Hanshu* 77.3258).

50. See *Hou Hanshu* 32.1119 for Fan Zhong, a man of property, who was the grandfather of Guangwudi; *Hou Hanshu* 32.1133 for the Yin family; *Hou Hanshu* 42.1431 for Liu Kang; *Hanshu* 81.3349 for the family of Zhang Yu; *Hou Hanshu* 70.2257 gives the "private land" (*si tian* 私田) of Zheng Tai 鄭太 (fl. 170–190), at 400 *qing*. On a lower scale, delighted at the discovery of his half-sister who had been living in obscurity in Changling 長陵, Wudi presented her with a sum of cash, a number of slaves, and 100 *qing* of *gong tian* 公田 (for this term, see section IV; also *Shiji* 49.1982, *Hanshu* 97A.3948).

51. For the average figure of 65–70 *mu*, see Patricia Ebrey, "Economic and Social History," 624. For a different estimate, see the Appendix above.

52. *Shiji* 6.224, 10.432, 15.751, 30.1419.

53. *Hou Hanshu* 80A.2598.

54. *Hanshu* 24A.1136, 1137, and 24B.1168.

55. *Shiji* 112.2961. Note 2 to *Hou Hanshu* 40A.1339 cites the *Qian shu yin yi* 前書音義 for three types of prominent persons who were moved to the settlements at the imperial tombs, i.e., high-ranking officials, men of wealth and the magnates, and the occupiers of great estates. For the movement of such persons to Maoling, see Loewe, "Displaced Persons in Han China."

56. *Shiji* 30.1425, *Hanshu* 24B.1163.

57. *Hanshu* 24B.1180; Swann, *Food and Money*, 336.

58. *Hou Hanshu* 28A.958.

59. *Hou Hanshu* 46.1553.

60. The *wu* were formed of the members of five households who were responsible for reporting to officials if they suspected any member of the group of criminal activities. See Loewe, "Social Distinctions, Groups and Privileges," 304.

61. See Zhongchang Tong's essay "Sun yi pian" 損益篇, in *Hou Hanshu* 49.1651. Elsewhere, he expatiates on the theme of the high style of living of the magnates; see *Hou Hanshu* 49.1648, in his essay "Li luan" 理亂; Balazs, *Chinese Civilization and Bureaucracy*, 219.

62. *Hanshu* 24A.1137.

63. See sections I, VII, for allocations in Han times. We have no definite record for such allocations during the Qin kingdom or empire.

64. *Hanshu* 72.3075; see also *Hanshu* 24B.1176.

65. It is far from clear in what ways and with what authority land could be given away. For an example in which Xuan Bing 宣秉 (d. 30), appointed Deputy to the Chancellor (*da situ sizhi* 大司徒司直), did this, for charitable purposes, see *Hou Hanshu* 27.928.

66. See section VII on vagrants and the item dated 49 BCE in the Appendix.

67. *Hanshu* 9.279 and 24A.1139.

68. *Hanshu* 69.2986.

69. *Hanshu* 97A.3948.

70. *Hanshu* 54.2467.

71. *Hanshu* 12.354.

72. *Hanshu* 8.248–49.

73. Around 67 BCE, Huo Shan 霍山 was observing that the poll tax (*fu* 賦) collected from those on these lands was being given to the poor; see *Hanshu* 68.2954.

74. *Hanshu* 24A.1139.

75. *Hanshu* 19B.823–24.

76. *Hanshu* 99B.4111.

77. *Hou Hanshu* 22.112.

78. *Hou Hanshu* 5.206 and 32.1128.

79. *Hou Hanshu* 3.145.

80. *Hou Hanshu* 29.1033–34.

81. *Shiji* 30.1430, *Hanshu* 24B.1167.

82. *Hanshu* 11.336–37, 24A.1143. Ru Chun 如淳 (fl. 221–265) takes these lands to be within the *guo* that were made over to the kings and nobles to administer and tax; the *ming tian* that were situated there were to be for their personal use. In his note to *Hanshu bu zhu* 11.3a, Wang Xianqian 王先謙 (1842–1918) identifies or explains *ming tian* as *zhan tian*, for which see the *Suoyin* note to *Shiji* 30.1430 and that of Yan Shigu 顏師古 (581–645) to *Hanshu* 24A.1137.

83. Ode no. 212; *Shi jing* 14 (1).16b (*Xiao ya* 小雅: "Da Tian" 大田); Swann, *Food and Money*, 133.

84. *Hanshu* 78.3276; Xiao Wangzhi (Imperial Counsellor 59–56 BCE) and Li Qiang 李彊 (then Commissioner for state visits, Da honglu 大鴻臚) were arguing against the proposal that remission from punishment should be granted to certain criminals in acknowledgement of their delivery of grain to the north west. For Zuo Xiong, Director of the Secretariat (*Shangshu ling* 尚書令), see *Hou Hanshu* 61.2016.

85. *Hanshu* 27B (1).1368 and 85.3479.

86. *Hou Hanshu* 5.222.

87. *Hou Hanshu* 42.1431.

88. For the lack of a regularity with which vagrancy was reported, see Loewe, "Displaced Persons in Han China," esp. p. 110.

89. See Loewe, ibid.

90. The formula reads *liu min wu ming shu yu zhan zhe* 流民無名數欲占者.

91. See *Hou Hanshu* 2.96 (57 CE); 2.106 (60); 2.114 (69); 2.121 (74); 2.123 (75); 3.136 (78); 3.137 (79); 4.181 (96); 4.186 (100); 4.194 (105); 5.212 (109); 5.220 (114); 6.252 (126); 6.256 (129); 6.259 (132).

92. *Hou Hanshu* 4.186 (100 CE; attempts to alleviate hunger); 4.190 (102; relief in Zhangye, Juyan, Wuyuan, Hanyang, and Kuaiji); 4.191 (103; food supplies for *liu min* who were en route for a return to their farms); and 5.209 (108).

93. *Hou Hanshu* 21.756.

94. *Hou Hanshu* 76.2459.

95. *Hou Hanshu* 41.1403, 73.2354.

96. *Hou Hanshu* 54.1784, 57.1849.

97. See "Jiangsu Yizheng Xupu 101 hao Xi Han mu"; Zhang, *Zhongguo lidai qiyue huibian kaoshi*, 27; Hinsch, "Women, Kinship, and Property." For an interpretation of the wording of the will, see Chen and Wang, "Yizheng Xupu 101 hao Xi Han mu 'Xian ling quan shu' chu kao."

98. Zhang Zhuanxi takes this as *po tian* 阪田.

99. See *Shiji* 1.34, where *yu* is used in a descending series of *xi* 賜, *yu* 與, and *yu* 予.

100. For the *san lao*, rendered as Thrice Venerable, and the *you zhi*, rendered as Petty Official with Rank, of the county and the district, see Bielenstein, *Bureaucracy of Han Times*, 101–3. For a different interpretation of the text, see Chen and Wang, "Yizheng Xupu 101 hao Xi Han mu 'Xian ling quan shu' chu kao," 21.

101. For the *wu*, or responsibility group, see note 60 above.

102. For a general consideration of these documents, see Li, "Ye tan 'di quan' di jianbie," and Wu, "Handai mai di quan kao."

103. See Zhangjiashan Statutes, strips nos. 367–68; Barbieri-Low and Yates, *Law, State, and Society* 2, 855.

104. These orders are numbered as for the list of the twenty, not including that of *gong zu*, which appears in the documents from Zhangjiashan but not in the *Hanshu*.

105. Zhangjiashan Statutes, strips nos. 366–67; Barbieri-Low and Yates, *Law, State, and Society* 2, 855.

106. Zhangjiashan Statutes, strips nos. 386–87; Barbieri-Low and Yates, *Law, State, and Society* 2, 863. See also strips nos. 379–80, for inheritance of the household in the absence of a son; this goes to the deceased person's father, mother, widow, daughter, grandchild, great-grandchild, grandfather, grandmother, and collateral. See Barbieri-Low and Yates, *Law, State, and Society* 2, 859.

107. See the Appendix for sale of plots of ten and five *mu* (in 179 and 184) and four *qing* (in 245).

108. Hong, *Bei jiang shi hua*, 6.50; the text gives, as an example, a deed found during the period 1522–66, dated to 284.

109. See Lao, *Juyan Han jian kaozheng*, 6, and Loewe, *Records of Han Administration* I, 116. For a contract for the sale of a camel, excavated at Astana, dated between 314 and 376, see *Xinjiang chutu wenwu*, no. 49. See Zhang, *Zhongguo lidai qiyue huibian kaoshi*, 32–68, for the texts of documents that may be described as deeds of purchase for various types of goods, dating from 73 BCE to 198 CE. These include the items listed in the Appendix above for 81, 169, 171, 176, 178, 179, 182, 184, and 188. In addition, he includes the texts of deeds that he suspects to be false, for the years 49 BCE (cf. 56, 125, and 198, given above), and items for 140 and 138 BCE (168, 173, and 188).

110. Juyan strip no. 26.1, dated 37 BCE, for the purchase of a fur coat for 750 *qian*; see *Juyan Han jian jia bian*, no. 187; *Juyan Han jian*, 78. See Ma, "Tulufan di 'Boque yuan nian yiwu quan,'" for a deed for the sale of clothing.

111. Juyan strip no. 262.29; Lao, *Juyan Han jian tuban zhi bu*, 354; *Juyan Han jian, jia pian*, no. 1373.

112. Juyan strip no. 557.4. Lao, *Juyan Han jian tuban zhi bu*, 488; *Juyan Han jian jia pian*, no. 2544A,B.

113. The exception is dated after Han, in 245, and concerned an area of four *qing* and fifty *mu* at a value of 3,500,000 *qian*; see "Anhui Nanling xian Maqiao Dong Wu mu," 978.

114. Exceptionally, perhaps, one deed of 176 CE calls for the same compliance as for the statutes and ordinances; Jiang, "Yangzhou Ganquanshan chutu Dong Han Liu Yuantai mai di zhuan quan," 57.

115. This text is given in Fang, "Cong Xu Sheng mai 'di quan' lun Handai 'di quan' di jianbie," 52. We may note: (1) Years are not identified by the *ganzhi* term in Western Han; (2) The first day of the fifth month of Huanglong 1 was *renyin*; so far from being the eighth day to follow, *yihai* preceded *renyin*; (3) In fact, personal names of two characters are somewhat unusual in Western and Eastern Han. It is not clear whether the adjoining land of He Fang and others denoted land under use or their graves. Identification of the site's position follows that of Lu, "Handai Xu Sheng mai di qianquan jianjie" (61); (4) Possibly the year Huanglong 1 denotes 229, in the kingdom of Wu. NB: Taken together, the foregoing points cast some doubt on the historicity of the land contract.

116. For a major study of this material, see Niida, *Chūgoku hōsei shi kenkyū* II, 400–61. For a discussion of whether these documents were valid or forged (*wei* 偽), see Wu, "Handai mai di quan kao," and Fang, "Cong Xu Sheng mai 'di quan' lun Han dai 'di quan' di jianbie." In particular, some scholars regard as genuine the document that concerns Sun Cheng 孫成 in 171 CE, while those that concern Zhuge Jing 諸葛敬 (49 BCE), Xu Sheng 徐勝 (56 CE), and Li De 李德 (125 CE) are cast as forgeries; see Lu, "Handai Xu Sheng mai di qianquan jianjie," and Fang, "Cong Xu Sheng mai 'di quan' lun Handai 'di quan' di jianbie." It would seem to the present writer that while in the one case the compiler had been able to produce an accurate version of a date, those of the other three inscriptions were neither as knowledgeable nor as careful.

Land Tenure and the Decline of Imperial Government | 95

117. See Morohashi no. 1966.

118. For illustration, see Dangtu xian wenwu guanli suo, "Dangtu xian faxian Dong Wu wanqi di quan," 92.

119. See the item dated 184 in the Appendix.

120. See the item for 125 in the Appendix.

121. For the situation of tombs in the same site, see, e.g., *Wenwu* (1989.1), p. 21 (a site in Shandong; nine graves); *Kaogu* (2009.10), p. 885 (fifty graves, in Shandong); *Wenwu* (2011.10), p. 51 (one hundred graves, in Shandong); *Kaogu* 1993.8, p. 741 (forty-seven graves, in Henan).

122. For the question of whether salaries were paid in cash or grain, see Nunome, "Hansen hankoku ron."

123. *Hanshu* 19A.743; *Hanshu bu zhu* 19A.31a; Loewe, *Men Who Governed*, 70.

124. Bielenstein, *Bureaucracy of Han Times*, 4–5; exceptions included 10,000 *shi* for the *san gong* 三公 (from 8 BCE) and salaries calculated in *dou* (1/10th of the *shi*) at the lowest end of the scale.

125. *Hanshu* 28B.1640.

126. These figures may have been given simply as examples of the exceptionally large size of some counties, e.g., Wan 宛 with its iron works (47,547 households), Chengdu 成都 as a commercial center (76,256 households), and Maoling 茂陵, a haven for the rich (61,087 households, 277,277 individuals). For Chang'an, with its 30,000 officials, *Hanshu* 26A.1543 gives the round estimates of 80,800 and 246,200. As against this bare figure, Professor Nylan writes, "By mid-to late Western Han, the immediate Chang'an metropolitan area boasted well over a million residents; greater metropolitan Chang'an and the immediate suburban counties, an estimated two million"; Nylan and Vankeerberghen, *Chang'an 26 BCE*, 18; see also 188, 195. Other counties for which we have figures include Luoyang 洛陽, Yangdi 陽翟, Yanling 傿陵, Lu 魯, Pengcheng 彭城, and Changling 長陵.

127. E.g., the commanderies of Shuju, Donghai, and Zhangye.

128. Bielenstein, "The Census of China," 139–41, 144; Bielenstein, "Wang Mang," 240. For the north, Bielenstein takes "China north of the Ch'in-ling Mountains, Huai Mountains and Yangtze estuary." He gives the figures of registered persons at 44 million for the north as against 13.7 million for the south for 1 CE, and 26 million for the north as against 22 million for the south in 140.

129. For further details, see the Appendix.

130. Cui Shi "Zheng lun" 政論 in Yan, *Quan Hou Han wen*, 46.10b.11a; Ebrey, "Economic and Social History," 622.

131. See the notes attached to *Xu Han zhi* 23.3534, *Hou Hanshu jijie* (tr.) 23B.31a–32a. These quote figures from the *Han guan yi* 漢官儀 of Ying Shao 應卲 (ca. 140–before 204) and the *Di wang shi ji* 帝王世紀 [or 紀] of Huangfu Mi 皇甫謐 (215–282), with favorable comments on the accuracy of Huangfu Mi's account. For Fu Wuji's work as an historian in the time of Shundi and Huandi, and the relation of his work to the *Dong guan Han ji*, see Crespigny, *Biographical Dictionary*, 233.

132. *Tong dian* 7, p. 144.

133. These figures are as corrected by Bielenstein, on the basis of the total of those given for the 103 administrative units of Western Han, and as against the total figures of 59,594,978 and 12,233,062 that the *Hanshu* 28B.1640 gives. In "Census of China," 139, he writes of "9,455,609 households with 48 million individuals" for 140 CE.

134. See Frier, "Demography," 812, 814.

135. Xun Yue 荀悅, [*Qian*] *Hanji* 前漢紀, *pian* 5 (comments appended to Huidi's Basic Annals).

136. *Hanji*, *pian* 11 (comments appended to Wudi's Basic Annals).

137. *Hanji*, *pian* 16 (comments appended to Zhaodi's Basic Annals).

138. *Hanji*, *pian* 18 (comments appended to Xuandi's Basic Annals).

139. *Hanji*, *pian* 28 (comments appended to Aidi's Basic Annals).

140. For Zuo Xiang, see *Hou Hanshu* 61.2015–19.

141. *Hou Hanshu* 30B.1068–74.

142. *Hou Hanshu* 61.2026.

143. *Hou Hanshu* 6.278.

144. E.g., in 40 CE some ten governors of commanderies and Bao Yong 鮑永, administrator of Lu, were brought up on a charge of sending in false measurements; *Hou Hanshu* 1B.66, 29.1020.

Asian-language Bibliography

"Anhui Nanling xian Maqiao Dong Wu mu" 安徽南陵縣麻橋東吳墓. Edited by Anhui sheng wenwu gongzuodui 安徽省文物工作隊. *Kaogu* 11 (1984): 974–78, 1020.

Bei jiang shi hua 北江詩話. See Hong Liangji.

Chen Ping 陳平, and Wang Qinjing 王勤金. "Yizheng Xupu 101 hao Xi Han mu 'Xian ling quan shu' chu kao." 儀征胥浦 101 號西漢墓 『先令券書』 初考 *Wenwu* (1987.1): 20–25, 36.

Chen Zhi 陳直. *Liang Han jingji shiliao luncong* 兩漢經濟史料論叢. Xi'an: Shaanxi remin chubanshe, 1958.

Dangtu xian wenwu guanli suo 當涂縣文物管理所. "Dangtu xian faxian Dong Wu wanqi di quan" 當涂縣發現東吳晚期地券. *Wenwu* (1987.4): 92.

Du You 杜佑. *Tong dian* 通典. References are to the punctuated edition. Beijing: Zhonghua shuju, 1988.

Fan Ye 范曄 (398–446) *Hou Hanshu* 後漢書, and Sima Biao 司馬彪 (d. 306) *Xu Han zhi* 續漢. References are to the punctuated edition. Beijing: Zhonghua shuju, 1965. See also Wang Xianqian.

Fang Shiming 方詩銘. "Cong Xu Sheng mai di quan lun Handai 'di quan' di jianbie" 從徐勝買地券論漢代地券的鑒別. *Wenwu* (1973.5): 52–55.

Fang Xuanling 房玄齡 (compiler, 578–648). *Jinshu* 晉書. References are to the punctuated edition. Beijing: Zhonghua shuju, 1982.

Gu shi bao shou lu 古石抱守錄. Compiled by Zou An 鄒安. Taipei: Xinwenfeng chuban gongsi, 1986.

Guoyu 國語. 2 vols. Shanghai: Shanghai guji chubanshe, 1978, with standard text divisions.

Hanshu 漢書. Compiled by Ban Gu 班固 (32–92) et al. References are to the punctuated edition. Beijing: Zhonghua shuju, 1962. To be read in conjunction with Wang Xianqian, *Hanshu bu zhu*.

Hanshu bu zhu. See Wang Xianqian.

Hong Liangji 洪亮吉 (1746–1809). *Bei jiang shi hua* 北江詩話. References are to the *Congshu jicheng* print.

Hou Hanshu. See Fan Ye.

Jiang Hua 蔣華. "Yangzhou Ganquanshan chutu Dong Han Liu Yuantai mai di zhuan quan" 揚州甘泉山出土東漢劉元臺買地磚券 *Wenwu* (1980.6): 57–58.

"Jiangsu Yizheng Xupu 101 hao Xi Han mu" 江蘇儀征胥浦101號西漢墓. Edited by Yangzhou bowuguan 揚州博物館. *Wenwu* (1987.1): 1–13.

Jinshi xubian. See Lu Zengxiang.

Jinshu. See Fang Xuanling.

Juyan Han jian 居延漢簡. Edited by Jiandu zhengli xiaozu 簡牘整理小組. 2 vols. Taipei: Zhongyang yanjiu yuan lishi yuyan yanjiu suo 中央研究院歷史語言研究所, 2014–2015.

Juyan Han jian jia bian 居延漢簡甲編. Beijing: Zhongguo kexueyuan kaogu yanjiusuo, 1959.

Kaogu 考古: As many early essays are anonymous, citations include all necessary information, except for essays with known authors, which are listed by authors.

Lao Gan 勞榦. *Juyan Han jian kaozheng* 居延漢簡考證. Taipei: Zhongyang yanjiu yuan lishi yuyan yanjiu suo, 1960.

———. *Juyan Han jian tuban zhi bu* 居延漢簡圖之部. Taipei: Zhongyang yanjiu yuan lishi yuyan yanjiu suo, 1957.

Li Shougang 李壽岡. "Ye tan 'di quan' di jianbie" 也談地券的鉴别. *Wenwu* (1978.7): 79–80.

Liu Tizhi 劉體智. *Xiao jiao jing ge jinwen tuoben mulu* 小校經閣金文拓本目錄, 1935; in *Sandai jinwen cunqi yicanzhao mulu: fu Xiao jiao jing ge jin wen* 三代吉金文存器影參照目錄：附小校經閣金文拓本目錄. Taipei: Taiwan xuesheng shuju, 1971.

Lu Po 魯波. "Handai Xu Sheng mai di qian quan jianjie" 漢代徐勝買地鉛券簡介 *Wenwu* (1972.5): 60–62.

Lu Yaoyu 陸耀遹. *Jinshi xubian* 金石續編. Edited by Lu Zengxiang 陸增祥 (1833–1874).

Lu Yuesen 魯岳森. "Mai diquan yanjiu san ti" 买地券研究三题, *Sichuan wenwu* 四川文物 (2001.1): 8–11.

Lu Zengxiang. See Lu Yaoyu.

Luo Zhenyu 羅振玉. *Zhensongtang jigu yiwen* 貞松堂集古遺文. In *Luo Xuetang xiansheng quanji* 羅雪堂先生全集. 1930.

Ma Yong 馬雍. "Tulufan di 'Boque yuan nian yiwu quan'" 吐魯番的白雀元年衣物券. *Wenwu* (1973.10): 61–65, 72.

Niida Noboru 仁井田陞. *Chūgoku hōsei shi kenkyū* 中國法制史研究. 4 vols. Tokyo: Tōkyō daigaku, 1964.

Nunome Chōfū 布目潮渢. "Hansen hankoku ron" 半錢半穀論. *Ritsumeikan bungaku* 148 (1967): 633–53.

Sima Biao. See under Fan Ye.

Shiji 史記. Compiled by Sima Qian 司馬遷 (?145–?86 BCE) et al. References are to the punctuated edition. Beijing: Zhonghua shuju, 1959.

Tong dian. See Du You.

Wangdu er hao Han mu 望都二號漢墓. Beijing: Wenwu chubanshe, 1959.

Wang Xianqian 王先謙. *Hanshu bu zhu* 漢書補注. Changsha (preface 1900). Reprint, Taipei: Yiwen chubanshe, 1955; Beijing: Zhonghua shuju, 1983. Also available in the *Basic Sinological Series*.

———. *Hou Hanshu jijie* 後漢書集解. Changsha (preface 1924). Reprint, Taipei: Yiwen chubanshe, 1955; Beijing: Zhonghua shuju, 1983. Also available in the *Basic Sinological Series*.

Wenwu 文物: As many early essays are anonymous, citations include all necessary information, except for essays with known authors, which are listed by author.

Wu Tianying 吳天穎. "Handai mai di quan kao" 漢代買地券考. *Kaogu xuebao* 考古學報 (1982.1): 15–34.

Xiao jiao jing ge jin wen. See Liu Tizhi.

Xinjiang chutu wenwu 新疆出土文物. Edited by Xinjiang Weiwuer zizhi qu bowuguan 新疆維吾爾自治區博物館. Beijing: Wenwu chubanshe, 1975.

Xu Han zhi. See under Fan Ye.

Xun Yue 荀悅. [*Qian*] *Hanji* 前漢紀. Taipei: Taiwan Shangwu yinshu guan, 1983.

Yang Lien-shang 楊聯陞. "Dong Han di hao zu" 東漢的豪族. *Ch'ing-hua hsüeh-pao* 清華學報 11, no. 4 (1936): 1007–63.

Yan Kejun 嚴可均. *Quan Hou Han wen* 全後漢文. In *Quan shanggu san dai Qin Han san guo liu chao wen* 全上古三代秦漢三國六朝文. 1887–93.

Yinwan Han mu jiandu 尹灣漢墓簡牘. Edited by Lianyun gang shi bowuyuan and Zhongguo wenwu yanjiu suo. Beijing: Zhonghua shuju, 1997.

Zhangjiashan Han mu zhujian 張家山漢墓竹簡. Edited by Zhangjiashan ersiqi hao Han mu zhujian zhengli xiaozu 張家山二四七號漢墓竹簡整理小組. Beijing: Wenwu chubanshe, 2001.

Zhang Zhuanxi 張傳璽. *Zhongguo lidai qiyue huibian kaoshi* 中國歷代契約彙編考釋. Beijing: Beijing daxue, 1995.

Zhensongtang jigu yiwen. See Luo Zhenyu.

Zhou Zhenhe 周振鶴. *Xi Han zhengqu dili* 西漢政區地理. Beijing: Renmin chubanshe, 1987.
Zhu Jiang 朱江. "Sijian meiyou fabiaoguo de di quan" 四件没有发表过的地券. *Wenwu* 文物 (1964.1): 61–64.

Western-language Bibliography

Balazs, Etienne. *Chinese Civilization and Bureaucracy: Variations on a Theme*. New Haven: Yale University Press, 1964.

Barbieri-Low, Anthony J., and Robin D. S. Yates. *Law, State, and Society in Early Imperial China: A Study with Critical Edition and Translation of the Legal Texts from Zhangjiashan Tomb no. 247*. 2 vols. Leiden: Brill, 2015.

Bielenstein, Hans. *The Bureaucracy of Han Times*. Cambridge: Cambridge University Press, 1980.

———. "The Census of China during the Period 2–742 A.D." *Bulletin of the Museum of Far Eastern Antiquities* 19 (1947): 125–63.

———. "Wang Mang, the Restoration of the Han Dynasty, and Later Han." In *The Cambridge History of China, Volume I: The Ch'in and Han Empires, 221 B.C.–A.D. 220*, edited by Denis Twitchett and Michael Loewe, 223–90. Cambridge: Cambridge University Press, 1986.

Bray, Francesca. *Science and Civilisation in China*. Vol. 6, *Biology and Biological Technology, Part II, Agriculture*. Cambridge: Cambridge University Press, 1984.

Crespigny, Rafe de. *A Biographical Dictionary of Later Han to the Three Kingdoms (23–220 AD)*. Leiden: Brill, 2007.

Ebrey, Patricia. "The Economic and Social History of Later Han." In *The Cambridge History of China, Volume I: The Ch'in and Han Empires, 221 B.C.–A.D. 220*, edited by Denis Twitchett and Michael Loewe, 608–48. Cambridge: Cambridge University Press, 1986.

Frier, Bruce W. "Demography." In *The Cambridge Ancient History, Volume XI: The High Empire A.D. 70–192*, edited by Alan K. Bowman, Peter Garnsey, and Dominic Rathbone, 787–816. Cambridge: Cambridge University Press, 2000.

Harris, W. V. *Rome's Imperial Economy*. Oxford: Oxford University Press, 2011.

Hinsch, Bret. "Women, Kinship, and Property as Seen in a Han Dynasty Will." *T'oung Pao* 84 (1998): 11–20.

Höllmann Festschrift. See under Loewe, "Displaced Persons in Han China and the So-called 'Mausoleum Towns.'"

Hsu, Cho-yun. *Han Agriculture: The Formation of Early Chinese Agrarian Economy (206 B.C.–A.D. 220)*. Seattle and London: University of Washington Press, 1980.

Loewe, Michael. "Displaced Persons in Han China and the So-called 'Mausoleum Towns.'" In *Vergangener Alltag: Neue archäologische, historische, philosophische *und völkerkundliche Betrachtungen aus alten Kulturen. Festscrift aus Anlase des 65. Geburtstag von Thomas O. Höllmann*, edited by Armin Selbitschka and Shing Müller, 107–20. Wiesbaden: Harrassowitz, 2017.

———. *Problems of Han Administration: Ancestral Rites, Weights and Measures, and the Means of Protest*. Leiden: Brill, 2016.

———. *Records of Han Administration*. 2 vols. Cambridge: Cambridge University Press, 1967.

———. "Social Distinctions, Groups and Privileges." In *China's Early Empires, supplement to The Cambridge History of China, vol. 1, Ch'in and Han*, edited by Michael Nylan and Michael Loewe, 296–307. Cambridge: Cambridge University Press, 2010.

———. *The Men Who Governed Han China*. Companion to *A Biographical Dictionary of the Qin, Former Han and Xin Periods*. Leiden: Brill, 2004.

———. "The Orders of Aristocratic Rank of Han China." *T'oung Pao* 48, nos. 1–3 (1960): 97–174.

Nishijima Sadao. "The Economic and Social History of Former Han." In *The Cambridge History of China, Volume I: The Ch'in and Han Empires, 221 B.C.–A.D. 220*, edited by Denis Twitchett and Michael Loewe, 545–607. Cambridge: Cambridge University Press, 1986.

Nylan, Michael, and Griet Vankeerberghen, eds. *Chang'an 26 BCE: An Augustan Age in China*. Seattle: University of Washington Press, 2015.

Swann, Nancy Lee. *Food and Money in Ancient China: The Earliest Economic History of China to A.D. 25, Han shu 24*. Princeton: Princeton University Press, 1950.

Yang Lien-shang 楊聯陞. "Great Families of Eastern Han." In *Chinese Social History: Translations of Selected Studies*, edited by E-tu Zen Sun and John de Francis, 103–34. Washington, DC: American Council of Learned Societies, 1956.

2

Water Control and Policy-Making in the *Shiji* and *Hanshu*

LUKE HABBERSTAD 何祿凱

> 水者, 地之血氣
>
> Water is the blood and *qi* of land.
>
> —Guanzi 管子

We can hardly overestimate the importance of water control during the Qin 秦 (221–207 BCE), Western Han 西漢 (206 BCE–9 CE), and Eastern Han 東漢 (23–220 CE).[1] Under all three empires, complex questions about manipulating riverine systems, lakes, and other sources of water dogged officials at all levels of government. Such an impression, at least, emerges from even a cursory overview of administrative positions related to water. During the Western Han, for instance, "managers of water" (*dushui* 都水) served under six different central government ministries with otherwise disparate duties.[2] Excavated manuscripts and seals further demonstrate that the lower ranks of officialdom were populated by a stunning array of water officials, including "Inspector of Canals" (Jianqu 監渠) and "Manager of Water Flow with Concurrent Duties over Boat Traffic" (Xingshui jian xingchuan 行水兼興船).[3]

By no means have students of early China ignored water. Studying water control to explain larger sociopolitical patterns has a long tradition in the scholarly literature, with historians, sociologists, and political scientists analyzing water engineering projects to categorize pre-imperial and early imperial institutions in comparative terms.⁴ Much of the "data" that such studies have relied upon, however, come from two sources with partly convergent but still distinct aims: the *Shiji* 史記 "Treatise on the Yellow River and Canals" (Hequ shu 河渠書) and the *Hanshu* 漢書 "Treatise on Ditches and Canals" (Gouxu zhi 溝洫志). Up to now, most scholars seem to have assumed that both treatises are important primarily because they provide technical information about advancements in early hydraulic engineering, with no significant differences between the two sources.⁵ (For summaries of the content of the treatises, see the two appendices to this chapter).

This chapter adopts a new approach for these treatises, however, by describing their distinct rhetorical patterns and outlining their starkly different models of decision-making and policy implementation, even as it presumes that rulers and their legitimacy always lay at the heart of imperial water control.⁶ Such an approach emphasizes the curious and underappreciated fact that many of the water-control projects mentioned in the *Shiji* and *Hanshu* treatises either failed or were never implemented in the first place. The treatises thus present not a series of case studies illustrating exemplary water-control techniques, but rather narratives designed to answer a question: what is required for a given engineering project to succeed?

The *Shiji* and *Hanshu* were hardly the only early sources to ponder such a question, as we will soon see, when we turn to a story about water control from the appropriately titled chapter "Taking Pleasure in Success" (Le cheng 樂成) from the *Annals of Lü Buwei* (*Lüshi chunqiu* 呂氏春秋, comp. 235 BCE). The story in the *Lüshi chunqiu*, a pre-imperial text well known to the compilers of the *Shiji* and *Hanshu*, emphasizes that the complexity and difficulty of water-control projects will inevitably draw strong local opposition. Such projects, then, were only feasible when directed by loyal (*zhong* 忠) officials and a supportive ruler who allowed his officials to "choose the best course" (*jue shan* 決善) and stick to it in the face of local opposition.

The motif of the loyal official supported by his ruler is hardly absent from the *Shiji* and *Hanshu* treatises, but they describe the two figures in different terms. The *Shiji* treatise tends to sketch a collaborative model between official and ruler, like the *Lüshi chunqiu*, though the *Shiji* puts more emphasis on the interventions by the emperor. Moreover, the *Shiji* shows an emperor not so much "choosing the best course" (because the treatise implies that the correct course is obvious), but actively securing divine sanction at the very site of the construction efforts to ensure its success. By contrast, the *Hanshu* shifts attention toward processes of group consultation that guided decision-making at court. The treatise abounds with detailed memorials and water-control proposals that reflect a bewildering array of incommensurate perspectives: soil composition, computations and budgeting, regularities or *shu* 數, *yin/yang* 陰陽, classical texts (including the Classics and masterworks), local surveys, and so on. In consequence, the end result in the *Hanshu* is less a celebration of the empire's ability to act, in the manner of the *Shiji*, than a meditation on effective methods to determine the single "best" (*shan* 善) proposal. Through a creative rereading of a passage from the *Analects* (*Lunyu* 論語), the *Hanshu* argued that the "best" proposal could be achieved only if there was assembled a group of advisors with superior cultivation of knowledge *and* virtues. Effective water control, and policy-making at court more generally, thus required not the most thoroughly skilled and knowledgeable technocrats who could produce the best proposal, but rather the right group of cultivated, virtuous advisors.

I. The *Annals* Story: The Zhang River, Loyalty, and Choosing the Best

As a compendium of statecraft designed to be comprehensive, and completed under the direction of one of the state of Qin's most powerful officials, the *Lüshi chunqiu* exerted an enormous impact on Western Han.[7] The compilers of both the *Shiji* and *Hanshu* knew the text, and each responded to one lengthy story about water control included in the "Taking Pleasure in Success" (Le cheng) chapter of the *Lüshi chunqiu*.[8] The narrative begins with the pre-unification King of Wei 魏 holding a banquet. After asking his attending officials to speak their minds, the king praised one Ximen Bao 西門豹, the magistrate of Ye 鄴, as a model official whom all should emulate. A banquet guest named Shi Qi 史起 immediately scoffed at the king's suggestion, noting with disapproval that land in Ye was dispensed in 200 *mu* 畝 allotments, compared to 100 *mu* elsewhere in Wei, "because the land [of Ye] was bad [not fertile]" (是田惡也). Shi Qi argued that the land could be improved by drawing irrigation water from the Zhang River 漳水. The fact that Ximen Bao had not used this method in his administration of Ye was due either to witlessness (*yu* 愚) or disloyalty (*bu zhong* 不忠), according to Shi Qi.

The king later summoned Shi Qi and asked him if an irrigation project at the Zhang River could be completed. Shi Qi responded in the affirmative, but said he feared the king would "be unable to do it" (*bu neng wei* 不能為). In response, the king asked for further details regarding Shi Qi's proposal. After giving his "solemn promise" (*jing nuo* 敬諾) to do his best, Shi Qi warned the king that the locals would probably hate him for initiating such a labor-intensive irrigation project. They would, in fact, try to coerce him into stopping or perhaps even threaten his life, in which case Shi Qi hoped that the king would send an immediate replacement to complete the project. The king responded with his own "promise" (*nuo* 諾), agreeing to the terms stipulated by Shi Qi, and so Shi Qi commenced diverting water from the Zhang River. As predicted, the people of Ye were incensed and duly threatened Shi Qi, who was forced to remain holed up in his residence. The king, staying true to his word, sent a second official to complete the project, which brought the expected benefits to Ye, so that the locals afterward celebrated Shi Qi's achievement in a short ditty.

The Zhang River tale in the *Lüshi chunqiu* concludes with a rumination on the minister-ruler relationship, juxtaposing the determined Shi Qi and Wei king with the witless and reactive masses. Worthy rulers and loyal ministers had to lead the ignorant, lest the subjects suffer. Because Shi Qi "understood transformations" (*zhi hua* 知化), in particular, how the locals would react to his project, he was able to serve his ruler with loyalty. The king of Wei was determined to both *jue* 決 "carve a channel" and *jue* 決 "decide" on the best course, and thus he served as an excellent example of a decision-maker who "would not change even when the masses raised a ruckus" (眾雖誼譁而弗為變). The *Lüshi chunqiu* thus highlights the necessary cooperation between knowledgeable ministers and discerning rulers who stuck to their principles. Note that the chief obstacle was human: ignorant and obstinate subjects who resented being conscripted for such a labor-intensive project. The success of the Zhang River diversion project, then, is ensured not by technical expertise, but rather by the collaboration of good ministers and rulers in the face of local opposition. Even though the *Shiji* alludes but briefly to the story of the Zhang River diversion, its discussion of water-control success reflects the same emphasis on minister-ruler collaboration that is found in the *Lüshi chunqiu*. As we will see, however, the *Shiji* also develops its own rhetorical agenda.

II. The *Shiji* Treatise

The "Treatise on the Yellow River and Canals" from the *Shiji* was the first attempt to provide a comprehensive narrative of efforts to build dikes and canals from high antiquity to the Han period. By the time Sima Qian 司

馬遷 (?145–?86 BCE) compiled the treatise, in addition to pre-imperial (even mythical) efforts from the ancient past, several Han emperors had attempted complex water-control projects. The treatise opens with the legendary sage-king Yu 禹, credited with taming the floodwaters and carving the channels of the Yellow River (see Appendix 1).[9] It then follows with a broad description of Western Zhou and Chunqiu-era water-control efforts and river systems before describing two Zhanguo water projects.[10] The first of these is the Zhang River diversion (Appendix 1, no. 3), but the treatise simply notes the successful completion of the diversion and its provision of abundant water for irrigation in Wei without mentioning either ruler or minister, let alone their collaborative efforts, which figured so prominently in the *Lüshi chunqiu*.[11] The second Zhanguo project mentioned in the *Shiji* treatise is the famous Zhengguo 鄭國 canal undertaken by pre-dynastic Qin (Appendix 1, no. 4). According to the treatise, the canal project was secretly promoted by the realm of Han 韓 to divert Qin's resources from the battlefield, but the canal project upon completion turned the Qin heartland of the Guanzhong basin into a highly productive farming region that was integral to Qin's eventual triumph.[12]

Notably, the passages describing these pre-Han water projects are brief, while more than two-thirds of Sima Qian's treatise focuses on water control under his reigning dynasty of Western Han, with a particular focus on the reign of Wudi 武帝 (r. 141–87 BCE), a period that witnessed ambitious projects to construct canals, connect waterways, and strengthen dikes (see Appendix 1, nos. 6–12). At the same time, many of the projects described in the treatise either were never executed or were failures, for which Sima Qian usually provides explicit reasons. For instance, the early days of Wudi's reign saw a major breach (*jue* 決) in the Yellow River dike at Huzi 瓠子 (Appendix 1, no. 6). After the first repair efforts failed, Wudi's Chancellor Tian Fen 田蚡 (d. ca. 130 BCE), who was Wudi's uncle and one of the most powerful men at court in the early years of Wudi's reign, intervened to stop the repair work. As the treatise tells us, Tian Fen held estates at Yu 鄃, to the northwest of the Yellow River.[13] Because the breached dike was on the river's southern bank, it diverted floodwaters in the opposite direction from Tian's estates. With flood damage no longer a danger, the estates enjoyed improved harvests. The wily Tian thus advised the emperor:

江河之決皆天事，未易以人力為彊塞，塞之未必應天。

Breaches in the Yangtze and Yellow Rivers are always a matter of Heaven. It has never been easy to marshal human labor to create strong enough barriers, and such barriers would not necessarily accord with Heaven.¹⁴

Wudi followed Tian's advice and did not attempt to repair the dike for decades. Given that the *Shiji* treatise highlights the economic benefit Tian reaped from the breach, readers must conclude that Sima Qian doubted Tian's rationale and suspected the motives of those who supported Tian Fen's proposal, a group of experts "who observed the *qi* to take advantage of the predictable regularities" (*wang qi yong shu zhe* 望氣用數者). With Tian Fen's economic interests couched in his invocation of cosmic patterns, supported by technical experts making questionable connections to the natural world, Sima Qian indicates that Tian and his supporters used disingenuous smoke screens to hide selfish calculations.

The story does not entirely end there, however, because Sima Qian immediately contrasts the Huzi dike breach with the construction of a canal. We read that one of the men initially sent to repair the dike before Tian Fen's intervention, Zheng Dangshi 鄭當時, Superintendent of Agriculture (*Da Sinong* 大司農), proposed the construction of a canal between Chang'an 長安 and the confluence of the Wei and Yellow Rivers (Appendix 1, no. 7). Such a canal, Zheng reasoned, would allow goods to be shipped to the capital both more quickly and more cheaply, because it would shorten the water route to Chang'an 長安 by some 600 *li*. At the same time, people could easily draw upon the water to irrigate vast amounts of land abutting the canal.¹⁵ Wudi approved the plan, and under the direction of a water engineer (*shui gong* 水工), the canal was completed in three years. It proved to be extremely successful, dramatically increasing irrigation and the volume of goods transported to the metropolitan region.¹⁶ By showing that Zheng Dangshi had been a failure at Huzi, but a success with the Wei River canal, Sima Qian drew further attention to Tian Fen's disloyalty, insofar as Tian blocked capable officials from completing projects that would benefit the empire. In this way, the *Shiji* treatise reflects a similar preoccupation with loyal officials analogous to that found in the Shi Qi story in the *Lüshi chunqiu*.

The contrast between Tian Fen and Zheng Dangshi (disloyal versus loyal), however, does not illuminate the ruler's role, which the *Lüshi chunqiu* had stressed. True, Wudi eventually approved Zheng Dangshi's plan for the canal, but three immediately subsequent stories in the treatise show Wudi approving water-control plans that either failed outright or achieved fewer benefits than promised (Appendix 1, nos. 8–10). In these cases, the *Shiji* informs readers, the emperor merely approved the projects without taking a personal interest in them.[17] That judgment is reiterated by a counterexample, a stunningly successful final effort to repair the Huzi breach (Appendix 1, no. 11), twenty years after Tian Fen blocked repairs. The *Shiji* description of the project, which Sima Qian or his father may have witnessed firsthand, forms the clear climax of the entire *Shiji* water treatise, and *finally* we see Wudi assuming a more direct role. According to Sima Qian, the emperor visited Huzi to oversee reconstruction efforts just one year after he had completed the *Feng* 封 and *Shan* 禪 sacrifices at Mt. Tai, which, according to age-old traditions, signaled a ruler's legitimacy.[18] The most important moment in the narrative comes when Wudi drowns a white horse and sinks a jade disc (*bi* 璧) in the Yellow River. After offering these valuable items, Wudi personally composes two impromptu songs (*ge* 歌) to "lament that the work was not yet successfully completed" (*diao gong zhi bucheng* 悼功之不成).[19] While a detailed analysis must remain outside the bounds of this chapter, the key point is that Sima Qian portrayed the songs as the turning point in the reconstruction project.[20] After Wudi sang them, we read, the local workers finally succeeded in plugging the breach in the Huzi dike, for which achievement the emperor constructed in gratitude a Xuanfang Temple 宣房宮 at the site. Sima Qian thus described a dramatic scene of labor mobilization paired with the emperor's efficacious ritual acts, contributions that were rightly commemorated in the newly erected temple.[21]

In its description of Wudi's role in the second Huzi incident, then, the *Shiji* treatise applauds the emperor's personal presence at the site and his ritual interventions, which led to the project's success. This role for the ruler goes far beyond the King of Wei showing resolute support for Shi Qi, described in the *Lüshi chunqiu*. Recall that in that story, the *Lüshi chunqiu* praised the king of Wei's ability to "decide on the best course" (*jue shan*) when the loyal Shi Qi had informed him of the true conditions. By contrast, in the *Shiji* treatise Wudi relies on an entire team of hard-working officials to execute the project, but the project's success evidently depended upon Wudi's reverent performance of sacrifices to the river god and singing of two laments that he composed. No real question of different "choices" obtrudes in the narrative, since the treatise makes clear that the dike's reconstruction is the only appropriate option. The scene thus invites readers to contemplate the replacement of a young and inept Wudi by an older, wiser, and more commanding Wudi. At the same time, the treatise achieves an even grander rhetorical aim in its closing appraisal, which offers nothing less than an exaltation of imperial achievement:

太史公曰：余南登廬山，觀禹疏九江，遂至于會稽太湟，上姑蘇，望五湖；東闚洛汭、大邳，迎河，行淮、泗、濟、漯洛渠；西瞻蜀之岷山及離碓；北自龍門至于朔方。曰：甚哉，水之為利害也！余從負薪塞宣房，悲瓠子之詩而作河渠書。

The Senior Director of the Archives says: I have traveled south and ascended Mt. Lu to view the nine rivers that Yu channeled.[22] Then I went to Kuaiji and Taihuang, ascending Gusu to survey the five lakes there. I traveled east to view the Luo's confluence[23] at Dapei, meeting the Yellow River. I traveled along the Huai, Si, Ji, Ta, and the canals of the Luo. To the west I looked upon Mt. Min and Li Ridge in Shu. In the north I traveled all the way from Longmen to Shuofang. And so I say, "Ah! These waters bring so much benefit and so much harm! I was among those in attendance upon the emperor who shouldered wood to plug the dike at Xuanfang. Moved by the imperial laments at Huzi, I composed this 'Treatise on the Yellow River and Canals.'"[24]

This appraisal verges on the romantic in meditating on the vast stretch of the Han imperial domain, threaded by riverine systems, which leads to imperial efforts to control the waters (not least because water could bring much "benefit" but also "harm"). By citing his own contribution at Huzi and his emotional reaction to Wudi's song, Sima Qian himself emerges as the most dutiful subject, insofar as he was able to support and commemorate the propitious achievements of his emperor. The mature Wudi deserves praise as a ruler finally able to lead loyal workers and secure divine blessings, both of which were indispensable factors in the success of the dike repairs.

In other words, whereas the *Lüshi chunqiu* tended to emphasize the conjoined labors of worthy rulers and loyal subjects, the *Shiji* treatise gave pride of place to the mature Wudi, with his loyal servant Sima Qian playing a supporting role.[25] Wudi had to overcome a selfish minister (Tian Fen) and incompetent water managers, rather than thoughtless objections by his mass of subjects. Only when Wudi as agent intervened in the repairs at Huzi were success and the greater good achieved. The treatise is not concerned with

Wudi's "deciding the best course" (*jue shan*) among competing proposals put forward by technical experts. As we will see, the *Hanshu* water treatise, by contrast, places the decision-making process and technical expertise at the center of its discussion.

III. The *Hanshu* Treatise

As noted above, many scholars have assumed that the *Hanshu* "Treatise on Ditches and Canals" simply reproduced and continued the *Shiji* "Treatise on the Yellow River and Canals." Ban Gu 班固 (32–92 CE), compiler of the *Hanshu*, certainly copied most of the *Shiji* treatise to serve as the first half of his treatise, with one significant exception (see below). The similarities stop there, not least because the *Hanshu*, in picking up the story of water control after Wudi's repairs at Huzi, ends the story with the downfall of Western Han (see Appendix 2). This second half of the *Hanshu* treatise evinces an entirely different narrative style from the *Shiji* copy that comprises the first half, because the second half quotes multiple memorials and proposals (some offering contradictory advice), which become increasingly detailed over the course of the treatise. In this second part, the *Hanshu* highlights the technical quandaries that plagued decision-making in water-control projects, so that the *Hanshu* treatise revives the "choosing the best course" theme from the *Lüshi chunqiu*, even as it modifies the earlier treatment of the theme. Specifically, through a creative reinterpretation of an *Analects* quotation that Ban Gu wove into his narrative, the *Hanshu* treatise presents a model of rulership in which rulers (and perhaps also their high ministers?)[26] are not direct actors but rather master-selectors. In particular, they are to be master-selectors of the right men, whose cultivation and virtuous behavior, rather than technical knowledge, will ensure selection of the best possible policy.

Even if the first half of the *Hanshu* treatise largely replicates the *Shiji* in its rhetorical moves, it still contains one significant departure: a detailed description of the Zhang River diversion (see appendix 2, no. 3). As readers will recall, while the *Lüshi chunqiu* included its own rendition of the story, one emphasizing a loyal Shi Qi who enjoyed the support and trust of the King of Wei, the *Shiji* treatise had mentioned the incident only briefly, without reference to Shi Qi. By contrast, the *Hanshu* treatise cleaves more closely to the *Lüshi chunqiu* story, while changing key parts.

Like the *Lüshi chunqiu*, the *Hanshu* version begins with a banquet scene at the court of Wei. The treatise, however, claims that Ximen Bao, magistrate of Ye, "had a fine reputation" (有令名), a statement absent from the *Lüshi chunqiu*.²⁷ When the Wei king wished his ministers well and expressed his hope that they would all follow the example of Ximen Bao, Shi Qi registered a protest. However, while the *Lüshi chunqiu* version had Shi Qi attributing Ximen Bao's decision not to divert the Zhang River water to his "witlessness" (*yu*) or "disloyalty" (*bu zhong*), the *Hanshu* treatise has Shi Qi asserting that Ximen lacked either "practical wisdom" (*zhi* 智) or "humaneness" (*ren* 仁). After the Wei ruler appointed Shi Qi magistrate, Shi rechanneled the Zhang and provided irrigation water to the people of Ye, who end up singing their ditty in praise of Shi Qi. Instead of replicating the focus on Shi Qi's loyalty in the *Lüshi chunqiu*, the *Hanshu* recasts Shi Qi as an upright official whose wisdom and humaneness trumps Ximen Bao's undeserved reputation. Indeed, loyalty to the ruler, let alone collaboration with him, are not mentioned, and the king himself fades into the background, as do the Ye locals who in the *Lüshi chunqiu* narrative forced the king to act. In the *Hanshu*, the drama lies in the story of official against official; it suffices for the king to have chosen Shi Qi, whose knowledge and wisdom sees the project through.

This message foreshadows themes that emerge more plainly in the second half of the *Hanshu* treatise, which consists of a series of memorials and debates that trade in ever more complicated and detailed discussions of technical knowledge, even as high waters and bursting dikes cause increasingly severe damage to areas along the Yellow River. For instance, prior to the floods of 29 BCE, Feng Qun 馮逡, Commandant (*Duwei* 都尉) of Qinghe 清河 commandery, submits a memorial about flood danger in the Yellow River delta, where Qinghe was located (Appendix 2, no. 17). According to Feng Qun, the soil in Qinghe was "friable and easily disturbed" (輕脆易傷). With a branch river upstream blocked and another branch silting up, the main Yellow River "simultaneously received the load of several rivers" (兼受數河之任), a situation that always threatened flooding.²⁸ Feng Qun therefore proposed several dredging and channeling projects to relieve the situation. The matter was referred to Xu Shang 許商, a renowned expert in the *Documents* classic (*Shangshu* 尚書) who "excelled in calculations and was able to estimate expenses for projects" (善為算, 能度功用).²⁹ Dispatched to observe the rivers mentioned in Feng Qun's memorial, Xu Shang recommended against Feng's plan, arguing that one of the branch rivers "could handle the high water" (盈溢所為) and, besides, the court's funds did not suffice to complete the project.[30] The *Hanshu* thus pits the local knowledge of Feng Qun against Xu Shang's expertise, not to mention the court's fiscal considerations.[31] Perhaps not surprisingly, the court's priorities won the argument. But notice that the dramatis personae rehearse the arguments pro and con at the court, a kind of scene that repeatedly occurs in the *Hanshu* treatise but is comparatively absent from the *Shiji*.

Unfortunately, Feng Qun's warning proved the wiser, for the Yellow River overflowed just three years later in 29 BCE, inundating land in thirty-two counties spread across four commanderies (Appendix 2, no. 18).[32] After Wang Yanshi 王延世 oversaw the repairs to the dike, Chengdi 成帝 (r. 33–7 BCE) inaugurated the new reign era, "River Pacified" (Heping 河平), and awarded Wang a court position, a high order of merit (*jue* 爵), and a hefty cash reward of 100 catties of gold. Nonetheless, just two years later, in 27 BCE, the river overflowed its banks once again (appendix 2, no. 19), and court officials of the highest rank began to question Wang Yanshi's competency in water-control matters. Specifically, Du Qin 杜欽 told Wang Feng 王鳳 (d. 22 BCE), Chengdi's uncle and de facto head of government, that Wang Yanshi (no kin to Wang Feng) had exaggerated his skills as a water engineer and that others were just as talented in handling floods, if not more so.[33] As Du continued:

且水勢各異，不博議利害而任一人，如使不及今冬成，來春桃華水盛，必羨溢.

> Moreover, the propensities of floods differ each time. When we do not broadly debate the benefits and drawbacks [of a proposal] but instead rely on the opinion of just one person, if he does not complete work before this winter, then when spring comes, and the peach trees blossom and the waters surge, they will certainly crest and overflow. . . .[34]

Du Qin advised Wang Feng to dispatch four people to the site of the flood so that they could "combine and initiate [different tasks]" (*za zuo* 雜作), sharing the work. Since Wang Yanshi and a second expert were skilled at flood control, Du Qin argued they would "deeply discuss the affordability [of their proposals] and try to best each other down to the last detail" (深論便宜，以相難極). The other two recommended emissaries, meanwhile, were "brilliant in making calculations" (明計算) so they could ascertain the most cost effective plans and thus "select the best policy to follow" (擇其善而從之).³⁵ Notably, Du Qin did not outline a specific role for Wang Feng or his emperor in this decision-making process; that may well have been too risky.³⁶ He seemed to argue strongly that delegating authority to specialists well versed in different techniques would increase the chances of success for any initiative, because the experts would be able to debate among themselves the best course of action. Wang Feng followed Du Qin's advice, and the dike was duly completed.³⁷

As a result, the *Hanshu* treatise pits hasty action, prompted by an avoidable crisis and possible corruption, against competitive yet collaborative decision-making by experts with technical knowledge in multiple fields, with the latter resulting in more cost-effective results over time.³⁸ The success of Du Qin's scheme was measured by the nine years that the Yellow River did not flood, not a very long period of time, perhaps, but certainly more successful than the two years provided by the discredited Wang Yanshi. In 17 BCE, however, floods again inundated some thirty-one counties and estates.³⁹ Two experts gave conflicting advice about how to address the crisis (Appendix 2, no. 20), and even though one of their proposals was accepted, two other officials subsequently criticized the sheer proliferation of water-control plans offered at court (no. 21). The two critics specifically condemned the "crowds of people who saw the weighty rewards given Wang Yanshi, and so they compete with expedient and crafty proposals, which cannot be employed."⁴⁰ To guard against such careerists avid for gain, the critics advocated halting all projects and adopting a more minimalist approach, one focused on monitoring water levels and allowing the river to shift course more naturally. Their proposal was finally adopted, so the remainder of Chengdi's reign saw no major water-control efforts.

The situation changed when Aidi 哀帝 (r. 7–1 BCE) came to the throne. The *Hanshu* treatise includes detailed memorials and proposals submitted during and immediately after Aidi's reign (nos. 22–27). The most striking feature of these memorials is their propensity to supply technical details across a range of specialties, surpassing previous memorials in the treatise (not to mention the contents of the *Shiji*). For example, whereas the claim early in Wudi's reign was that the Wei River canal would "reduce conscripted labor" (*sheng zu* 省卒),⁴¹ after 17 BCE one proposal submitted by two experts argued that a drainage canal would "reduce the officers and conscripted laborers who fix the dikes to provide flood relief by an annual

30,000-plus people" (又省吏卒治隄救水, 歲三萬人以上).[42] Of course, we need not assume that this number is "accurate."[43] Nonetheless, by invoking a specific number, the memorial gestures toward an archive of technical knowledge (perhaps the government's corvée labor registers) to bolster its argument. For our purposes, the statement suggests a shift toward more detailed technical rhetoric, at least compared with Wudi's reign.

A similar contrast could be made between Feng Qun's discussion of "soil that is friable and easily disturbed" (土壤輕脆易傷) and statements made in later memorials discussing soil conditions in even more precise terms. In a memorial submitted around 7 BCE, for instance (see no. 23), Jia Rang 賈讓 wrote that some forty years earlier, the Yellow River was several tens of paces from the base of a dike. For Jia, that the water had only reached the foot of the dike by his own day indicated that the earth at the base of the dike was quite solid and would provide a safe foundation for a heavier stone dike.[44] Meanwhile, Zhang Rong 張戎 wrote in a post–2 CE memorial (no. 25) about silting conditions with reference to a specific ratio: "one *shi* of water contains six *dou* of silt" (一石水而六斗泥).[45] Note the difference in the type of knowledge invoked: whereas Feng Qun's characterization of the soil could have been based on a onetime observation or interview, Jia Rang's memorial cited the changing water levels over time, which were necessarily based either on written observations, or perhaps interviews with old-timers among the locals, while Zhang Rong's statement in theory would have been based on some sort of observation or experiment that involved filtering silt out of a precise measure of river water. Again, our concern is not whether such detailed measurements were accurate; our concern is the rhetorical function performed by claims of access to technical knowledge. Similar observations could be made from multiple later water-control memorials, which refer to more detailed information from other realms of knowledge, including classical texts[46] and heavenly (i.e., astronomical) patterns.[47]

In short, the *Hanshu* treatise presents a wealth of memorials that marshal different forms of expertise that no single person could have hoped to master in a lifetime. This characterization is particularly true of the series of memorials submitted after 2 CE, which collectively reveal a complicated debate about water control at the imperial court during a period when Wang Mang 王莽 (46 BCE–23 CE) began to command ever-greater power. How and who could decide the most advantageous course of action? This seems to have been an ongoing question at the late Western Han court. Indeed, the *Hanshu* treatise says as much in its concluding appraisal, which discusses the role that Huan Tan 桓譚 (43 BCE–28 CE) played in gathering and judging these memorials:

> 沛郡桓譚為司空掾, 典其議, 為甄豐言: 凡此數者, 必有一是。宜詳考驗, 皆可豫見, 計定然後舉事, 費不過數億萬, 亦可以事諸浮食無產業民.

> Huan Tan, from Pei commandery, served as an Assistant to the Imperial Counsellor and put these discussions into order. He told (Imperial Counsellor) Zhen Feng: "Among these many proposals, there must be one that is correct. It would be proper to thoroughly investigate and verify them so that we can predict all outcomes. After we determine the best plan, we can proceed. Our expenses will not then run into millions upon millions, and we can meanwhile work for all the people who lack food and a livelihood."[48]

According to Ban Gu, Huan Tan was deemed the right judge to assess the water-control debates in the closing years of the Western Han and the beginning of Wang Mang's Xin dynasty. Huan's stated aim, in the treatise, was to evaluate the outcomes of all the competing proposals by a more rigorous method. Through such a process, the best proposal could be ascertained.

Fortunately, portions of the debate that Huan Tan analyzed are found in fragments ascribed to Huan's *New Treatise* (*Xin lun* 新論).[49] While the fragmentary nature of the text precludes arriving at definitive conclusions about Huan Tan's judgments, this much is clear: Huan praised the expertise shown by several specialists who had participated in the water-control debates in a fashion not seen in the *Hanshu* treatise. For instance, Huan Tan stated that Guan Bing's 關並 "talent and wisdom were penetrating" (材智通達), Zhang Rong was "well practiced in irrigation" (習溉灌事), and Han Mu 韓牧 "excelled at water management" (善水事).[50] In articulating the particular forms of expertise commanded by a range of water-control specialists, Huan Tan's decision-making process recalls that of Du Qin, as described above, for Du Qin likewise had advocated sending a group of experts with specific skill sets (water engineering and budgetary calculations) to respond to a flood in 27 BCE. Note, however, that the similarities between Du and Huan end there. Du Qin suggested that the relevant experts should correct and compete with one another so that the "best" proposal would become apparent without further intervention by the court. By contrast, Huan Tan assumed that he and the Imperial Counsellor together commanded sufficient ability to select the best policy via a comparison of various participants' contributions to the water-control debates. The *Hanshu* treatise thus presents two models of decision-making. The first, advocated by Du Qin, delegated significant authority to on-site technical experts, with only a limited role for an imperial court (perhaps so that it could escape blame for unwise decisions). The second, advocated by Huan Tan, reserved an active role for court ministers to select the best policy by set criteria: the particular technical competency of the expert and his specific policy content.

Evidently, Ban Gu dismissed both these models toward the end of his treatise, as we can see from his rejection of the post–2 CE water-control debate. By his description of events, "during the time of Wang Mang, esteem was given only to empty [abstract] theories and nothing of use was carried out" (王莽時, 但崇空語, 無施行者).[51] The phrase "empty theories" (*kong yu* 空語) is more important than it may look, for several Han sources contrast "empty theorizing" with more knowledge-based rhetoric drawn from authoritative sources, such as cosmic patterns, eyewitness accounts, classical texts, and so on.[52] Ban Gu, by raising the notion of "empty theories," thus characterized the water-control debate as one that relied on untrustworthy sources. What and who should be trusted, then? In this connection, the evidence from Huan Tan becomes important, for even if the relationship between Huan Tan, as author of the *New Treatise*, and Ban Gu, compiler of the *Hanshu* treatise, remains unclear,[53] Ban Gu clearly thought it unnecessary or inappropriate to include any description of the various specialties of the debate participants. Proposal after proposal, many of them contradictory, appear in the *Hanshu* treatise, but Ban offers no pat resolution to any water-control problems. Ban Gu's editorial decision echoes his earlier revision to the story of Shi Qi's Zhang River diversion (see above). As readers will recall, Ban Gu had Shi Qi dismiss Ximen Bao's "good reputation" as meaningless, because Ximen was sadly lacking in wisdom (*zhi*) and humaneness (*ren*). Presumably, Shi Qi, who criticized him, embodied both qualities, for we know that Shi Qi put forth an effective plan. The "empty" (i.e., undeserved) reputations of lesser men (e.g., Ximen Bao) could not compare with the sterling characters of superior men (e.g., Shi Qi). By Ban's account, rulers such as the King of Wei had only to pick such men of sterling character to realize a successful project.

This perspective comes through quite clearly in the final appraisal at the end of the *Hanshu* treatise, which cites, but recasts, a passage from the *Analects*. Ban Gu begins with a faint echo of the praise meted out to the sage-king Yu's work in the *Shiji* appraisal for the water-control chapter (see above), but then the *Hanshu* treatise changes course:

古人有言：「微禹之功, 吾其魚乎！」
中國川原以百數, 莫著於四瀆, 而河為宗。
孔子曰：「多聞而志之, 知之次也。」
國之利害, 故備論其事。

> The ancients had a saying:
> "If not for the accomplishments of Yu, we would be but fish!"[54]
>
> In the Central States, the headwaters of rivers are counted in the hundreds, but none are more important than the four great rivers, and the Yellow River is their lineage head.
>
> Kongzi (Confucius) said: "Hearing much before setting one's mind [to a decision] is a second-best form of knowledge."
>
> Since it brings profit and damage to the realm, I laid out this treatise in detail relating to the matter [of rivers].[55]

The second quotation ascribed to Kongzi actually alters a citation found in *Analects* 7.28, reproduced in full here:

子曰：蓋有不知而作之者, 我無是也。多聞擇其善者而從之, 多見而識之, 知之次也。

> The master said: "No doubt there are those who can act without knowledge, but I am not one of them. Hearing much, picking out the best, and then going on with that, as well as seeing much before claiming to know something—these are [but] secondary forms of knowledge."[56]

The passage, however difficult to parse, does seem to contrast an intuitive and instinctual knowledge with the "second best" (*ci* 次) forms of knowledge based on broad learning and wide-ranging observations, which allow the person to "pick the best." It can hardly be coincidental that the speech attributed to Du Qin cites the same *Analects* passage, when Du urges Wang Feng to "pick the best." There is a key difference, however, between Du Qin's speech and the *Hanshu* appraisal: the *Hanshu* rates technical expertise as a good, but a good that is distinctly inferior to sagely understanding. While Du Qin advocates "picking the best," the *Hanshu* advocates "hearing much" (*duo wen* 多聞) and "setting minds" (*zhi* 志) to a course of action without any discussion of choosing anything; the *Hanshu* quite openly deletes mention of "picking the best" from the *Lunyu* passage. In contrast to Du Qin's speech, then, selecting the technically "best" *proposal* is less important than gathering at court the right men with the right virtues. The emperor's men should be learned, cultivated, and totally dedicated to "setting their minds" to getting the job done.

Presumably, assembling such men was the emperor's responsibility, though the *Hanshu* does not highlight this point. Very likely, such a role for the emperor could be taken for granted. On a more speculative note, however, it is possible that Ban left it intentionally ambiguous: important to remember here is that Ban was responding to Du Qin and Huan Tan's models of decision-making, which explicitly understood decision-making to involve group consultation among court officials. By leaving the role of the emperor unstated, perhaps Ban left a role for virtuous men at court to recommend similarly cultivated peers for court office. In any case, if my reading is correct, the entire *Hanshu* "Treatise on Ditches and Canals" can be read simultaneously as an emendation to *Analects* 7.28, a statement alluding to the criteria for proper decision-making at court, and an argument about the value of carefully cultivated virtues over mere technical expertise.[57]

IV. Water Control Redux

Water control takes trouble and expense. Successful water-control projects are well-nigh miracles, and it is hard to predict how long the project will work. Careful consideration of the *Shiji* and *Hanshu* water treatises forces these undeniable facts on the readers' attention, not least because both treatises present many schemes that turn out to be entire or partial failures. While contemporary social scientists ponder the political, economic, and social ramifications of spending so much time and money on water control, the early sources reflect different concerns and questions: How are such resources to be expended most effectively, so as to achieve water success? And what, indeed, constitutes "success"? The story of the Zhang River diversion shows us that such questions commanded attention even in pre-imperial times, because the *Lüshi chunqiu* emphasized collaboration between loyal officials and steadfast rulers who "choose the best course" and remain unmoved by loud complaints from the "masses." Even if such stories acknowledge the heavy investment that water-control projects require, the *Lüshi chunqiu* narrative treats success as if it were mainly reliant on relationships of trust between ministers and rulers.

The *Shiji* "Treatise on the Yellow River and Canals" similarly emphasizes the importance of the human actors, but in the *Shiji*, it is officials collaborating with the ruler, as the "masses" have dropped out of sight; compared with the *Lüshi chunqiu* story, the ruler receives considerably more attention. To be sure, the selfish designs of ministers like Tian Fen could prevent consideration of even the most beneficial water-control efforts, and a level of technical expertise on the part of those proposing and executing a given project was necessary. Above all, however, the most impressive water-control projects owed their success to the active intervention of the emperor. For Sima Qian, the model is the repairs at Huzi, where the emperor completed sacrifices to the Yellow River, composed poetic laments extemporaneously, and constructed a commemorative temple for the reconstructed dike. With the help of his laborers and officials (including Sima Qian himself), Wudi ensured maximum efficacy by securing divine support for the human endeavor. Notably, there is no significant concern in the *Shiji* treatise about choosing the right course of action, as the dike obviously needs to be repaired. The treatise thus emphasizes that the emperor's active engagement is crucial to major water-control projects, which demonstrate the glorious potential of a centralized empire that by Sima Qian's time commanded unprecedented territory and power.

Such a theme is by no means absent from the *Hanshu* "Treatise on Ditches and Canals" because most of the first half of the *Hanshu* reproduces the earlier *Shiji* treatise. But, as we have seen, the *Hanshu* treatise introduces new issues while laying bare the dizzying array of technical specialties involved in water-control proposals and projects. The second half of the treatise adds the observation that certain officials were presenting water-control proposals to the late Western Han court *primarily* to gain favor, rather than to serve their ruler. Perhaps that explains why each of the competing proposals is more complicated than the last, with appeals to ever more specialized bodies of knowledge. Also according to the *Hanshu* treatise, the success of a particular water engineering project did not turn on the trust within minister-ruler relations, but rather on the court's ability to "choose the best" from a wide range of alternatives. (As readers will recall, the theme of "choosing the best" appeared in the *Lüshi chunqiu* but was absent from the *Shiji*.)

Unlike the *Lüshi chunqiu*, the *Hanshu* treatise does not assume that the precise meaning of "the best" was self-evident; indeed, it shows that "the best" was a source of contention, so that senior advisors at the late Western Han court sought to devise the more reliable means for selecting the best proposal. The treatises show us two competing selection methods, though both of them, it is worth emphasizing, required group consultation of experts and court officials. The first (epitomized by Du Qin) had the court delegating significant authority to experts, who were to debate the finer points among themselves. The second selection method (epitomized by Huan Tan) was to have a single master, who combined expertise in the Classics with expertise in technical matters, review all the proposals. Huan Tan's preferred method retained a significant role for high-ranking advisors, as it was one of their number who was to verify and predict the probable outcomes for various plans, always with reference to the reputation and knowledge of the experts who first proposed it. Unlike the *Shiji*, the *Hanshu* did not focus primarily on the emperor's role in water-control matters; one may speculate that the *Hanshu* tried to distance the throne from the frequent failures in water control. (Note, for instance, how Ban Gu deemphasized the king of Wei in the Zhang River story).

In any case, with support from a creative rereading of an *Analects* passage, Ban Gu championed yet a third variation of the model for decision- and policy-making (though one that still seems to have assumed group consultation). For Ban, a group of perfectly cultivated and broadly learned advisors should play the central role, "consulting widely" (*duo wen*) with the other members of the court before "setting their mind" (*zhi*) to a given measure.[58] Technical competency, selecting the most effective proposal, or even predicting eventual outcomes—these were not Ban's central concerns, it seems. Rather, Ban advocated choosing the right men, with the right attitude toward learning, who would commit to getting the job done. The emperor was ultimately responsible for assembling such men at his court, and he, of course, risked bearing the blame for botched water projects, but Ban's treatise does not belabor these points. Indeed, as speculated above, by not specifically mentioning the emperor, Ban Gu not only gave the throne distance from failures, but also perhaps left a role for officials to recommend men of virtuous character to court.

In a way, Ban Gu's rhetoric conforms to stereotypical understandings of Chinese government as being driven by officials focused on self-cultivation and virtuous administration. That the *Hanshu* treatise pushes this argument so hard suggests, contrary to the stereotype, that the idea was by no means universally accepted and that such officials were not easily found. Ban's method was clearly not the only model of good governance available at the time. In the treatise, Ban Gu worked hard to convince his contemporaries that technocratic decision-making alone did not suffice in important matters of state.[59] Even if the technical complexities of water control threatened to overwhelm other considerations, the policy-making process still had to be driven by the virtues of high-quality men. The broad-minded knowledge and steadfast commitment of such men, among whose ranks Ban Gu no doubt included himself, took precedence over technical expertise. Water, in other words, was primarily a problem of human character and cultivation, not a technical problem, and excessive emphasis at court on the details of hydraulic engineering projects had to be avoided.

This rhetorical pattern comes into more focus when we compare it with evidence from the Song (960–1279 CE), especially the second half of the eleventh century, a period that infamously witnessed many floods on the Yellow River plain along with simultaneous attempts at government renewal and reform. As Christian de Pee has recently demonstrated, Song officials, many of them well versed in the technical complexities of water control, argued above all that water-control problems were best solved by cultivated men schooled in the Classics who understood the connection between material aspects of imperial management (the "body physical," in de Pee's terms) and the larger health and sustainability of the body politic. Technical mastery, then, could only be effective if executed by "talented men with true learning" who could manifest the "immanent pattern of the moral Way" in government administration.[60] This assumption was shared across different political factions, even while they clashed on policy details. By the twelfth century, however, the consensus broke down, because the continuing failure of Song water infrastructure caused officials to withdraw from government life, despairing that any amount of learning and cultivation could successfully re-create the perfect administration of the ancient sage kings by connecting the body physical to the body politic. While the emphasis on choosing cultivated men is easily discernible in the *Hanshu* treatise, it is this final step of withdrawal that is entirely missing from the "Treatise on Ditches and Canals." Ban Gu was busy making the apparently still-novel argument that the selection of men of broad learning and cultivation should be the primary concern of the court and the basis for successful decisions about technical policies. In other words, this idea did not constitute a shared consensus among all advisors and officials at court, in contrast to the Song.

On a final, more speculative note, Ban Gu's claim that the details of water-control proposals should be subordinated to questions of virtue and character might have had a long-term impact on imperial historiography, if we can take as evidence the complete absence of water-control treatises from imperial historiography after the *Hanshu*. The dominant explanation for the disappearance of such treatises until the *History of the Song (Song shi* 宋史; compiled 1343 CE) is that there were simply fewer floods along the Yellow River in the post-Han period. As the story goes, the collapse of the Eastern Han, its eventual replacement by a ruling house rooted in the nomadic steppe culture of the north, and increased migration of Chinese sedentary agriculturalists to the south brought less intensive development along the middle reaches of the Yellow River and thus less soil erosion. With a corresponding reduction in silt buildup along the lower reaches, the river breached its dikes much less frequently. It was thus not until the late Tang and especially the Song period that flooding along the lower reaches of the Yellow River increased in frequency and intensity.[61]

The above analysis of the *Shiji* and *Hanshu* treatises does not subvert this narrative, and certainly the evidence marshaled by de Pee confirms that there was a clear uptick in anxiety about the Yellow River during the Song, when concern about floods roiled court debates. The evidence from the *Shiji* and *Hanshu* reminds us, however, that the imperial histories cannot be transparently mined for facts about ecological history. If one of Ban Gu's arguments was that the imperial court and the officials working therein should not get too bogged down in the technical particulars of hydraulic engineering, then one compelling response by his successors would be to stop writing treatises about them. More research is necessary to explore this hypothesis, which in any case does not at all support a conclusion that officials ceased worrying about hydraulic engineering after Ban Gu; the issue continued to be a problem even in the immediate post-Han period.[62] It does suggest the possibility, however, that a change in the status of water control, spurred on by the complex debates and polemical representations of the subject in the two Han courts, inevitably affected its treatment (or lack thereof) in later texts. It would be abundantly clear to any reader of the treatises that water-control projects were difficult and dangerous, with unpredictable long-term consequences, so we might forgive later authors for concluding that it was best to omit the topic entirely.

Appendix 1: Summary, *Shiji* "Treatise on the Yellow River and Canals"

Time Period	Summary	
High antiquity	1)	Yu 禹 tames the floods, creates Nine Provinces (*jiu zhou* 九州), and establishes the course of the Yellow River.
"After that" (*zi shi zhi hou* 自是之後)	2)	Description of the waterways and canals connecting different Zhou-era realms.
Zhanguo period	3)	Ximen Bao 西門豹 [not Shi Qi] diverts the Zhang River 漳水 to irrigate Ye 鄴.
	4)	Qin 秦 constructs the Zhengguo Canal 鄭國渠. Initially, the effort was spurred by a Han 韓 plan to distract Qin and force it to expend resources, but the plan backfires when the canal ends up vastly increasing Qin's agricultural production.
Western Han, to reign of Wudi (eight stories)		*Reign of Wendi* 文帝 (r. 180–57 BCE)
	5)	Yellow River bursts banks at Suanzao 酸棗, 167 BCE, and a large number of conscript laborers are mustered from Dong 東 commandery to repair the breach.
		Reign of Wudi 武帝 (r. 141–87 BCE)
	6)	During the Yuanguang 元光 reign period (134-29 BCE), the Yellow River bursts its banks at Huzi 瓠子. Tian Fen 田蚡 prevents repairs.
	7)	On the advice of Zheng Dangshi 鄭當時, the Wei River 渭水 canal is constructed, making grain transport in capital region easier.
	8)	The governor of Hedong 河東 commandery gains approval for constructing irrigation canals east of the narrows in the Yellow River at Dizhu 砥柱, but the canals fail after the Yellow River changes course. Eventually, the newly irrigated lands are abandoned.
	9)	Some people gain approval for constructing roads between the Bao 襃 and Mian 沔 Rivers to connect Shu 蜀 (Sichuan) more directly to the capital. The road proves to be useful, and yet the Bao and Mian Rivers are not navigable, and so they cannot transport grain.

Water Control and Policy-Making in the *Shiji* and *Hanshu* | 123

Time Period	Summary
	10) Zhuang Xiongpi 莊熊羆 gains approval for constructing a canal from the Luo River 洛水 to irrigate Chongquan 重泉. The work is much more difficult than anticipated, with workers forced to dig and then connect together a series of deep wells in order to create an underground canal. In the end, however, the canal does not create much irrigation benefit.
	11) Wudi leads a large work party to finally repair the dikes at Huzi. He composes and sings songs about the difficulty of the work. The workers succeed and construct a temple on the new dike (109 BCE).
	12) Short mention is made of several water control projects proposed in the wake of Huzi's success.

Appendix 2: Summary, *Hanshu* "Treatise on Ditches and Canals"

Time Period	Summary
High antiquity	1) Yu 禹 tames the floods, creates Nine Provinces (*jiu zhou* 九州), and establishes the course of the Yellow River.
"After that" (*zi shi zhi hou* 自是之後)	2) Description of the waterways and canals connecting different Zhou-era realms.
Zhanguo period	3) Shi Qi 史起 [not Ximen Bao] diverts the Zhang River 漳水 to irrigate Ye 鄴.
	4) Qin 秦 constructs the Zhengguo Canal 鄭國渠. Initially, the effort was spurred by a Han 韓 plan to distract Qin and force it to expend resources, but the plan backfires when the canal ends up vastly increasing Qin's agricultural production.
Western Han through beginning of Xin, during reign of Wang Mang	*Reign of Wendi* 文帝 (r. 180–57 BCE) 5) Yellow River bursts banks at Suanzao 酸棗, 167 BCE, and a large number of conscript laborers are mustered from Dong 東 commandery to repair.

continued

Time Period	Summary
	Reign of Wudi 武帝 (r. 141–87 BCE)
6)	Yellow River bursts its banks at Huzi 瓠子. Tian Fen 田蚡 prevents repairs (134–29 BCE).
7)	On the advice of Zheng Dangshi 鄭當時, the Wei River 渭水 canal is constructed, making grain transport in capital region easier.
8)	The governor of Hedong 河東 commandery gains approval for constructing irrigation canals east of the narrows in the Yellow River at Dizhu 砥柱, but the canals fail after the Yellow River changes course. Eventually, the newly irrigated lands are abandoned.
9)	Some people gain approval for constructing roads between the Bao 褒 and Mian 沔 rivers to connect Shu 蜀 (Sichuan) more directly to the capital. The road is useful but the Bao and Ye end up being unable to transport grain.
10)	Zhuang Xiongpi 莊熊羆 gains approval for constructing a canal from the Luo River 洛水 to irrigate Chongquan 重泉. The work is much more difficult than anticipated, with workers forced to dig and then connect together a series of deep wells in order to create an underground canal. In the end, however, the canal does not create much irrigation benefit.
11)	Wudi leads a large work party to finally repair the dikes at Huzi. He composes and sings songs about the difficulty of the work. The workers succeed and construct a temple on the new dike (109 BCE).
12)	Short mention is made of several water control projects proposed in the wake of Huzi's success.
13)	Ni Kuan 兒寬 proposes further canals in the capital region to support the Zhengguo Canal. His proposal is approved by Wudi (111 BCE).
14)	An otherwise unknown figure by the name of Lord Bai 白公 repeatedly proposes dredging a canal that would draw from the Jing River and enter the Wei River at Yueyang. This Bai Canal is successful, and the people sing its praises (95 BCE).

Time Period	Summary
	15) A man named Yannian 延年 proposes leveling the Yellow River and possibly diverting it. The proposal is rejected.
	Reign of Xuandi (r. 74–49 BCE)
	16) After the Yellow River bursts its banks and carves a new river bed, called Tunshi 屯氏, Guo Chang 郭昌 surveys the northern bend in this new channel, constructing release canals at three points in order to reduce the water speed and danger of flooding (69–66 BCE). Later, in 39 BCE, during the reign of Yuandi 元帝, the Yellow River breached its dikes elsewhere, and the Tunshi is cut off.
	Reign of Chengdi (33–7 BCE)
	17) Feng Qun 馮逡 proposes dredging to reconnect the Tunshi 屯氏 River and thus relieve pressure on the Yellow River banks. The proposal is rejected on the advice of Xu Shang (ca. 33 BCE).
	18) After a catastrophic flood at Guantao and Metal Embankment, several officials lead rescue efforts, and Wang Yanshi 王延世 reconstructs the dike. Chengdi starts new reign era "River Pacified" (Heping 河平) and rewards Wang (29 BCE).
	19) Another flood occurs at a breached dike. On the advice of Du Qin 杜欽, Wang Feng 王鳳 dispatches Wang Yanshi, Yang Yan 楊焉, and two other officials to repair the dike (27 BCE).
	20) After the Yellow River overflows its banks at three separate points, Sun Jin 孫禁 and Xu Shang 許商 give contradictory advice about the best solution. Xu Shang's proposal is accepted (17 BCE).
	21) Li Xun 李尋 and Xie Guang 解光 criticize the flurry of water control plans at court and successfully manage to stop all water control work (17 BCE).
	Reign of Aidi (7–1 BCE)
	22) Ping Dang 平當 proposes more dredging and channeling, but Inspectors and Governors are unable to find enough people to implement his plan (7 BCE).

continued

Time Period	Summary
	23) Jia Rang 賈讓 submits a long and detailed memorial proposing three plans, ranked by order of effectiveness, for alleviating the problem of flooding and burst dikes on the Yellow River, but no action is taken (7 BCE or immediately after).
	Reign of Pingdi (1 BCE–6 CE) or later (Wang Mang in power)
	24) Guan Bing 關並 proposes that no residences or government offices be built in a roughly 180 *li*-wide band extending north and south of the Yellow River, but no action is taken.
	25) Zhang Rong 張戎 proposes reducing irrigation from the Yellow River, since during the dry months it causes lower water levels and increased sedimentation, ultimately increasing the risk of breaches and floods during the rainy season. No action is taken.
	26) Han Mu 韓牧 proposes following evidence from the "Tribute of Yu" (Yu gong) chapter of the *Documents* (*Shangshu*) to carve out channels of the ancient "Nine Rivers" near the mouth of the Yellow River. No action is taken.
	27) Wang Heng 王橫 criticizes Han Mu's plan and argues that channels should be opened up elsewhere. No action is taken.

Notes

1. Thank you to the editors and anonymous reviewers, as well as Jesse Chapman, Garrett Olberding, and Michelle H. Wang, all of whom provided valuable comments on earlier drafts of this chapter.

Early texts use "water control" (*shui li* 水利), a term that seems to have first gained currency during Western Han, to refer to three different kinds of projects: flood control, transport canals, and irrigation works. See Fujita, "Kan dai ni okeru suiri jigyō tenkai."

2. The ministries were Ceremonial (*Taichang* 太常), Agriculture (*Da sinong* 大司農), Lesser Treasury (*Shaofu* 少府), Waterways and Parks (*Shuiheng duwei* 水衡都尉), Metropolitan Superintendent (*Neishi* 內使), and Commandant, Orders of

Honor (*Zhujue duwei* 主爵都尉). See *Hanshu* 19a.726, 731, 735, 736. As Hans Bielenstein showed long ago, the devastating floods of 11 CE very possibly led to the collapse of the Xin dynasty. See "Wang Mang," 241–44.

3. For an overview, see Guo, "Handai shuili zhiguan kao lun," 1–5 and 3 for the two mentioned titles.

4. The paradigmatic example is the idea of "Oriental despotism," reviewed in Minuti, "Oriental Despotism." The concept received its most influential modern articulation in Wittfogel, *Oriental Despotism*, which argued that a primarily Asian and especially Chinese dedication to elaborate hydraulic projects led to the development of centralized bureaucracies and "despotic" political systems. For an early and particularly strident critique, see Joseph Needham's review of the book (Wittfogel and Needham's disagreements were partly rooted in political disagreements). Implicit and explicit critiques of the hydraulic empire thesis continue, with recent authors tending rather to emphasize the prolonged military conflicts of the Zhanguo period as the driving factor in the development of state institutions in pre-imperial China. See, e.g., Fukuyama, *The Origins of Political Order*; Zhao, *The Confucian-Legalist State*, 204–7 and passim.

5. For instance, Dorofeeva-Lichtmann, "Where Is the Yellow River Source?," 11, stated that there is "little difference in the content of the two [*Shiji* and *Hanshu*] treatises." In their exhaustive work on the history of Chinese hydraulic engineering, Joseph Needham, Wang Ling, and Lu Gwei-djen referred regularly to the "Treatise on Ditches and Canals" and even included a full translation of Jia Rang's 賈讓 long memorial (Appendix 2, number 23) detailing three possible water-control measures. See Needham et al., *Science and Civilisation in China*, vol. 4, part III, 211–378.

6. Note that the *Hanshu* "Treatise on Ditches and Canals," along with the "Treatise on Geography" (Dili zhi 地理志), are the *only* treatises in the *Hanshu* that do not correspond to a directly related bibliographic section in the "Treatise on Classics and Writings" (Yiwen zhi 藝文志), which listed and categorized the texts contained in the imperial library. Most texts dealing with the examination of geographic features are categorized under "Forms and Models" (*xing fa* 刑法), which includes texts addressing everything from physiognomy to the assessment of swords and other blades. In the "Treatise on Classics and Writings," description of the land and water pairs with other ways of evaluating external appearances and "forms" to determine intrinsic, hidden patterns, often, apparently, for purposes of divination. The fact that the "Treatise on Ditches and Canals" evinces little concern for such matters suggests that, from the earliest days of imperial historiography, water control was conceptualized and understood in terms of imperial administration.

7. For a detailed discussion of the compilation of the *Lüshi chunqiu*, see Knoblock and Riegel, *Annals of Lü Buwei*, 12–14 and 27.

8. *Lüshi chunqiu* 16/5; Knoblock and Riegel, *Annals of Lü Buwei*, 391–93.

9. *Shiji* 29.1405.

10. *Shiji* 29.1406.

11. *Shiji* 29.1408. The treatise states only that Ximen Bao successfully diverted the Zhang River. At the end of the "Accounts of the Crafty Talkers" (Guji liezhuan 滑稽列傳) chapter in the *Shiji*, Chu Shaosun 褚少孫 (?103–?30 BCE) appended accounts of six other men Chu characterized as *guji*, including Ximen Bao. Chu Shaosun's account casts Ximen Bao as a water-control hero, because he managed both to stop a local Ye custom of throwing young women into the river as sacrificial wives of the River Lord (He bo 河伯) and to construct an irrigation system that continued to be used in the Han period (*Shiji* 66.3211–13). The relationship between Sima Qian's laconic statement about Ximen Bao's success in the treatise and Chu Shaosun's elaborate story is unclear. In a detailed exploration of different stories about Ximen Bao and Shi Qi, Hamakawa, "Shōsuikyo no kenzōsha" argues that a) statements suggest Ximen Bao likely served at the Wei court substantially earlier than Shi Qi; b) the relatively later date of stories about Shi Qi completing the Zhang River project probably reflect the fact that salinization had rendered Ximen Bao's earlier efforts ineffective, and so there was a later effort to repair the Zhang River irrigation works; c) and the *Hanshu* inclusion of the story of Shi Qi was designed to refute the outlandish claims of Chu Shaosun. Hamakawa, however, does not explore differences between the *Lüshi Chunqiu* and *Hanshu* versions of the Shi Qi story (explored further below), nor does he discuss the stories within the overall rhetorical program of the *Shiji* and *Hanshu* treatises.

12. Ibid. The famous story is perhaps the most dramatic example of a water-control project bringing unforeseen long-term consequences. The story did not end there, of course, because maintenance obligations persisted long after the Han, even up to the present, as detailed in Will, "Clear Waters vs. Muddy Waters."

13. Yu county bordered the Yellow River in Qinghe 清河 commandery, some 250 km to the northwest of Huzi. See Tan, *Zhongguo lishi ditu ji*, vol. 2, 26.

14. *Shiji* 29.1409.

15. *Shiji* 29.1410.

16. The canal became the centerpiece of a network of transport canals that, by the end of Western Han, carried enormous shipments of grain throughout the Guanzhong region. See Li, "Lun Xihan Guanzhong pingyuan de shuiyun jiaotong."

17. After the Zheng Dangshi proposal and two of the three failed proposals, the treatise states, "the Son of Heaven held it to be as described" (*Tianzi yi wei ran* 天子以為然). See *Shiji* 29.1410–11. Passive approval of the emperor, then, was a necessary but not sufficient condition for water-control success.

18. *Shiji* 29.1412.

19. *Shiji* 29.1413.

20. We cannot overestimate the centrality of these songs to the treatise, because in his appraisal Sima wrote that he composed the treatise after being "moved by the poem at Huzi" (悲瓠子之詩) (*Shiji* 29.1415).

21. Note that in the treatise's appraisal (see below), Sima Qian referred to stopping up the dike at Xuanfang. Of course, Xuanfang did not exist prior to

Wudi's efforts at the dike, so the temple itself became the means by which the reconstruction effort was remembered.

22. Presumably, being part of the Yangtze River system, these nine rivers (*jiu jiang* 九江) differ from the Nine Rivers (*jiu he* 九河) associated with the Yellow River delta.

23. There is a second reading for 汭, one of which reads this as the Rui River, following the "Yu gong" chapter of the *Documents*.

24. *Shiji* 29.1414.

25. Other than Sima Qian, the treatise only briefly mentions two other officials in relation to the Huzi repairs: Ji Ren 汲仁 and Guo Chang 郭昌, both ordered by Wudi to raise several tens of thousands of corvée laborers. Sima Qian's praise for Wudi's efforts at Huzi contrast with passages striking a more critical note about Wudi's interventionist acts, particularly in matters relating to foreign wars and coinage manipulation. See Hans van Ess, "Emperor Wu of the Han."

26. The *Hanshu* treatise is not clear on this point, as discussed in further detail below.

27. *Hanshu* 29.1677. According to Yan Shigu, this statement meant that "he had a reputation for excellence in governance" (有善政之稱), a perhaps excessively positive interpretation of Ban Gu's choice of words.

28. *Hanshu* 29.1687.

29. *Hanshu* 29.1688. The "Yiwen zhi" 藝文志 of the *Hanshu* lists a *Wuxing zhuan ji* in one *pian* by Xu Shang in the *Documents* section, plus a *Xu Shang suan shi* in the calendrical writings section (*li pu* 曆譜). As the *Hanshu* treatise and the story of Xu Shang's role in water-control policy demonstrate, expertise in calculations was applied across different domains of knowledge.

30. *Hanshu* 29.1688. Xu appears to suggest that even though the Tunshi River is blocked off, it could still receive overflow during high water flow on the Yellow River. The descriptions in the *Hanshu* treatise and included memorials evince some disagreement, or even confusion, about the status of the Tunshi River.

31. As readers may recall, expertise in the *Documents* often meant expertise in predicting the future.

32. *Hanshu* 29.1688. The treatise seems to validate Feng Qun's memorial, because it states that the river "as a result overflowed" (果決) immediately after describing Xu Shang's rejection of the proposal.

33. *Hanshu* 29.1689. As is typical of many of his comments in the *Hanshu*, Du Qin, who served as Wang Feng's advisor, supposedly offered this opinion to Wang Feng in private, raising ultimately unanswerable questions about Ban Gu's sources. At the very least, and as we discuss in more detail below, Ban Gu's quotation of Du Qin here is clearly meant to engage with his larger discussion in the treatise about proper policy-making and rulership.

34. *Hanshu* 29.1689.

35. This statement is an almost verbatim quote from *Analects* 7.28. Ban Gu also cites this same *Analects* passage, but transforms it (see below).

36. Du Qin certainly got his position as advisor largely because of his status and family background. But Du Qin is constantly giving Wang Feng advice on all kinds of problems.

37. In the treatise, accordingly Wang Feng comes out looking quite a bit better than his nephew Wang Mang (see below).

38. Recall that Feng Qun's proposal prior to 29 BCE was rejected after just one person, the budget and numbers expert Xu Shang, deemed the project too expensive. Du Qin's plan thus hints that allowing budgetary concerns to trump technical considerations is not wise, even financially wise, in the long term. A collaborative process, Du Qin seems to suggest, will help prevent short-term concerns about money from having an upper hand in policy debates.

39. *Hanshu* 29.1690.

40. 眾庶見王延世蒙重賞, 競言便巧, 不可用 (*Hanshu* 29.1691). The two men were Li Xun 李尋 and Xie Guang 解光, who further argued that water-control efforts were ineffective, because floods and other changes in water level were caused by the rhythmic flux of yin *qi* and the "Way of Heaven" (*tian dao* 天道). See n. 47 below.

41. *Hanshu* 29.1679.

42. *Hanshu* 29.1690.

43. For a detailed discussion of the role of truthfulness and accuracy in court debates, see Olberding, *Dubious Facts*.

44. *Hanshu* 29.1695.

45. *Hanshu* 29.1697.

46. Whereas Feng Qun's pre-29 BCE flood memorial stated that the so-called "Nine Rivers" (*Jiu he* 九河) dredged by Yu 禹 to drain the Yellow River delta region into the sea were "difficult to identify" (難明) (*Hanshu* 29.1687), a memorial from Xu Shang after the 17 BCE flood stated that "ancient explanations regarding the names of the Nine Rivers mention Tuhai, Gusu, and Gejin, which today are found within the borders of Chengping, Dongguang, and Ge counties" (古說九河之名, 有徒駭、胡蘇、鬲津, 今見在成平、東光、鬲界中) (*Hanshu* 29.1690). See also the discussion of Han Mu 韓牧 and Wang Heng 王橫 below.

47. Li Xun and Xie Guang advocated a laissez-faire approach to flood management, using a logic based on natural patterns that is completely absent from earlier memorials. As their memorial stated, "when the *qi* of Yin flourishes then water responds by increasing" (陰氣盛則水為之長), and "even if water usually stays in low land, like the sun and the moon changing on the first and fifteenth days of the month, it is clear that the Way of Heaven has factors that give rise to actions" (猶日月變見於朔望, 明天道有因而作也) (*Hanshu* 29.1691). In other words, flooding was the result of rhythms that could not be changed, and dikes were ultimately useless against the unceasing patterns of low, then high water brought about by Heaven.

48. *Hanshu* 29.1697.

49. The passage in question is from the "Encountering Affairs" (Li shi 離事) chapter. See Pokora, *Hsin-lun (New Treatise)*, 112–13.

50. My translation follows, with some changes, Pokora, *Hsin-lun (New Treatise)*, 112–13.

51. *Hanshu* 29.1697.

52. The most famous example comes in the *Shiji*, when Hu Sui asks Sima Qian why Kongzi 孔子 (Confucius) compiled the *Annals* (*Chunqiu* 春秋). Sima Qian responds by quoting Kongzi to the effect that "empty words" (*kong yan*) cannot compare to the "profundity and brilliance of actions and affairs [of the past]" (行事之深切著明) (*Shiji* 130.3297).

53. Regarding Zhang Rong's memorial, Pokora wrote that the *Hanshu* text was "undoubtedly based on the *Hsin-lun*," but "more detailed and somewhat different." See Pokora, *Hsin-lun (New Treatise)*, 128n12.

54. As Yan Shigu pointed out in his commentary, the statement comes from the *Zuozhuan*. See *Zuozhuan*, Lord Zhao 1.5; Yang, *Chunqiu Zuozhuan zhu*, 4.1210; Durrant et al., *Zuo Tradition*, 3.1315.

55. *Hanshu* 29.1698.

56. With some alterations, my translation follows that found in Leys, *Analects of Confucius*, 20.

57. Note that the title of the treatise itself, "Gou xu zhi," is a reference to *Analects* 8.21:

子曰：「禹，吾無間然矣。菲飲食而致孝乎鬼神，惡衣服而致美乎黻冕，卑宮室而盡力乎溝洫。禹，吾無間然矣。」

The Master said: "In Yu, I find no flaw. He drank and ate a frugal fare, but displayed utter devotion in his offerings to the ghosts and spirits; he wore coarse cloth, but his liturgical vestments were magnificent; his dwelling was modest, and he spent his energy on ditches and canals (*gou xu*). In Yu, I find no flaw."

Translation follows Leys, *Analects of Confucius*, 23. Interestingly, however, whereas *Analects* 8.21 praises the direct physical efforts of Yu, *Analects* 7.28 (or, at least, the *Hanshu* interpretation thereof) praises not direct effort rooted in sagely knowledge but rather the broad accumulation of knowledge to assist in decision-making. So, while the "Treatise on Ditches and Canals" title might lead us to believe that the treatise is an implicit paean to Yu and his rule, the treatise itself seems to imply that a good ruler probably cannot hope to have Yu's prescience, and so must find an alternative mode of decision-making.

58. Garrett Olberding has translated *zhi* 志 as "the truth of the matter" or "vision." His point is that *zhi* is not a necessarily positive moral "intention" cultivated internally: "*Zhi* . . . [is] more than simply an intended objective but less than a strictly moral point of view. It refers to an objective to which one is personally committed and in which the actor believes there is something essential about the

world ... captured in it" (*Dubious Facts*, 37). A full assessment of Olberding's many interesting arguments must remain outside the bounds of this chapter, but his characterization of *zhi* seems to fit Ban Gu's use of the term in the "Treatise on Ditches and Canals," even if Ban clearly understood *zhi* to be morally positive. At the same time, other evidence from the *Hanshu* treatise provides a counterpoint to Olberding's argument that in early imperial debates the "correctness of the facts ... were less the point than the general propositions being forwarded" (Ibid, 174). While this is certainly true in many instances, the *Hanshu* treatise also shows us that at least some people within the late Western Han imperial court (e.g., Du Qin and Huan Tan) took technical facts and considerations very seriously and tried to devise strategies that would allow such facts to drive decision-making processes.

59. In his chapter in this volume, Jesse Chapman analyzes a different *Hanshu* treatise, the "Treatise on Astronomy" (Tian wen zhi 天文志), demonstrating that the treatise called for a new kind of authority (classical learning) to supplement or even supplant technical knowledge. Though the claims and rhetoric in the "Treatise on Ditches and Canals" are somewhat different, the arguments in this chapter are broadly congruent with those of Chapman.

60. De Pee, *Lost in the City* (forthcoming), ms. p. 210.

61. The most famous and influential version of this argument comes from Tan, "He yi Huanghe zai Dong Han yihou hui chuxian yi ge changqi anliu de jumian?" See also Elvin, *Retreat of the Elephants*, 25. Note that some scholars have disputed Tan Qixiang's theory, arguing that a larger shift toward a drier climate was also a factor in the post-Han reduction in flooding. For an overview of the debate, with citations and a more moderate take emphasizing that Tan was neither entirely "right" nor entirely "wrong," see two essays by Hamakawa Sakae: "Shimin shichijū man jin to kōdo kōgen" and "Kan Tō aida no kasai no genshō to sono genin."

62. To wit, in 278 CE, a figure no less important than Du Yu 杜預 (222–84) opined on dammed lakes (*bei* 陂) in the plain stretching between the Yellow and Huai rivers. For a detailed discussion, see Muramatsu, *Chūgoku kodai kankyōshi no kenkyū*, 185–212.

Asian-language Bibliography

Fujita Katsuhisa 藤田勝久. "Kan dai ni okeru suiri jigyō tenkai" 漢代における水利事業展開. *Rekishi gaku kenkyū* 歴史学研究 521 (1983.10): 1–16, 61.

Guo Junran 郭俊然. "Handai shuili zhiguan kaolun" 漢代水利職官考論. *Jiangsu keji daxue xuebao (Shehui kexue ban)* 江蘇科技大學學報 (社會科學版) 12, no. 2 (June 2012): 1–5.

Hamakawa Sakae 浜川栄. "Kan Tō aida no kasai no genshō to sono genin – Tan Kino setsu o meguru saikin no giron ni yosete" 漢唐間の河災の減少とその原

因--譚其驤説をめぐる最近の議論によせて. In *Chūgoku kodai shakai to Kōga* 中国古代社会と黄河, 323–63. Tokyo: Waseda daigaku shuppanbu, 2009.

———. "Shimin shichijū man jin to kōdo kōgen–Zen Kan Butei ki ni okeru kōdo kōgen no kankyō to kaihatsu" 徙民七十万と黄土高原—前漢武帝期における黄土高原の環境と開発. *Tōyō bunka kenkyū* 東洋文化研究 6 (2004.3): 32–61.

———. "Shōsuikyo no kenzōsha o meguru ni setsu ni tsuite" 漳水渠の建造者をめぐる二説について. In *Chūgoku kodai shakai to Kōga* 中国古代社会と黄河, 60–80. Tokyo: Waseda daigaku shuppanbu, 2009.

Hanshu 漢書. Compiled by Ban Gu 班固, 32–92, Ban Zhao 班, 48–ca. 116, et al. 12 vols. All refs. to the punctuated edition. Edited by Yan Shigu 顏師古. Beijing: Zhonghua shuju, 1962.

Li Lingfu 李令福. "Lun Xihan Guanzhong pingyuan de shuiyun jiaotong" 論西漢關中平原的水運交通. *Tangdu xue kan* 28, no. 2 (2012): 5–14.

Muramatsu Kōichi 村松弘一. *Chūgoku kodai kankyōshi no kenkyū* 中国古代環境史の研究. Tōkyō: Kyūko shoin 汲古書院, 2016.

Shiji 史記. Compiled by Sima Qian 司馬遷 et al. 12 vols. All refs. to the punctuated edition. Beijing: Zhonghua shuju, 1972.

Tan Qixiang 譚其驤. "He yi Huanghe zai Dong Han yihou hui chuxian yi ge changqi anliu de jumian?" 何以黃河在東漢以後會出現一個長期安流的局面. *Xueshu yue kan* 學術月刊. 1962 (2): 23–35.

———. *Zhongguo lishi ditu ji* 中國歷史地圖集. Shanghai: Ditu chuban she: Xihua shudian, 1982–1987.

Yang Bojun 楊伯峻 (editor and annotator). 1981 (2009). *Chunqiu Zuozhuan zhu* 春秋左傳注, vols. 1–4. Beijing: Zhonghua shuju.

Western-language Bibliography

Bielenstein, Hans. "Wang Mang, the Restoration of the Han Dynasty, and Later Han." In *The Cambridge History of China, Volume I: The Ch'in and Han Empires, 221 B.C.–A.D. 220*, edited by Denis Twitchett and Michael Loewe, 223–90. Cambridge: Cambridge University Press, 1986.

De Pee, Christian. *Lost in the City*. Forthcoming.

Dorofeeva-Lichtmann, Vera. "Where is the Yellow River Source? A Controversial Question in Early Chinese Historiography." *Oriens Extremus* 45 (2005/06): 68–90.

Durrant, Stephen, Wai-yee Li, and David Schaberg, trans. *Zuo Tradition, Zuozhuan,* 左傳: *Commentary on the "Spring and Autumn Annals*, vols. 1–3. Seattle: University of Washington Press, 2016.

Elvin, Mark. *The Retreat of the Elephants: An Environmental History of China*. New Haven, CT: Yale University Press, 2004.

Fukuyama, Francis. *The Origins of Political Order: From Prehuman Times to the French Revolution*. New York: Farrar, Straus, and Giroux, 2011.

Knoblock, John, and Jeffrey Riegel. *The Annals of Lü Buwei: A Complete Translation and Study*. Stanford, CA: Stanford University Press, 2000.

Leys, Simon, trans. *The Analects of Confucius*. Edited by Michael Nylan. New York: Norton, 2014.

Minuti, Orlando. "Oriental Despotism." *Europäische Geschichte Online (EGO)*, published by the Leibniz Institute of European History, 2012-05-03. http://www.ieg-ego.eu/minutir-2012-en. URN: urn:nbn:de:0159-2012050313 [08-07-2020].

Needham, Joseph. "Review of *Oriental Despotism: A Comparative Study of Total Power*, by Karl Wittfogel." *Science and Society* 23, no. 1 (1959): 58–65.

Needham, Joseph, Wang Ling, and Lu Gwei-djen. *Science and Civilisation in China*. Volume 4, *Physics and Physical Technology, Part III, Civil Engineering and Nautics*. Cambridge: Cambridge University Press, 1971.

Olberding, Garrett. *Dubious Facts: The Evidence of Early Chinese Historiography*. Albany, NY: State University of New York Press, 2012.

Pokora, Timoteus. *Hsin-lun (New Treatise), and Other Writings by Huan T'an (43 B.C.–28 A.D.): An Annotated Translation with Index*. Ann Arbor, MI: Center for Chinese Studies, 1975.

van Ess, Hans. "Emperor Wu of the Han and the First August Emperor of Qin in Sima Qian's *Shiji*." In *Birth of an Empire: The State of Qin Revisited*, edited by Yuri Pines, Gideon Shelach, Lothar von Falkenhausen, and Robin D. S. Yates, 239–57. Berkeley, CA: University of California Press, 2013.

Will, Pierre-Étienne. "Clear Waters vs. Muddy Waters: The Zheng-Bai Irrigation System of Shaanxi Province in the Late-Imperial Period." In *Sediments of Time: Environment and Society in Chinese History*, edited by Mark Elvin and Liu Ts'ui-jung, 283–343. Cambridge: Cambridge University Press, 1998.

Zhao, Dingxin. *The Confucian-Legalist State: A New Theory of Chinese History*. New York: Oxford University Press, 2015.

3

The *Hanshu* Geographic Treatise on the Eastern Capital

Lee Chi-hsiang 李紀祥

Translated by Michael Nylan

Preliminaries

The "contextualizing and recontextualizing of history" has become an important part of the recent work done by historians and archaeologists.[1] This chapter contributes to that ongoing effort with respect to early China in general and the *Hanshu* geographic treatise ("Dili zhi" 地理志) in particular, in light of earlier contributions to scholarship.[2] The chapter's several parts treat distinct and complex problems, but they are designed to build upon each other to achieve a much larger payoff, namely, the insight that Ban Gu's 班固 (32–92) geographic treatise posits a series of anachronistic equations to tie together a highly idealized vision of early Western Zhou, equipped with a secondary capital supposedly dating more than a millennium before Ban Gu's era. Accordingly, Ban Gu's geographic treatise by various means ties the capital of Western Zhou to Ban Gu's own capital of Luoyang in early Eastern Han. Ban Gu's imagined world, rooted in an understanding of the early Western Zhou as the best of all governmental orders, may have been inspired by a Postface to the *Documents* classic, known to Ban but now lost. In any case, Ban Gu's account quickly became established "fact" for scholars down through the ages and even until today.

To reconstruct what happened in remote antiquity for readers likely to be new to these materials, the chapter first summarizes what we know about the production of the current version of the *Documents* classic, which has been in circulation since the early fourth century CE (called by many scholars the pseudo-Kong *Documents*). It then highlights a conflation between two graphs 雒/洛 that occurs within the two authoritative texts, the *Documents* classic and the *Hanshu* (but in neither case consistently), because this particular conflation has had significant consequences for all readings of the pre-imperial past after Ban Gu.³ Again, by way of background, this chapter then proceeds to acquaint readers with several contradictory descriptions given by the *Documents* classic for the construction of a new city or settlement at Luo (雒/洛邑). Only after setting out these three logically separate problems does the chapter turn to its real focus: the relevant passage in Ban Gu's geographic treatise devoted to the capital of Luoyang. The chapter's final aim is to show how Ban Gu's deployment of a disparate range of sources advances a seemingly quiet but potentially explosive set of claims about Luoyang in his own day: to wit, that the Eastern Han capital of Luoyang could trace its heritage as capital back to the early Western Zhou and the city-building activities of its founding hero-kings.

Given the complexity not only of the multilayered *Documents* traditions but also of the relation between Ban Gu's *Hanshu* and its earlier counterpart the *Shiji* 史記, compiled nearly two hundred years earlier, the reader will need more than a dollop of forbearance to correlate these insights, even if this chapter tries to simplify and summarize many arguments. Not surprisingly, perhaps, the archaeological evidence does not offer much help to moderns dutifully wading through the sources with the firm intention of resolving the outstanding controversies. However, patience yields just rewards, for by the end of the chapter, readers will gain a far better understanding not only of Ban's methods in compiling his standard history, but also of how, intentionally or not, historical "facts" can sometimes be made up out of virtually nothing,⁴ and simply accepted because they are so often repeated. (For a similar effort that reaches similar conclusions, see J. Holmgren's illuminating essay "Myth, Fantasy or Scholarship: Images of the Status of Women in Traditional China," published in 1981).

I. On the *Documents* Postface/Chapter Prefaces (*xu* 序)

As all students of Chinese history soon learn, the *Shiji* describes Kongzi/Confucius (551–479 BCE) as the original compiler of a pre-unification collection titled the *Documents*, on the presumption that Kongzi must have assembled in that compilation the archaic pre-unification seal script writings available to him. According to the Han legends, at the time of Han Wudi (ca. 100 BCE), a distant descendant of Kongzi, Kong Anguo 孔安國, came into possession of part of that pre-Qin version of the *Documents* classic. Kong Anguo transcribed its archaic script into "modern" script (i.e., the clerical script used in Kong Anguo's mid-Western Han), so that he and the members of his family could produce a more user-friendly version for the throne, even as they made it the "family business" to teach the newly transcribed *Documents*. Han legends tell us, in addition, that the version taught by the Kong family was ten-odd *pian* (literally, bamboo bundles) longer than the version taught by Fu Sheng, the first court-sponsored Academician in Qin and early Han said to specialize in the *Documents* classic. Sima Qian adds, "In all probability, the [early] *Documents* was longer than this [meaning, the combined total of twenty-eight or twenty-nine court-sponsored chapters plus ten-odd more]."[5] Assuming that later editors did not interpolate this assertion into the *Shiji*, we note that by ca. 100 BCE there were already pronounced differences between a Kong family transmission and that associated with the Academician Fu Sheng and his disciples. However, the *Shiji* makes no reference to either a Postface to the Fu Sheng version of the *Documents* classic or to a Postface written by either Kongzi or Kong Anguo for the Kong family version of the *Documents*.[6] There is no early evidence, in other words, that any such Postface existed.

Indeed, the first references to a Postface appear roughly two hundred years later, in several passages in Ban Gu's *Hanshu,* including Ban's bibliographic treatise, where it is dubbed the "Hundred-*pian Documents* Postface" 百篇書序 and ascribed to the Supreme Sage Kongzi himself.[7] Evidently, Ban Gu believed that such a piece of writing had once existed, but, judging from the extant sources at our disposal, no lengthy *Documents* Postface describing a hundred chapters ever surfaced in either Western or Eastern Han.[8] Ban Gu was also the first to provide additional information about the *Documents* classic taught by the Kong family. Ban claimed that sixteen *pian* of manuscripts had been found in the walls of the Kong family home [in what we below call the "wall find"], so that the Kong family version of the *Documents* contained fully forty-five *pian*, rather than Fu Sheng's twenty-eight or twenty-nine *pian.* (NB: Other early sources deny that all of the sixteen *pian* found in the wall find were *Documents* materials.) Ban's bibliographic treatise says that "several ten" *pian* discovered in the wall find were once or in Ban Gu's time—the phrase lends itself to two entirely different constructions—transcribed into "archaic script," meaning the seal script in use in pre-unification times (凡數十篇, 皆古字也).⁹ Then Ban explains away the oddity that no one at court thought to sponsor this lengthier version. Because of the chaos engendered by the *wugu* 巫蠱 (witchcraft) affair of 91 BCE, Ban says, no Academician was ever appointed for this version of the *Documents* (未列於學官), despite its supposedly impeccable provenance within the family of the Supreme Sage himself.¹⁰ The *Hanshu* biography of King Yuan of Chu contains a letter by Liu Xin 劉歆 (d. 23 CE) complaining about the current crop of court Academicians, and Liu Xin's letter seems to corroborate Ban Gu's later account, insofar as it, too, states that sixteen *pian* were found in the wall find for the *Documents*, but somehow these chapters never went into general circulation among members of the governing elite.

Plainly, major discrepancies exist between the *Shiji* and *Hanshu* accounts, most importantly, that only the *Hanshu* talks of a "Hundred-*pian Documents* classic," which Ban Gu thinks must have once existed, covering the reigns from the sage-king Yao down to the Qin. For such a lengthy *Documents* classic, Ban Gu then imagines a hundred-part Postface designed to elucidate the circumstances and intentions behind the composition of each chapter supposedly known to Kongzi.¹¹ In early Tang, Yan Shigu 顏師古 (d. 645) wrote a commentary for these lines in the *Hanshu* bibliographic treatise, claiming that Liu Xiang 劉向 (*not* Liu Xin, but Liu Xin's father) believed that the *Yi Zhoushu* (*Remnant Zhou Writings*) corresponded to the "extra chapters" of the *Documents* known to Kongzi/Confucius but not included in Fu Sheng's version of the *Documents*],

> These Proclamations, Orations, and Edicts [in the *Yi Zhoushu*] *probably* were left over from the 100-*pian* version of the *Documents* referred to in Kongzi's teachings.¹²

No matter that none of Kongzi's teachings refer to this version of the *Documents* classic.

In all likelihood, then, talk of Kongzi's "Hundred-*pian Documents*" and its Postface arose with Liu Xiang (d. 7 BCE) and was perpetuated via his son's very influential writings. Note that Ban Gu's bibliographic treatise, we think, was based substantially on Liu Xin's "Seven Summaries" (Qi lue 七略); also that Ban Gu's treatise on the pitch standards cites two works by Liu Xin, Xin's *Classic of the Ages* (*Shi jing* 世經) and his "Triple Concordances Calendar" 三統曆, and that same treatise supplies dates for certain events that rely on a Postface for the *Documents* that Ban Gu, ca. 80

CE, either has in hand or knows of.¹³ As experts in the *Documents* classic frequently note, the only known Han-era *Documents* Postface apparently surfaced in connection with Chengdi's palace library collecting, which Liu Xiang supervised during the years 26–7 BCE, the last twenty years of his life, and Liu Xin for one year after his father's demise. It is almost certain, then, that talk of the "Hundred-*pian Documents*" arose during Chengdi's reign, and this is *not* an ancient tradition going back to Kongzi, contra the *Hanshu* assertions.¹⁴

Sadly, there has been no way to determine the truth of the matter, because for more than 1,500 years we have had but two main avenues by which to approach the Han-era *Documents*: the early fourth-century *Shangshu* (often confusingly identified as the "Archaic Script" *Documents* or the pseudo-Kong *Documents*) and the genuine Xiping Stone Classics 熹平石經 (carved in 175–183), now in fragments. Plainly, the early fourth-century version of the *Documents* comes from a different world than that related in Sima Qian's writings.¹⁵ So, not surprisingly, from at least the time of Zhu Xi 朱熹 (1130–1200), scholars have expressed doubts about whether Kongzi wrote a Postface to the *Documents* or not.¹⁶ And while some, if not most, scholars deem this "pseudo-Kong" version an outright forgery, I would emphasize this: the fourth-century version submitted to the Eastern Jin court by Mei Ze 梅賾 or his associates was accepted as genuine from the early fourth century down to the seventeenth century, in mid-Qing, in Yan Ruoqu's era. Beginning in the Tang (618–907 CE), this early fourth-century version, enshrined in the Tang-era *Wujing zheng yi* (Correct Meaning of the Five Classics 五經正義), became the sole court-sponsored version, and it is thus only relatively recently, since mid-Qing, that scholars have come to think of it as "forged" in Eastern Jin. Long before the fourth century, however, there existed extensive commentaries on earlier *Documents* versions (now lost), written by such eminent scholars as Ma Rong 馬融, Zheng Xuan 鄭玄, and Wang Su 王肅, not to mention the Xiping Stone Classics transcribed in the late second century BCE (175–183).¹⁷ But those sources exist now only in fragments as well.

Recall meanwhile that, so far as we know, Fu Sheng's court-sponsored version from Qin and early Western Han had no Postface at all, but Eastern Han scholars two or three centuries later knew one or more versions of a Postface. (Chinese does not distinguish singular from plural.) In my view, once Ban Gu thought he had seen a Postface compiled by Kongzi, then he recorded that the *guwen Shangshu* he knew had forty-six chapters (*juan* 卷, or scrolls).¹⁸ At the same time, based on present evidence, when the *Hanshu* bibliographic treatise speaks of a *guwen*/Archaic *Documents* and specifies its length, we can only say that Ban Gu definitely had seen or knew of a one-*juan Documents* Postface. Without sufficient evidence (and indeed, with counter-evidence in hand), we cannot posit that the version of the *Documents* classic seen by Ban Gu was identical to the version of the classic known to Sima Qian,[19] let alone how closely related the version seen by Sima Qian or by Ban Gu is to the fourth-century *Documents* Postface ascribed to Kongzi. To add an additional layer of complexity: the Postface (*xu*) Ban Gu and his peers saw *cannot be identical* to the early fourth-century *xu*/Postface promoted by Mei Ze, if only because the court-sponsored *Documents* classic in Han had fewer chapters and a different arrangement of chapters. As postfaces in the early empires functioned as something like "signing statements," explaining to readers the traditions surrounding a given text's composition and provenance, it is small wonder that the world of classical learning is still riven by debates about multiple issues relating to the early history of the *Documents*.

That said, when we turn to consider a second piece of writing ascribed to Ban Gu, his *Hanshu* geographic treatise, it bears every sign that Ban Gu was profoundly influenced by the late Western Han notion that Kongzi had compiled a lengthy Postface for the *Documents*, and Ban did not realize—or did not find it convenient to acknowledge—that this notion was a comparatively late invention, possibly as late as Chengdi's reign.[20] We can see that Ban Gu in his *Hanshu* treatises takes directions from a late Western Han Postface—not from the hand of Kongzi, but alleged to be so. This is especially evident in the geographic treatise's treatment of the "Tribute of Yu" 禹貢 chapter from the *Documents* classic, for the geographic treatise offers (1) direct citations from the late Western Han Postface for the *Documents* and (2) less explicit citations drawing from the same Postface. As we cannot assume an identity between the Postface that Ban Gu saw and the fourth-century collection of Little Prefaces anachronistically traced to Kong Anguo, scholars want to know more about the precise relation between the late Western Han–era Postface and the fourth-century Postface and, specifically, what changes occurred between Ban Gu and Mei Ze.

We are seriously hampered in that often we can only use Ruan Yuan's 阮元 *Shangshu zhushu* 尚書注疏 (preface completed 1815) to learn a bit about the problem under examination.[21] That said, the goal of this chapter is nonetheless to forge one possible entryway into such knotty problems, by a close examination of a single passage in the geographic treatise Ban Gu wrote for the *Hanshu*. After all, Ban Gu wrote as if he believed there was a transparent relation between Kongzi's original transmission of the *Documents* classic in archaic script and the one-*juan* Postface for the *Documents* that Ban knew. Indeed, there is some reason to suspect that Ban's belief in a 100-*pian* Postface ascribed to Confucius exerted a profound influence on Ban's writings. Evidently Ban Gu accepts the idea put forward by Liu Xiang and Liu Xin, father and son, that Kongzi/Confucius compiled a 100-*pian Documents*, because Ban also made his own history a work in 100 *pian* (unlike Sima Qian's work, in 130 *pian*).[22]

Therefore, the remainder of this chapter examines clues gathered from a single passage of the *Hanshu*'s geographic treatise, which treats the capital site of Luoyang as a capital site common to three eras—Western Zhou, Eastern Zhou, and Ban Gu's own Eastern Han—to see how Ban's acceptance of anachronistic views of Kongzi's civilizing project helped him to establish a strong sense of moral equivalence between the halcyon days of Western Zhou and Ban Gu's own court.

II. What's in a Graph? Luo 雒 or Luo 洛

In old texts—we don't know precisely when—the two characters 雒/洛 began to be treated as loan characters that were freely interchangeable.[23] This situation has caused a great deal of confusion to this day, especially when scholars discuss the topic of classical learning in Han. Significantly, Ban Gu's geographic treatise consistently, *in every case*, uses 雒, with a bird radical 隹, to designate place names in Henan commandery (e.g., Luo city 雒邑 and Luoyang county 雒陽縣), but Luo 洛, with the water radical, to indicate the Luo River whose source is in modern-day Shaanxi. Like the *Documents* "Tribute of Yu" chapter, then, Ban Gu's geographic treatise distinguishes Henan located in Yuzhou 豫州 from Yongzhou 雍州, basically the Guanzhong basin.[24] In this, Ban's geographic treatise appears to follow precisely the same conventions as the *Shiji*, as well as the convention used in the "Tribute of Yu" chapter in the *Documents* classic.[25] In Ban Gu's time, it seems—or at least to Ban Gu himself—there was no possibility of mixing up the two sites.

At the same time, outside the geographic treatise, the rest of the current *Hanshu* is inconsistent in its usage, perhaps reflecting the activist editing perpetrated on the history after Ban Gu's day.[26] Analogously, outside the "Tribute of Yu" chapter, the fourth-century *Documents* classic *in every case* undermines the important distinction between the two graphs, writing Luo

洛 for Henan, as well as for Shaanxi, suggesting that the ready conflation of the two graphs occurred before the early fourth century. The *Documents* does this with Luo city 洛邑 and with the chapter title "Luo gao" 洛誥, for example.[27] These intratextual inconsistencies are consistent with one another, in other words.

That Ban Gu's geographic treatise writes Luo city 雒邑 for the capital city in Henan, always, is at once unmistakable and vitally important—and not only because it tells us that it was likely later scholars who conflated the two characters for Luo. Why is it so important? Mainly because Luo 雒, referring to Henan, can signify not only the Eastern Zhou capital, but also the Eastern Han capital that the founder Guangwu established and Ban Gu served in Luoyang county 雒陽縣, Henan commandery. This usage alerts us, in other words, to the existence of a huge problem, a conflation or a confusion that creates a major stumbling block whenever we try to research the historical events in early Western Zhou history, especially the building of an Eastern Capital or ritual center at Luo.

Certainly, *Documents* scholars have made much of this inconsistency within the *Documents* usage across chapters. Pi Xirui's *Jinwen Shangshu kaozheng* acknowledges the issue:

> The [present/received] *Great Commentary to the Documents* (*Shangshu dazhuan*) in all the versions cited writes "Luo 洛 gao," as does the *Shiji* . . . , but in other passages Luo city is sometimes written as Luo 雒 City. I suspect that the graphic form of Luo in the chapter title was emended by later people, as the Xiping Stone Classics, for the *Documents*' "Duo shi" 多士 chapter writes Luo 雒 to indicate that place [in Henan].[28]

Wang Xianqian's 王先謙 *Shangshu Kongzhuan canzheng* 尚書孔傳參正 (1904) notes the same problem in his commentary on the chapter title "Luo gao" (now written as 洛誥):

> "Luo" 洛 ought to be written as "Luo" 雒, as we can see plainly from the "Tribute of Yu" chapter in the *Documents*. The Xiping Stone Classics writes that graph Luo 雒,[29] and Pi and other scholars, while not daring to emend the character, have raised the issue. We therefore know that the old place name Luoyang 雒陽 was emended to Luoyang 洛陽.

Wang, following Duan Yucai 段玉裁,[30] believes that Cao Pi 曹丕 (186–226), first emperor of Wei, decided to emend the character Luo for cosmological reasons. Unsurprisingly, by this account, Cao chose to justify his decision by saying he was reversing an earlier Han precedent and returning to earlier usage.[31] Yan Shigu, in commentary on the *Hanshu* geographic treatise, cites a story told by the Three Kingdoms scholar Yu Huan 魚豢 and then remarks, "If we follow . . . the story, then it was after Guangwu, r. 25–57 [that the Eastern Han court] changed the character to become Luo 雒."[32]

In my opinion, Yan Shigu's account is not credible, even if it is based on a relatively early source. After all, Ban Gu lived in Eastern Han, and he served two courts after Guangwu. He was careful to preserve the distinction between the two graphs in some cases. Therefore, the surmise that "after Guangwu" we first see the change from Luo with the water radical to Luo with the bird radical is simply not tenable. It seems rather clear that from the Wei-Jin period the two characters were mixed up, insofar as Du Yu's 杜預 (222–285) *Chunqiu Zuoshi jingzhuan jijie* 春秋左氏經傳集解 and Yang Xuanzhi's 楊衒之 (d. 555) *Luoyang qielan ji* 洛陽伽藍記 both use Luo 洛 when speaking of the city of Luoyang, in Henan.[33] But no matter. For our purposes, it suffices to know that the *Hanshu* geographic treatises consistently take Luo 雒 to be the correct *ancient* form for the Henan site. Qing scholars in the Evidential Learning tradition ascertained, moreover, that the usage in Ban Gu's geographical treatise tallies with the usage in the late Eastern Han Xiping Stone Classics,[34] ensuring that the two Luos were not always deemed identical in Eastern Han.[35]

By this point in our inquiry, it may help to recapitulate several related facts:

1. The discussion of the Western Zhou Luo city differs in the *Shiji* and in the *Hanshu*, and only the *Hanshu* (compiled ca. 82 CE) seems to have been cognizant of an authoritative postface to guide its discussions about past history.

2. The only *Documents* Postface to surface in late Western Han and Eastern Han writings was one-*juan* long, and hence the only version Ban Gu could have seen was a standalone manuscript in itself. By contrast, the fourth-century counterpart to the Han-era Postface eventually came to be separated into chapter postfaces, then chapter prefaces (aka

the "Little Prefaces") that reflected a larger *Documents* corpus that incorporated not only the Han-era materials of Fu Sheng but also a set of additional chapters incorporated into the new Mei Ze version of the *Documents*.[36]

3. Both Ma Rong's and Zheng Xuan's commentaries treated the *Documents* Postface in a single, one-*juan* piece of writing, a fact Duan Yucai proved in his *Guwen Shangshu zhuan yi* 古文尚書撰異. Therefore, the fragments now ascribed to Ma Rong and Zheng Xuan must have undergone significant alterations to retrofit them to the new, longer fourth-century CE Mei Ze version of the *Documents* classic.[37]

4. The late Eastern Han Stone Classics (erected 175–183 in Luoyang) used the court-sponsored Modern Script versions, which can be traced back to Fu Sheng's teachings on the *Documents* and reproduce the twenty-nine *pian* transmission by the Ouyangs, who were Academicians at the Western Han court, i.e., twenty-eight chapters plus a "Tai shi" of dubious origin that was submitted to the court either under Han Wudi or Han Xuandi. The Xiping Stone Classic for the *Documents* distinguishes the two graphs for Luo.[38]

5. Aside from the "Tribute of Yu" chapter, the so-called Kong Anguo (aka Mei Ze) version of the *Documents* ignores the distinction between the two graphs for Luo, most importantly in the chapter title and contents of the "Luo gao."

The foregoing suggests that, gradually over time, *Documents* experts tended to conflate the two Luos, regardless of their preferred interpretive lineages, simply because of the importance ascribed to a fourth-century Postface. This fourth-century Postface (eventually chopped into chapter postfaces) possibly incorporates some lines from the Han-era Postface known to Ban Gu. Still, it had to be substantially new if it was to accompany the Mei Ze version. How the current *Documents* chapter postfaces derived from the fourth-century Mei Ze submission relate, in turn, to Ma Rong's and Zheng Xuan's commentaries is unknown, and there is no way to know this, absent future manuscript finds supplying new evidence. For this reason, I below argue that Ban Gu's geographic treatise preserves a distinction that should shape our own thinking about events in early history. For Ban clearly thought the proximity of the Western Zhou Luo city site conferred great legitimacy on the Eastern Han capital in Luoyang more than a thousand years later. To understand the subtle complexities underlying the case that Ban Gu commits to, and recognize the confusion that it has engendered, we have to go back in time once more in the following section.

III. On Siting Cities in the "Zhou Writings" 周書 *Documents*' Section

The *Documents* includes a great many records about the legendary Zhougong's move, early in Western Zhou, east to an area near a Luo River. They concern the building of a major site on the Luo, long construed as an eastern capital paired with the western capital in the Guanzhong basin, in Shaanxi (see below for the "two capitals" theory). For example, the "Proclamation for Kang" ("Kang gao" 康誥) chapter says that "Zhougong at the beginning built a 'great new city/settlement' 新大邑 in the eastern kingdoms, at Luo 洛."[39] Upon close reading, we learn that Zhougong first divined the new site, making it likely that Zhougong had built an entirely new city, with walls, and did not merely "renew" or repair an old city,[40] after which Zhougong, as the Western Zhou representative of the king, or the king himself, supposedly himself came to Luo.[41] Zhougong then offered a *she* 社 sacrifice at the "New City," which consisted of one ox, one sheep, and one pig, after which he gave the remnant Yin groups their orders. By the *Documents*' account, Shaogong, a senior Western Zhou advisor, also supposedly offered presents to Zhougong during the course of the ritual ceremonies. According to Han-era traditions, the city was duly completed in year 7, at which time Zhougong "returned the rule" to the young King Cheng. By some Han traditions, Zhougong also reported to King Wen and King Wu at a temple that his work had been completed. Several chapters in the "Zhou Writings" section of the *Documents* classic—but especially the "Luo gao" 洛誥 and "Shao gao" 召誥 chapters—recorded phases in what seems to be a single lengthy process of constructing this New City, even if the "Shao gao" makes Shaogong its principal speaker, whereas the "Luo gao" mainly records a dialogue between Zhougong and King Cheng.[42]

Several features are noteworthy in this "Zhou Writings" section in the *Documents* classic:

1. Zhougong reportedly made no fewer than three divinations about the site (not one), with the text specifying the sequence: 我卜、乃卜、又卜. Presumably the first divination was not auspicious, so they divined again, but the second round of divinations were all auspicious 洛食. (Note the use of that pesky Luo character again.)

2. Whereas four chapters—"Kang gao," Luo gao," "Duo shi," and "Shao gao"—mostly concern the construction of a city, they are inconsistent in the ways they describe the new city. The "Kang gao" uses the three-graph phrase, "New Great City/Settlement" 新大邑. The "Duo shi" and "Shao gao" chapters use a shorter, two-graph formula, "Great City/Settlement" 大邑. But the "Luo gao" speaks twice of the "New City/Settlement" 新邑. Meanwhile, the "Duo shi" speaks of the "New City at Luo" 新邑洛 and also uses the phrase "[at] this Luo" 茲洛. Notably, the *Documents* classic itself only one time calls the new city "Luo city" 洛[=雒]邑, in contrast to later sources referring to the *Documents,* including Sima Qian's *Shiji*.

3. Sima Qian's "Basic Annals of Zhou," by contrast, uses the term Luo city 雒邑, describing its construction as the fulfillment of the Zhou founder's wishes. Plainly, the *Shiji*, in speaking of Luo city 雒邑—never the New City/Settlement—merges several passages from different *Documents* chapters to describe a single ongoing event, the construction of Luo city.

4. *Shiji* contributes to later traditions in yet another seminal way, insofar as it alleges that Zhougong placed the Nine Tripods in a building, presumably a temple, at Luo city. Indeed, the *Shiji* makes that one ceremonial occasion, along with Zhougong's decision to "return the rule" to King Cheng, the real motivation underlying the composition of no fewer than four of the *Documents* chapters, "Shao gao," "Luo gao," "Duo shi," and "Wu yi" 無逸.[43]

5. Thus *Shiji* answers an important question about *when* the Nine Tripods were placed at the new site. At the same time, Sima Qian never states when the Zhou got the Nine Tripods, sign of Heaven's Mandate. The tripods suddenly appear,

shortly after the Zhou conquest.[44] But the *Shiji* states, clearly enough, that the planning or construction of an eastern city site to house those tripods began under King Wu, and Zhougong completed it to "follow King Wu's intention" 如武王意.[45] By Sima Qian's account, we must conclude, it was King Wu's express wish, recorded in his testamentary edict, that there be built an eastern metropolis, and Zhougong and his liege King Cheng set about to realize this earlier plan. By Sima Qian's reckoning, still later came the edicts recorded in several other *Documents* chapters ordering the local lords to gather the "remnant Yin men" who were to work on a construction site,[46] although Sima Qian's account is remarkably unclear on the question of where the remnant Yin are finally to be settled, at the same construction site or at a second construction site.

6. By contrast, Ban Gu's *Hanshu* geographical treatise, following statements in the *Zuo shi chunqiu*, precursor to today's *Zuozhuan/Zuo Traditions*, places most of these events during the short reign of the Western Zhou founder, King Wu, *before* Zhougong and King Cheng are in power.

7. Judging from other early extant sources, the date of the beginning of the legendary building process—whether it was during the reign of King Wu or after King Wu's death, during the reign of King Cheng, heir to King Wu—remained a bone of contention during Han times.[47]

Why should we even care at this remove about such seemingly trivial matters—a matter of one or more city sites, let alone a matter of a few years, more or less—in the historical retrojections that picture a time more than a millennium before the Eastern Han events? I would answer, because the various pictures and dates given for the same process may well have different implications, a fact that should not be overlooked. To take one example, it matters a great deal to Western Zhou history whether Zhougong built some sort of temple or a secondary capital at a certain location, and when, before or after, he returned the rule to King Cheng.

As readers may recall, Western Zhou was founded circa 1050 BCE, though traditional dates put the founding decades earlier. Building a new city at Luo would have had to happen not long afterward, within the first ten years of the new dynasty, judging from the *Documents* classic and its *Great Commentary* (*Shangshu dazhuan*). Roughly three centuries later, in 771 BCE, a badly weakened ruler of Western Zhou, King Ping, had to flee the Guanzhong basin for a Luo city in Henan, and roughly two centuries later, another king of Zhou, King Jing 景王 (r. 544–520 BCE), supposedly built a city wall around a site called "ChengZhou" (Zhou Completed/Perfected).[48] Many classical scholars who knew the *Documents* traditions assumed these last two major events happened at one and the same site, whose significance can be traced back to the "New City" at Luo that is described in the *Documents* classic.[49] However, when the *Shiji* speaks of King Ping "moving east" in 771,[50] its author, writing in the mid-Western Han capital in Chang'an, in the Guanzhong basin, never confused the king's enforced move to a "new city" (identical with Luo city and hence at or near Luoyang) with residence in a rightful Western Zhou capital. For Sima Qian, then, the "proper" or "rightful" Western Zhou capital was always at Hao 鎬, in the Guanzhong basin in Shaanxi, prior to the enforced move to Henan. But the situation had changed drastically for Ban Gu, residing some two hundred years later in the Eastern Han capital of Luoyang, in Henan. For Ban Gu made King Ping's flight to Luo city—dubbed the "King's City" first in Eastern Han—a glorious return to a hallowed secondary capital site with impeccable early Western Zhou credentials.[51] Put another way, the force of Sima Qian's vision, however strong, did not preclude later texts from envisioning a different model that I dub the "two capitals" model outlined below.[52]

Further Complications: ChengZhou, near the Old Luo City Site?

Over time, almost certainly in Eastern Han (i.e., after the Han capital was relocated in Luoyang), there came to be a theory insisting that the *Documents* classic described two capitals (Hao and Luo city) with different names at two locations. In truth, the *Documents* classic remains remarkably unclear on the question of whether early Western Zhou and its superheroes Zhougong, Shaogong, and King Cheng built one city or two, and where. Ban Gu's *Hanshu* bibliographic treatise and, still more so, his *Hanshu* geographic treatise played an important role in establishing or promoting the "two capitals" model, configuring Hao as the western capital and Luo as the eastern. I submit, as my working hypothesis, that Ban Gu supported this "two capitals" model by drawing heavily from the one-*juan* Postface he knew was appended to the Han-era *Documents* classic.[53]

In the current edition of the *Shangshu zhushu* (Ruan Yuan's edition of 1815), the fourth-century chapter preface appended to the "Shao gao" *Documents* chapter says,

> King Cheng was in Feng. He wanted to take up residence (or simply build?) at Luo city, and he sent Shaogong to oversee the location, and there was compiled the "Shao gao."[54]

But the preface currently appended to the "Luo gao" *Documents* chapter says, by contrast, that Zhougong built ChengZhou (Zhou Completed/Perfected), and then he composed the "Luo gao."[55] As one *Documents* preface speaks of Luo city and one of ChengZhou, we are left with a conundrum regarding the relation between the two sites. Were there, in fact, two Western Zhou capitals or only one, and two major cities in Henan or only one? If we then turn to compare these two chapter prefaces in the fourth-century Mei Ze *Documents* version with the chapter preface appended to the "Duo shi" chapter in the same *Documents* version, we read more discrepant language describing Zhougong's early efforts. That chapter preface says,

> Once ChengZhou was completed, they moved the recalcitrant people of Yin there. Zhougong, using a royal decree, made a proclamation there, and he compiled the "Duo shi" chapter.[56]

As we read on in the "Luo gao" and "Shao gao" chapters from the fourth-century *Documents*, we can, in fact, construe events in one of two contradictory ways: either Zhougong and Shaogong each built a single city, even if they were working in a cooperative mode, or Zhougong and Shaogong each had his respective tasks in building a single city. Yes, the "Duo shi" preface emphasizes the building of a ChengZhou, in connection with settling the remnant Yin peoples around the lands that were in or near modern Anyang, but because the "Luo gao" has Zhougong divining two sites 卜二地, perhaps he superintended the building of two discrete cities 營二邑. And whereas the fourth-century chapter prefaces seem to indicate or hint at such a possibility, the so-called "Kong Anguo commentary" and Kong Yingda's subcommentary make that possible scenario far more explicit.[57] In any case, the name or epithet applied to the new city, "Completed or Perfected Zhou," might (a) refer to Zhougong's recent conquest over the rebels; or it might (b) signify that the new city was designed to complement ZongZhou 宗周 (Ancestral Zhou), in the Feng 灃 river valley, essentially the old capital at Hao; or it might (c) by design simply praise good King Cheng for his excellent reign. Moderns will never know if one, two, or all three meanings were important to the early Western Zhou advisors to the court at the time.

But some things should be plain enough to modern historians. If we now step back and think a bit, we will quickly see how unlikely it is that King Cheng, given the transportation and communication systems of the time, would have made two inspection visits to the eastern lands, despite the language used in the fourth-century CE chapter prefaces.[58] Furthermore, nowhere does the *Documents* classic lend itself to talk of two major cities under construction in early Western Zhou. However, the precise language used in the so-called "Kong Anguo commentary" for the fourth-century "Duo shi" chapter does lend itself to the vision of two major city sites built at or near Luo:[59]

> The [former] Yin counsellors and men in service were not always in their hearts modeling themselves on the charismatic virtues, and so they were *moved near the king's capital* 徙近王都 so that they might be instructed.[60]

Going further, the fourth-century *Documents* chapter preface seems to imagine two cities of different size and significance at or near Luoyang 洛陽, one a "secondary or lower capital, a "king's capital," and the second a metropolis "near the king's capital."[61] Both the king's capital and ChengZhou were thought to be in or near Luoyang, as the subcommentary ascribed in early Tang to Kong Yingda makes plain.

We have no early evidence that there was a city by the name of Luoyang 雒陽 in existence in early Western Zhou. If I am correct, it was the fourth-century version of the *Documents* classic that starts talking about Luoyang (aka the New City) in relation to a second city site of ChengZhou, which is thought to be lesser in size and importance. Clearly, the fourth-century commentary incorrectly ascribed to the Han-era scholar Kong Anguo adopts relatively new phrases such as "king's capital" 王都, equated with Luo city, to cast one of the two Henan sites as a "secondary capital," second only to that at Hao, in the Guanzhong basin, in Shaanxi.[62] By my working hypothesis, then, the Kong Yingda subcommentary was probably wrong, but it drew inspiration from somewhere, most likely the fourth-century *Documents* "Lesser Prefaces." For the fourth-century *Documents* "Lesser Prefaces," the so-called "Kong Anguo commentary," and Kong Yingda subcommentary together represent a seamless tradition carefully stitched together that soon became the "common wisdom" among classical scholars. That seamless tradition drew support, I believe, from two unlikely sources: (1) the "Luo gao's" talk of these "divining two locations" 卜二地; and (2) the historical situation in Eastern Zhou after King Ping moved east to Luoyang, because we find in the *Annals*' traditions in Eastern Han similar talk about both a "King's City" and a "ChengZhou."[63]

We now have to ask whether that same "common wisdom" informed the treatises in Ban Gu's *Hanshu*. It is my belief that Han thinkers, when trying to reconstruct early Western Zhou events, at a remove of more than a millennium, gained some inspiration from the *Documents* classic itself, but their thinking was no less heavily influenced by what the *Annals* classic had to say about King Ping's momentous move of his capital east. This dual influence helps to explain why the Han classicists then compiled the Han-era *Documents* Postface in the way that they did, and then (mistakenly) ascribed that piece of writing to Confucius himself by the time of Liu Xiang and Liu Xin. Ban Gu's *Hanshu* geographic treatise, which was compiled, not coincidentally, during a time when the capital was at Luoyang, allows us a valuable and unique glimpse of this process, I would argue.

IV. The "Two Capitals" Model Promoted in the *Hanshu* Geographic Treatise

a. Overview of the Model

Scholars today hotly debate the historical accuracy of the dynastic histories composed during the early empires and, even more so, the historical accuracy of the few Classics and masterworks that seem to describe the Shang-Western Zhou transition. Ban Gu's geographic treatise speaks of early Western Zhou events in a single entry discussing Henan commandery and Luoyang county, the environs of the Eastern Han capital of Luoyang in Ban's day. It begins by telling us that Henan commandery, under the Qin dynasty (221–210 BCE), was once called "Three Rivers Commandery" (Sanchuan jun 三川郡). As Ban Gu himself comments, it was the Western Han founder who changed the name of the area to Henan commandery. Readers are then told that Luoyang county belongs to Henan commandery,[64] and the population figure for Luoyang city (not county) is given as 52,839 households. The main text also supplies other figures: the registered households in Henan commandery number 276,444, and the individual mouths number 1,740,279⁶⁵ in its twenty-two constituent counties.

From this quiet beginning, the text goes on to make major assertions,

> Luoyang is the place where Zhougong moved the Yin people. It became ChengZhou 雒陽, 周公遷殷民, 是為成周. In the *Annals* classic, for the thirty-second year of Lord Zhao, it says, "The state of Jin brought together the local lords at Diquan" 狄泉, since this land is the great city of ChengZhou, where King Jing of Zhou lived [in the sixth century BC]. . . .
>
> Henan [county] is the old land of Jiaru 郟鄏. King Wu of Zhou moved the Nine Tripods here. When Zhougong brought about the great peace, he built a metropolis there, which was to be the King's City, and it came to pass that King Ping lived there [at Luoyang centuries later, in the early eighth century].⁶⁶

Obviously enough, Ban Gu, in talking of the Han commandery of Henan, wants to trace the locale as far back as possible to its historical origins. In doing so, it is no less clear that Ban takes Luoyang to be essentially the early Western Zhou city of ChengZhou, that is, the "secondary capital" as opposed to the eastern capital at Hao, in close proximity to Ban's Eastern Han capital at Luoyang; also that Ban thinks this is the same ChengZhou of which the *Annals* classic spoke, when recounting events in 771–770 BCE, the time of the forced transition to Eastern Zhou and the transfer of some capital functions from Hao to Luoyang. Ban further equates the Han county of Henan with the Luo city built by Zhougong, that is, what the *Annals* calls "the King's City." This Ban locates at a specific site, and, no less importantly, Ban says that this is the very site where King Wu placed the Nine Tripods.⁶⁷ Moreover, he introduces a brand-new term, "King's City" 王城, in his geographic treatise (though whether that phrase should be lower- or upper-case, with the latter indicating a proper noun, is far from certain).⁶⁸ In light of the discrepancies in the stories about the Nine Tripods in the *Shiji* and *Hanshu*, not to mention the *Zuozhuan*,⁶⁹ we are reminded of how many rival traditions were in circulation during the early empires. Yet Ban Gu happily conflates a series of terms for important Western Zhou sites, including ChengZhou, Luo city, Luoyang, and King's City. To reiterate, in hopes of obviating further confusion: Ban Gu's "two capitals" model refers to Zongzhou/Hao, on the one hand, and Chengzhou/

Luoyang, on the other, with the epithet "King's City" reserved by Ban Gu for Chengzhou/Luoyang, as we have seen and will see below.

The *Hanshu* geographic treatise then continues,

> The [distant] forebear of the Qin ruling house was called Boyi 柏益.... At the time when King You was defeated by the Dog Rong, King Ping moved east to Luo city 雒邑. Lord Xiang led troops to save Zhou, and he was meritorious, so, as reward, he received the lands of Zhi 邹 and Feng 鄷, at which point he was ranked among the Zhou nobility.[70]

Ban Gu follows long-standing traditions in having King Ping moving the capital east to Luo city, certainly. But he then adds color and detail when he writes,

> Long, long ago, Zhougong constructed Luo city. As he took the site to be the center of the land, he thought that the local lords from the four quarters would [find it convenient] to assemble there [for court audiences], and so he set up a capital [in that place]. When it came about that King You was seduced by and dallied with Bao Si 褒姒, so that ZongZhou (i.e., Hao) was destroyed, his son Lord Ping moved the capital east to Luo city [in 771–70 BCE]. Later, the hegemons led the local lords to revere the Zhou house. Therefore, the Zhou was the longest lasting of the Three Dynasties [of legendary Xia, Shang, and Zhou]. For 800 years, it [the dynasty] lasted, until King Bao 赧, when their capital became a single administrative ward 里 under Qin.[71]

In a valiant attempt, perhaps, to reconcile the *Annals* traditions with those recorded in the *Documents* classic (see the Appendix), Ban Gu imagines Luo city and ZongZhou (Hao) to be two royal domains with their respective capitals, east and west, from the beginning of Western Zhou dynastic history. He continues,

> When the Han arose, it employed the Qin administrative system.... But during the reign of Han Wudi (r. 141–87 BCE), who pushed back the northern and southern barbarians and thereby greatly enlarged the empire's territory, the com-

mandery of Jiaozhi was set up in the south, and Shuofang in the north. . . . And at this time, Yong 雍 [i.e., Guanzhong basin, the old Zhou base territory] was called Liang 涼; and the older Liang 梁 territory was then named Yi 益 [as the provincial borders were adjusted to the new map of Han territories in order to retain the traditional Nine Provinces], making thirteen [new] commanderies, for which an Inspector was established.

The traces of the former kings are far distant, and the place names were also repeatedly changed, and so we have examined the old sayings, and looked for clues in the *Odes* and *Documents* in order to extrapolate the [correct] sites of the mountains and rivers, from the compiled "Tribute of Yu," the *Zhou guan*, and *Annals* classic, right down to the Zhanguo, Qin, and Han.

The foregoing passages not only reveal that Ban Gu's geographic treatise, in reconstructing early Western Zhou events, puts forward a "two capitals" model at Hao and at Luo, but that Ban always fashions his narrative about these two distinct seats of power in antiquity *by taking into account the much more recent histories of events taking place in those locations*. Of course, Ban Gu is not so crude as to draw a straight line from early Western Zhou down to the Western Han, or even from Eastern Zhou to Eastern Han. Still, he sketches connections, as the *Annals* traditions in particular allow him to do, between a reconstructed early Western Zhou "eastern" or "secondary" capital at Luoyang and the Han commandery-county system he knew. In all likelihood (but here I speculate), Ban was moved to write of the two Western Zhou capitals because his theory of "two capitals" legitimates the Eastern Han decision by the founder to move the capital from Chang'an (in Guanzhong, in Shaanxi) to Luoyang (in Henan) at the start of Eastern Han.

b. Later Influence of Ban Gu's "Two Capitals" Model

Ban Gu's reconstructed "two capitals" model has exerted an enormous influence on later scholars, who generally thought his account absolutely authoritative. We could begin with Zheng Xuan's (127–200) understanding of the King's City, as supplied in his commentary on the *Odes*, where Zheng writes,

> ZongZhou 宗周 is the capital Hao 鎬; it refers to Western Zhou. Talk of the Zhou "King's City" refers to Eastern Zhou. When

ZongZhou was destroyed, King Ping moved the capital east, and the government became weaker. . . .⁷² The term "King's City" refers to the Zhou's Eastern capital, whose domain is 600 *li* square; the fief was then in Yuzhou 豫州, according to the "Tribute of Yu" chapter. . . . In the beginning, when King Wu built his city at Hao, he renamed it ZongZhou, and made it the "Western capital". . . . In the fifth year of Zhougong's rule, King Cheng was in Feng, and he wanted to make a residence at Luo city, so he sent Shaogong to inspect the city. Once that was built, he called it the "King's City," and it became his Eastern Capital. This is present-day Henan 河南. Once Shaogong had inspected the site, Zhougong went to build ChengZhou, which is present-day Luoyang 洛/雒陽. King Cheng lived at Luo city, and he moved the recalcitrant Yin people to ChengZhou, after which he returned to his Western Capital [at Hao/ZongZhou].⁷³

Ban's influence (directly or indirectly, through Zheng Xuan) is also perfectly clear when we read Zhang Shoujie's 張守節 (fl. 725–735) *Shiji zhengyi* commentary, which equates Luo city with the "King's City."⁷⁴ Evidently, Du Yu's 杜預 (222–85) commentary on the *Zuozhuan* also borrows from Ban Gu's geographic treatise, even as it departs from it, insofar as Du, when recounting the reign of King Jing, has him saying that he wants to "restore the city of King Cheng" to its former glory, in hopes that this might settle the unrest of the local lords (今我欲 . . . 脩成王之城 . . . 諸侯用寧).⁷⁵ The fifth-century *Hou Hanshu* by Fan Ye (d. 445) equally attests Ban Gu's enormous influence, for Fan adds a record taken from Sima Biao's 司馬彪 (d. 306) "Treatise on the Commanderies and Kingdoms" (Jun guo zhi 郡國志), which says, in the entry for the Overseer of Henan 河南尹, "Luoyang 雒陽 in the Zhou period was named ChengZhou. . . . [In] Henan there was the walled city Luo city 雒邑 built by Zhougong; during the Chunqiu period, it was known as the "King's City."⁷⁶ Li Daoyuan's 酈道元 *Shuijing zhu* 水經注 (dated 525?) is still another work that hearkens back to Ban Gu's model in the *Hanshu* geographic treatise.⁷⁷

Turning to the modern period, Ban Gu's influence continues unabated, as is clear from reading Zhao Yi's 趙翼 (d. 1814) writings on the subject, which likewise try to reconcile all the materials that he knew, in particular, Ban Gu's geographic treatise, the *Annals* traditions, and the *Documents*' talk of divining two sites.⁷⁸ Similarly, the Qing scholar Liang Yusheng 梁玉繩 (1744–1819) follows Ban Gu and Zheng Xuan in the belief that they are absolutely authoritative when it comes to antiquity.[79] By the time we get to Takigawa Kametarō's 瀧川龜太郎 *Shiji huizhu kaozheng* 史記會注考證 (1932–34), we can hardly be surprised that Takigawa speaks confidently of Zhougong building a "secondary capital" in Henan.[80]

This sort of "two capitals" model should be highly suspect, however, chiefly because it assumes that there could be no changes in the urban landscape between early Western Zhou and the site many centuries later, in Eastern Zhou, or even Eastern Han. The identifications that have made Luo city the "King's City," in contrast to ChengZhou, are all deeply problematic as a result. We sense this if we carefully consider the work of Zhao Yiqing 趙一清 (1709–1784), who, in collating editions of the *Shuijing zhu*, writes under the entry for the Jian River 澗水 near Luoyang,

> When King Ling 靈 of Zhou dammed the Gu River 穀水 . . . the two rivers, the Gu and the Chan 瀍水, still did not flow into the Luoyang walled city 洛陽城. But when the Eastern Han made that city their capital, they drew water from 15 *li* away, at Qianjin jie 千金堨, so that it [the water] would travel around the city, as a moat. North and South then had the capacity for transport vessels to pass through, and the Chan River *for the first time*, along with the Gu River 穀, flowed east. In antiquity, certainly the Chan did not join up with the Jian. Nor did it pass through Luoyang county. It traveled southward but not as far as Yanshi 偃師.[81]

The desire to merge and reconcile disparate traditions (even relatively late traditions, such as we find in the *Annals* traditions) apparently surpassed any impulse to investigate the sources more closely.[82] It is my belief that the "two capitals" traditions came into existence only after Guangwu, in Eastern Han, made Luoyang his capital; it was Guangwu and his court who named Luoyang the "Eastern Capital," as he deemed Chang'an the "Western Capital." As both Ban Gu and Zheng Xuan lived and worked during Eastern Han, their work reflects the contemporary discourse that they knew.

V. Archaeological Evidence for Luoyang

Given the contributions of scientific archaeology, we must ask whether and how much it can contribute to elucidating events that Ban Gu could hardly hope to clarify, despite his best efforts, in view of his rhetorical duty to glorify the choice of Luoyang and Henan for the Eastern Han capital of his imperial patrons. Put another way, it is vital to consider whether two capitals with two names were built during early Western Zhou times, or whether talk of two capitals is an artifact of Eastern Zhou or even later, in Eastern Han. As scholars today continue to debate this problem, this section therefore reviews the archaeological evidence at hand relating to this important question. In truth, only archaeology can resolve such a question, even if today's archaeologists always pay too much deference to the literary evidence when deciding where to excavate.

While the current *Documents* classic talks of Zhougong divining potential sites in two places, we know that King Cheng's principal (only?) capital was at Hao. The extant literary or archaeological record gives no sign that King Cheng ever moved to an eastern capital outside Guanzhong. As we have seen above, thanks to Zhao Yiqing, the local rivers in the area have shifted course, which necessarily means their relation to the "King's City" and to Luoyang 雒陽 has itself shifted. If during Eastern Han people already recognized this, how could it be true that in early Western Zhou, in the time between King Ping's move east and King Jing's 敬 reign, there was no change at all in the courses of the same rivers? In fact, the old riverbed of the Chan River has already dried up; there is no longer any water in its riverbed. This alone suggests that the course of today's Luo River 洛水, while still in existence, has undergone many changes over the *durée*, and today's riverbed is not likely to coincide with the ancient riverbed. This needs to be kept in mind as we review my summary of the archaeological record below.

From 1954 to 1980 and beyond, archaeologists from the national-level Academy of Science and the Luoyang Archaeological Brigade have found and excavated multiple Western Zhou sites and Eastern Zhou sites in the environs of Luoyang. They have even found evidence of the old dams between the ancient riverbeds of the Chan River 瀍水 and the Jian 澗水, as well as evidence of the changes time has wrought. There is an ancient city wall site southeast of the present-day Chan riverbed, apparently of mid-Western Zhou date (see below), or even Eastern Zhou, after the move of the Zhou capital in 771.[83] An important discussion has been published under the title "Was there or not in early Zhou an eastern capital, or a King's City" 周初營建東都是否有王城.[84] No matter what the scholarly consensus, many reports and essays are investigating the range of possibilities, in large part because the ancient sites in Henan still have not yielded definitive proof of anything.

When we look at the archaeological results, several discussions relating to different archaeological strata should be kept in mind in relation to this chapter: (1) the excavation reports devoted to the early Western Zhou sites;

(2) the reports on sites dating to the middle and late Western Zhou; and (3) the reports on the Eastern Zhou "King's City." Everyone agrees that in 770 BCE, because of the threats from the Rong and Di, King Zhou moved the capital from Hao to Luo city or Luoyang, making it his "King's City" or "king's city" as descriptor. (Below we will consistently treat that as name of or epithet for the city, using capital letters, following many, but not all commentators.) But no evidence has been found to date that would allow us to decide whether the "King's City" and Western Zhou ChengZhou occupied one site or two nearby sites. In the 1950s, the Henan county wall and the Eastern Zhou city walls were discovered, and we now know, as a result, that the Eastern Zhou "King's City" was positioned at the confluence of the Jian River and the Luo River. But below that, in an earlier archaeological stratum, or nearby, there has been no discovery of a prior site with *hangtu* pounded-earth city walls, as the *Documents* traditions specify.[85] In other words, up to now, archaeological teams have not been able to confirm Ban Gu's "two capitals" model with respect to the extant Western Zhou sites. (Abundant evidence meanwhile exists for Hao and its environs.) In the Luoyang area, archaeologists have indeed located tombs that date to early Western Zhou, but they have found no city sites from that era at all, and this fact is key to the present analysis. (That there would be Western Zhou settlements hardly surprises, in that the Western Zhou forces had to oversee the administration of sites near the old Shang capital at present-day Anyang; also, the *Documents* speaks of enfeoffing a Prince Kang in that very area to undertake oversight responsibilities.)[86] Perhaps it is relevant that *up to now* our earliest extant reference to the "King's City" in both the *Annals* and *Zuozhuan* postdates King Ping's move east in 770 BCE.

What methods have historians, equipped with archaeological reports, used to devise their theories to date? As we have seen, according to the *Documents* traditions, the Luo River 雒水 region has a very complicated relation with its various cities, including Luo city 雒邑, ChengZhou, and whatever Xiadu 下都 (literally, capital below) means. (NB: The term Xiadu never appears in the *Documents* itself, only in commentaries discussing the *Documents*.) Among modern scholars, Qian Mu 錢穆 (d. 1990) was one to argue that Zhougong's earlier and later constructions were named Luo city and Xiadu, suggesting the expression ChengZhou (Zhou Completed/Perfected) is just a general descriptive or celebratory epithet applied to either or both sites.[87] Qian Mu refers to it somehow "coming about" that Xiadu was used to distinguish one of the sites from the main capital, and that time period when this "came about" must be the time of King Jing of Zhou. This seems to muddy the waters.

Much more recently, Peng Yushang's 彭裕商 essay "Investigation into the New City" 新邑考 assembled all the rival theories regarding Luo city before dividing them into three types: (1) theories claiming that there were two capitals; (2) the "one capital" theory; and (3) what Peng calls the "package" (*baoshe* 包攝) theory, wherein the early Western Zhou rulers really built two walled cities, but ChengZhou was used as the generic name for both, possibly because the term "King's City" referred only to the palace district in the one city. Peng names six major scholars who have embraced this "package" theory: Lü Zuqian 呂祖謙 (d. 1181), Tong Shuye 童書業 (1908–68), Li Xueqin 李學勤, the archaeologist Yang Kuan 楊寬, Wang Yuzhe 王玉哲, and Shi Weile 史為樂.[88] Obviously enough, Qian Mu's theory would also fit the "package" hypothesis, for Qian Mu believed that ChengZhou, in early Western Zhou, was not a specific name for a specific city, but a general name for new sites marking the Zhou achievements. But Qian Mu was also clearly influenced by the commentaries attached to the fourth-century CE chapter preface to the "Duo shi" chapter, which mention Xiadu, the "lower" or "lesser capital" (apparently a secondary capital).[89] So Qian Mu devised a new "one site, with two names" hypothesis, alluded to above, to identify ChengZhou with Luo city. All these scholars basically agreed that, after the move of the capital east in 770 CE, it is unlikely that Hao would ever have been identified in the contemporary or roughly contemporary sources as "the King's City." From 770 on, then, Luoyang would be the "King's City," aka the "capital" (*jingshi* 京師), the term favored in later eras. As for the term ChengZhou, the scholars speculate, it was probably originally meant simply to contrast with ZongZhou 宗周 (Ancestral Zhou), that is, Hao.

The modern scholar Huang Zhangjian 黃彰健 (d. 2009),[90] following Zheng Xuan's theory, cited the *Shangshu dazhuan*'s reference to ChengZhou being built in the fifth year of Zhougong's rule.[91] Huang then argued that the name ChengZhou was generated during Zhougong's regency, with the "Cheng" referring to the same situation that the posthumous name King Cheng does: the establishment of the Western Zhou peace after the conquest of the rebels.[92] Accordingly, the next *full* year count becomes the "first year of King Cheng," whose name reflects the peace, and the last year of Zhougong's rule is the year *before* the "first [full] year of King Cheng."[93] By Huang's ingenious theory, in the first full year of King Cheng's reign, King Cheng went to ChengZhou to carry out a great celebration, during which time he formally announced his success to High Heaven and the Zhou former kings at a ritual site. Huang Zhangjian reasoned this way: if ZongZhou as an epithet for the capital emphasized the model of the ancestors, ChengZhou is an epithet that emphasized the military and political victory, in Huang's view. But King Cheng would likely have remained behind in Hao, and the powerful force of the phrase ChengZhou would have gradually weakened until the point when ChengZhou was nothing more than a city with walls.

Even if Huang's hypothesis is correct, I find it puzzling that the fourth-century CE *Documents* classic has two chapters ("Luo gao" and "Duo shi") linking ChengZhou to the notion of settling "the recalcitrant leading men of the Yin,"[94] while the narrative setting of "Shao gao" specifies the site was at ChengZhou. Kong Yingda's "Correct Interpretation" commentary for the *Documents* (*Shangshu zhengyi* 正義) conflates this very ChengZhou with Xiadu (a secondary capital). But let us remember: it is not only Kong Yingda who does this. Kong is following the historical line articulated by Ban Gu, Zheng Xuan, Du Yu, and others, who say much the same thing: that the ChengZhou site that is known from the time of King Ping is the same as the site that Zhougong (nominally at the behest of King Cheng) built to lodge these conquered people, and, *for this reason*, Xiadu must be identical with ChengZhou, making two major cities in near proximity to one another.

Looked at with a critical eye, this sort of thinking about ChengZhou is completely illogical in one sense. For when King Cheng first fixed his capital at Hao, it was also called by the epithet ZongZhou. Although there were two names, there was only one city, which remained the capital city throughout the entire Western Zhou. (This is the capital that the *Odes* calls "Zhongguo" 中國, the present name for "China," by the way.)[95] It seems reasonable, then, to surmise that just as "Hao, the capital" and "ZongZhou" were names for the same site, Luo city and ChengZhou are also names for the same site, despite the firm belief articulated by Ban Gu in his geographic treatise that ChengZhou must be a second site specifically built to house the old rebel leaders among the conquered Yin. If Luo city and Cheng-Zhou are one city with two names, calling it Luo city may indeed refer to Zhougong building it with the intention that it would become an eastern locus of political power; perhaps Zhougong even had the idea that King Cheng might move his base there, and it might become his future capital. It is equally plausible, however, that Zhougong was planning to move the ancestral temple to the site where his camp had been, or that other scenarios unfolded, spurred by other motivations. Speculations aside, we reiterate:

King Cheng, after all, never moved his capital from Hao, and *perhaps* he or his representative Zhougong only conducted a series of ceremonies at a Luo city while still regarding Hao as the capital.⁹⁶

The conclusion that I draw from the foregoing is this: that the early Western Zhou ChengZhou city is not the same as the ChengZhou city known in connection with King Jing 敬 or with his predecessor, King Ping of Eastern Zhou. As these binomial epithets for cities gesture toward complex associations, they are complex signifiers (much as the word "capital" is a complex signifier) that need not be attached to a single city.⁹⁷ Similarly, I believe, the expression the "king's city" (I prefer the lowercase) just refers to the Eastern Zhou capital, with a focus on the king's palace residence there. Because the *Documents* chapters simply gesture toward King Cheng's reign and the peace achieved during that reign, whenever the chapters employ the phrase "ChengZhou," we do not have to assume any identity between it and the later city called "ChengZhou" centuries later, in Eastern Zhou. That this sort of anachronistic conflation has occurred so frequently has generated great difficulties down through time, when scholars try to reconstruct events in the remote Western Zhou and early Eastern Zhou worlds from the scanty evidence at hand. But American scholars should remember that Paris, Frankfurt, Versailles, and Athens are all cities in Kentucky, as in Europe, and great scholars of the early twentieth century (e.g., Fu Sinian) knew full well that city names over time moved around in kindred ways.

Conclusion

It is easy to see how the most careful classicists can go wrong when reconstructing the past, but why should this confusion matter to us today? If we give this question some thought, we will see that the era of King Cheng (and Zhougong, of course) is usually regarded as the "height" of the "Zhou Way," which in turn is said to represent one of the heights, if not *the* height, of Chinese civilization for all time. The decline of that Zhou Way is ascribed, by both Sima Qian and Ban Gu, to the time of the inept King Li, when, as Ban Gu wrote, "the rites and music, as well as the punitive campaigns, were initiated by the local lords" and not by the Zhou king himself.⁹⁸ (Kongzi's teachings were meant to remedy this lamentable situation, or so the traditions say.) That being the case, should we emphasize the long continuous sweep of history from the time of King Cheng down to Eastern Zhou, or should we read into that history at least one major interruption? Admittedly, the extant sources are limited. The *Documents*' account is downright confusing, and we cannot expect to derive greater illumination about early Western Zhou from the *Annals* traditions, the few contemporary bronze inscriptions found to date, or the archaeological reports produced so far.

Sima Qian has something interesting to say about this very problem, as it happens: that scholars were debating these very problems already in his day, if not earlier. He writes the following, as Senior Director of the Archives,

學者皆稱周伐紂，居洛邑，綜其實不然。武王營之，成王使召公卜居，居九鼎焉，而周復都豐、鎬。至犬戎敗幽王，周迺東遷于洛邑。

Those who study the matter all claim that when Zhou fought the last Shang king, [he, King Wu] made his residence at Luo city but, looking at all the sources I have gathered together, this was not the case. King Wu planned it [Luo city] out,⁹⁹ but it was King Cheng who sent Shaogong to divine the sites, and it was he who had the Nine Tripods placed there, but the Zhou [king] himself returned to [restore?] the capitals he had at Feng and Hao. *Only with the Dog Rong's defeat of King You did the Zhou move its capital east to Luo city.*¹⁰⁰

Obviously, Sima Qian does not fully accept an identity between Luo city 雒邑 and the later "eastern capital" theory outlined by Ban Gu, though possibly such a theory may have predated Ban Gu. Sima Qian may have believed that "an ordering of the *Documents*" was undertaken by Confucius himself,¹⁰¹ but he guessed that the records that Kongzi saw had so many lacunae in them that not even a Supreme Sage could extract a coherent narrative from them. As we have seen, it's also abundantly clear that Sima Qian doesn't have a *Documents* Postface of any length whatsoever in front of him, because he is struggling to find the proper order of chapters and of events—conundrums that any authoritative Postface would have been designed expressly to resolve. Modern scholars, based on extant sources, can't find a Postface in circulation until the reign of Han Chengdi (r. 33–7 BCE), yet somehow, by Ban Gu's era, more than a century later, there had arisen a strong belief that there once existed a *Documents* classic known to Confucius/Kongzi in 100-*pian* ("chapters").¹⁰² As for the current *xu* 序 (now a set of chapter prefaces) associated with Mei Ze's fourth-century version of the *Documents* classic, we moderns simply have no idea which parts of it reflect genuine traditions dating back to antiquity and which do not. Consequently, we must presume the whole version dates to Mei Ze's time.[103]

The point of this exercise, then, has been to trace a single line of transmission from a seemingly insignificant passage in Ban Gu's geographic treatise to show how the authority of ancient texts like Ban's continues to shape modern scholarship. Ban Gu's influence is indisputable, although it has often gone unrecognized. For Ban Gu's "two capitals" model, like the notion of "unchanging China," can be found not only in Ban's geographic treatise in the *Hanshu*, but also in the writings of Ma Rong, Zheng Xuan, Wang Su, and nearly every other *Documents* expert to the present time. Even Cai Yong 蔡邕, in a *fu* of frustration aimed at the late Eastern Han court, cannot seem to resist portraying wondrous Luoyang and its undeniable allure to all newcomers.[104]

Modern scholars are still wrestling with old ideas, and far too often they do not trouble to ascertain the sources of those ideas, never puzzling afresh over the complex insights registered in the older scholarship. But unless we have a good idea of where our assumptions come from, we will not be able to actively interrogate our views, or skillfully correlate history and archaeology, in light of presentist claims about remote antiquity. It seems likely that Ban Gu saw a portion of an Eastern Han Postface for the *Documents* mistakenly ascribed to Kongzi/Confucius in his day. If we care to look through this single window, we may begin to glimpse a bit more of the Eastern Han landscape, and then all the time and effort expended in compiling the evidence for this inquiry will not have been wasted. In addition, modern scholars sometimes place too much faith in archaeology, thinking it can resolve nearly all historical problems. What the naifs do not understand is that all levels of archaeological reasoning require constant extrapolation from and reference to the old texts and traditions.[105] Finally, I would end with an observation on manuscript culture: as we now know, even relatively short works such as Ban Gu's "Treatise on Earth's Patterns" (*Dili zhi*) are often composite texts composed of disparate strands of different origins. The next step in studying Ban's treatise should be to begin the study of other strands of potentially different origins to see what more we can find.

Appendix for Chapter 3

It may be helpful for readers of classical Chinese to see the passages on which part of my chapter relies. For that reason, the *Annals* passages are gathered here. All relevant *Documents* classic passages are cited in a new essay by Michael Nylan, which builds on my insights: "Cross-reading 'Kang gao,' 'Shao gao,' and 'Luo gao,'" *Journal of Asian History* 54:1 (summer, 2020), 1–62. The numbering refers to the new three-volume translation *Zuo Tradition* (University of Washington Press, 2016).

Lord Zhao, Year 22 (520 BCE, Year 25 in the reign of King Jing 景王 of Zhou)

The *Annals* classic 22.4–5 says:
夏四月乙丑，天王崩。[106]
六月，叔鞅如京師。[107]
葬景王。[108]
王室亂。[109]

The *Annals* classic 26.7 says: 秋，劉子、單子以王猛入於王城。[110]

The *Gongyang* commentary explains: 王城者何？西周也。其言入何？篡辭也。[111]

Lord Zhao, Year 23 (519 BCE)

The *Annals* classic 23.8 says:
秋。天王居于狄泉。[112] . . . . 尹氏立王子朝。[113]

He Xiu's 何休 subcommentary for the *Gongyang* explains: 時居王城邑，故號西周王。[114]

Today's *Zuozhuan* 23.6 explains: 其勉之，先君之力可濟也。周之亡也，其三川震。今西王之大臣亦震，天棄之矣！東王必大克。」[115]

Du Yu's 杜預 commentary for today's *Zuozhuan* 23.6 explains: 子朝在王城，故謂西王 . . .[116] 敬王居狄泉，在王城之東，故曰東王。」[117]

Lord Zhao, Year 26 (516 BCE, Year 4 in the reign of King Jing 敬王 of Zhou)

The *Gongyang* commentary explains: 成周者何？東周也。[118]

Kong Yingda's 孔穎達 "Zhengyi" 正義 subcommentary explains: 此時始得入于成周，遂以成周為都，來告，故特書之。[119]

Lord Zhao, Year 32 (510 BCE, Year 10 in the reign of King Jing 敬王 of Zhou)

The *Annals* classic 32.3c says: 韓不信如京師，合諸侯大夫於狄泉，尋盟，且令城成周。

The *Gongyang* commentary continues: 仲孫何忌會晉韓不信、齊高張、宋仲幾、衛世叔申、鄭國參、曹人、莒人、薛人、杞人、小邾人，城成周。[120]

Today's *Zuozhuan* 32.3a–b explains:
> 秋八月，王使富辛與石張如晉，請城成周。[121]
> 天子曰：天降禍於周，俾我兄弟並有亂心，以為伯父憂。….
> 昔成王合諸侯，城成周以為東都，崇文德焉。今我欲徹福，假
> 靈於成王，脩成王之城，俾戍人無勤，諸侯用寧，蠻賊遠屏，
> 晉之力也。[122] 而又焉從事。魏獻子曰：善。使伯音對曰：天子有命，
> 敢不奉 承以奔告於諸侯，遲速衰序，於是焉在。[123]

Du Yu's commentary for today's *Zuozhuan* explains:
> 子朝之亂，其餘黨多在王城，敬王畏之，徙都成周，成周狹小，故請城
> 之。[124]

The phrase "king's city" appears in *Gongyang*, *Guliang*, and *Zuozhuan*, e.g., as follows: 晉侯辭秦師而下，三月甲辰，次於陽樊，右師圍溫，左師逆王。夏四月丁巳，王入于王城。[125]

* * *

For the backstory, see also the *Annals* classic:
1. Lord Yin (r. 723–712), Year 11
> 夏，揚拒泉皋伊雒之戎，同伐京師，入王城，焚東門，王子帶召之
> 也，秦晉伐戎以救周。秋，晉侯平戎于王。[126]
2. Lord Huan (r. 712–694), Year 9
> 京師者何？天子之居也。京者何？大也，師者何？眾也。天子之
> 居，必以眾大之辭言之。[127]

2. *Zuozhuan* for Lord Zhuang (r. 694–662), Year 28:
Today's *Zuozhuan* says,
> 凡邑有宗廟先君之主曰都，無曰邑；邑曰築，都曰城。[128]

Kong Yingda's 孔穎達 subcommentary then explains:
> 此傳所發，乃為小邑發例。大者皆名都，都則悉書曰城，小邑
> 有宗廟，則雖小曰都，無乃為邑。邑則曰築，都則曰城。為尊
> 宗廟故，小邑與大都同名。[129]

Notes

1. The translator would like to thank Vanessa Davies, the Egyptologist, for her insightful remarks on this chapter, and also Luke Habberstad for his sensitive reading. See Pauketat, *Chiefdoms and other Archaeological Delusions*, 199.

2. For example, Ma's *Liang Han Shangshu xue yanjiu*, 317, cites a number of changes that he believes Ban Gu introduced into the record, based on what Ma thinks is *guwen*. Zhou, one of China's foremost historical geographers, discussed the dating and compilation of the *Hanshu* treatise in his *Xi Han zhengqu dili*. Zhou's exhaustive study concluded that that treatise, nominally ascribed to Ban Gu, essentially represents court documents from late Western Han, specifically the courts of Chengdi, Aidi, and Pingdi (reigns 33 BCE–6 CE), and that aside from the brief framing device at the beginning of the treatise, no substantial part of the received work reflects an Eastern Han context a full century later in time. I prefer Qian's work on geography, *Gu shi dili luncong*, but Qian does not focus there on many problems in Western Han place names. For more on Qian, see below. NB: Ban Gu himself notes that the place names have been repeatedly changed over time (*Hanshu* 28A.1543: 地名又數改易).

3. Sometimes Ban Gu's readings even change territorial borders.

4. See the Appendix.

5. *Shiji* 121.3125. *Hanshu* 88.3607 uses the identical phrasing. In some cases one wonders whether the *Hanshu* phrasing has been interpolated into the *Shiji* phrasing, but at this remove this is impossible to ascertain. I believe that the punctuation given in the Zhonghua shuju edition of the *Hanshu* is incorrect and should be changed to the *Shiji* punctuation.

6. If such a Postface existed in Sima Qian's time, it must have been stored away in the imperial palace archives, out of sight. But Sima Qian, as Senior Director of the Archives, *should* have been familiar with Kong Anguo's transmission, if it existed in the palace archives, and also with the collated edition in Modern Script that was associated with the Kong family mansion. That he evinces no knowledge of this therefore speaks volumes. So whereas the phrase 孔子序書 exists in the *Shiji*, as well as in the *Hanshu* (where it is 孔子為百篇書序), in the two different histories, it does not bear the same meaning. In the *Shiji*, the word *xu* means "to put in proper order," whereas in the *Hanshu* it refers to a Postface.

7. Translator's note: strictly speaking, the "Yiwen zhi" doesn't have this precise term (it ascribes the Postface to Kongzi in the description of the *shu* section of the "Yiwen zhi"). For Chengdi's reign (33–7 BCE), when a Postface known to be a forgery or pastiche was introduced to the court by one Zhang Ba 張霸, we again have only Ban Gu's testimony.

8. A much shorter one-*juan* Postface (almost certainly dated to late Western Han) did circulate in the circle of classicists familiar with the Ban family. See note 13 below for further information.

9. Even this wording is odd, as *shu shi* would more likely mean "twenty or thirty" than "ten-odd." The translator wonders whether the *Hanshu* text has been emended in the fourth century CE or afterward.

10. *Hanshu* 30.1706.

11. *Hanshu* 36.1969–70.

12. *Hanshu* 30.1706.

13. Liu Xin justifies the dating in his calendrical classic no fewer than three times on the basis of what he regards as an antique Postface to the *Documents*. Notably, Liu Xin cites one chapter, "Yin xun" 伊訓, that is not in the Western Han court-sponsored version. *Hanshu* 21.1014: 故《書序》曰：成湯既沒，太甲元年。使伊尹作《伊訓》。《伊訓》篇曰. . . . Cf. ibid. 21.1015, citing the Postface to the *Documents* (twice): 文王受命九年而崩，再期，在大祥而伐紂，故《書序》曰：惟十有一年，武王伐紂，(作)《太誓》. . .八百諸侯會。. . . 故《書序》曰：「武王克殷，以箕子歸，作《洪範》」。

14. For a review of the arguments about the Postface, see He and Nylan, "On the Han-era Postface (*xu* 序)," an essay inspired, in part, by Professor Lee's work.

15. Translator's note: here Professor Lee specifically mentions the biographies found in the *Shiji* chapter "Forest of Classicists" (Rulin 儒林).

16. Many have argued that any such work must have been a product of the pre-unification classicists writing hundreds of years after Kongzi. In late Qing, Cui Shu 崔述, Liao Ping 廖平, and Kang Youwei 康有為 all began to insist that such a Postface did indeed exist in Ban Gu's era, it having been forged by Liu Xin. Gu Jiegang's 顧頡剛 *Gushi bian* 古史辨 continued a similar line of thinking, because Gu based his work on Yan Ruoqu's 閻若璩 (1636–1704) *Evidential Analysis of the Archaic Script Chapters of the Documents* (*Shangshu guwen shuzheng* 尚書古文疏證). Accordingly, Gu assumed that the received *Shangshu* could only be "false" (*fei* 非), either the product of unspecified "later classicists" or of Liu Xin. At this point, it is important to remember that the *Documents* version seen by Zhu Xi, by the Qing scholars, by Gu Jiegang, and by us is one and the same *Shangshu*, and that version is based on the court-sponsored *Shangshu zhengyi* 尚書正義 (Correct Meaning of the *Documents*, compiled between 637–653), which in turn is based on an early fourth-century CE version of the *Documents*. This is likewise the very same version that Sun Xingyan and Pi Xirui and others have used to "reconstruct" the Modern Script version ascribed to Fu Sheng in early Western Han.

17. Translator's note: the current commentary ascribed to Zheng Xuan must contain later interpolations and emendations, for Zheng's original commentary had to be retrofitted to account for Mei Ze's version. The same is probably true of that for Ma Rong and Wang Su.

18. If my hypothesis is correct, then it is more than likely, too, that Fu Sheng's transmitted chapters had no Postface, contrary to the claims made by many scholars from Tang through Qing. We should note meanwhile, in connection with memorials by Kuang Heng 匡衡 and Zhang Tan 張譚 regarding the suburban sacrifices in the north and south of Han Chang'an, that there is a quotation of the "Tai shi" *pian*, in *Hanshu* 25B.1255. That means that at that time, a "Tai shi" (apparently not identical with the pre-Qin "Tai shi") perhaps was included in the *pian* count ascribed to the late Western Han *Documents*, making twenty-nine chapters in all. Again, Qing classicists expended much effort on ascertaining the precise number of

chapters in the Han-era version, to no avail. See, e.g., Wang's *Jingyi shuwen*, which thinks the one extra *pian* should be the "Tai shi"; Chen's *Zuohai jing bian*, which thinks it should be a one-*pian* or one-*juan* Postface; Gong's *Da shi da wen*, which together with Pi's *Jingxue lishi*, proposes that the one-*juan* addition reflects a split between the "Gu ming" and "Kang wang zhi gao" chapters. In the modern period, all New Text adherents assume that Fu Sheng's version had a Postface (see Duan Yucai, Pi Xirui, Wang Xianqian, etc.), with the result that many of those scholars (with Pi Xirui an exception) believe there were two Postfaces (one for the *jinwen* and one for the *guwen* texts of the *Shangshu*).

19. Translator's note: it is clear that the *Shiji* uses *guwen Shangshu* to refer only to the antiquity of the *Documents* classic, as the chapters it mentions as *guwen* (old writings) are indeed what we would call *jinwen* (so-called "Modern Script" chapters) today. This has been noted by many scholars.

20. Translator's note: the material between the dashes has been added by the translator, Michael Nylan, and is not in the original Chinese text by Professor Lee.

21. Recall that Ruan's compilation was based on the early Tang *Shangshu zhengyi* 尚書正義, which was itself based on the fourth-century *Documents* materials presented to the Eastern Jin throne by Mei Ze 梅賾.

22. Ban Gu, in his own personal narration (chapter 100 of the *Hanshu*), says that he intends "to bring together conduct and deeds, and to get at the heart of the Five Classics, so that above and below communicate well, and to this end, to make an investigation like that found in the *Annals* 為春秋考. To that end, Ban compiled Basic Annals, Tables, Treatises, and Biographies, in 100 *pian* altogether." See *Hanshu* 100B.4235.

23. Several recent essays have tried to ascertain when this change was made, but the evidence is far from conclusive, although most scholars follow Wang Xianqian and Duan Yucai in ascribing the change to Cao Pi (see below). The *Shiji suoyin* by Sima Zhen (679–732) identifies the Luo 雒 River with the Luo 洛 character, as does the Tang-era *Wujing zhengyi*.

24. Contrast *Hanshu* 28A.1530 with *Hanshu* 25A.1206, *Hanshu* 28A.1541.

25. In the main, Ban's distinction aligns with the *Shiji* practice and although present editions of the *Shiji* occasionally write "Luo city" using the character 洛 (see, e.g., *Shiji* 4.133), this appears to be a later emendation. As readers may recall, the original *Documents* classic ascribes the activity of making the four rivers enter into the Yellow River; it is not clear whether Ban Gu accepts this or not, but Ban's use of the particle *ji* ("once" 既) suggests that he does.

26. For example, see *Hanshu* 1B.28 (Zhonghua shuju ed. of 1962), which talks of Luoyang 洛陽 in Henan.

27. Likewise, the Song Danshu 單疏 edition of Kong Yingda's commentaries 尚書疏 uses 洛 consistently.

28. Pi, *Jinwen Shangshu kaozheng*, 343. Pi mistakenly says that the *Hanshu* also fails to observe the distinction; hence the ellipsis here, to forestall reader confusion.

29. Wang, *Shangshu Kongzhuan canzheng*, juan 22, 721.

30. Wang writes, "The key to this lay in the Three Kingdoms (Sanguo 三國) period, in Wei, and it involved the Five Powers cyclical theory, as we can see in the "Wei lue" 魏略. Duan Yucai says, "According to the *Wei zhi* 魏志, 'In the first year of Huangchu (220 CE), the emperor visited Luoyang.' " Pei Yin's 裴駰 (active 438) commentary on this entry is: "The imperial edict said, when using the Fire Phase of the [Eastern] Han dynasty, Fire dreads Water. Therefore, they deleted from the character for Luoyang the water radical and added a bird. The Wei dynasty rules by the next Phase, which is Earth. Earth is Shepherded by Water, and when Water gets Earth, then it flows, and when Earth gets Water, then it is soft. Therefore, our [Wei] dynasty should delete the bird radical and add the water radical, changing Luo 雒 to Luo 洛."

31. Duan, "Yi Luo guzi but zuo Luo kao," *juan* 1, 883; cf. Tan, *Qingren wenji dili lei huibian* 清人文集地理類类汇编, 403. Some surmise that the Eastern Han rulers wanted to avoid the Water Phase in naming their capital of Luoyang, lest too much Water prove detrimental to their dynastic fortunes.

32. *Hanshu* 28A.1555. Editors' note: One may compare *Taiping yulan* 17A. Bielenstein and Chavannes went into this question thoroughly. See Bielenstein, "Lo-yang in Later Han Times," 103n2, which agrees with Chavannes's *Les mémoires historiques de Se-ma Tsien*, II, n. 2 (lines 287–88), that from earliest times the name of the river (and hence that of the city) was written as 雒. They speculated that the form 洛 was only adopted in 220 CE by the new empire of Wei. For more on Luoyang, see de Crespigny's *Fire over Luoyang*. The fish story appears in many Han and post-Han apocryphal texts. Some, but not all, of these stories appear in the *Bohu tong*, a compilation ascribed to Ban Gu, which evinces interest in the policy question of how to site a capital properly.

33. Based on this evidence, Wang Niansun felt he had to say that the two characters for Luo were interchangeable. Because of Wang Niansun, the late editions of *Shuijing zhu* 水經注 were all emended as well, and the text improbably described "the Southern and Northern branches of the Luo River" to distinguish the Luo 洛 River in Yongzhou from the Luo 雒 in Yuzhou. See n. 35 immediately below. Tan Qixiang, as lead editor for the historical map collection, *Zhongguo lishi ditu ji*, followed this practice of using the Luo character with the water radical for both Luo Rivers, although the Guanzhong basin river empties into the Wei River and the Henan river empties into the Yellow River. See Tan, vol. 1, maps on pp. 15–16, 17–18, 19, 20–21, 22–23. Tan's motivation may have been to use simplified characters, however.

34. Regarding the Xiping Stone Classics, recently Liu (d. 2012) used Hong, *Li shi* (before 1184) and Ma, *Han Shijing jicun*, plus a stone fragment "found" in 1949 (reported in *Kaogu xuebao* 考古學報), to demonstrate that the Tang-era Stone Classics, unlike the Han-era Xiping Stone Classics, used the character Luo 洛 for the *Documents* "Duo shi" chapter, presumably under the court's mandate or with

its acceptance, at least. From this, we learn why other Tang books, including Kong Yingda's *Shangshu zhengyi* and other authoritative writings produced by the Tang court, employed the character Luo 洛. Liu Qiyu thought that explained why the Song editions also used the same form of Luo 洛, which by then had become customary.

35. Wang, *Dushu zazhi*, 4:6, 248, correctly stated that Luo 洛 and the Luo River were in Yongzhou (i.e., Shaanxi) and Luo 雒 in Yuzhou (i.e., Henan). Wang believed that Cao Pi had changed the character, and Wei Bao 衛包 had done so for the *Documents*, confounding the Luo character that properly belonged to Yongzhou. Duan Yucai's commentary for the *Shuowen* 說文, under the entry for Luo 洛, likewise correctly identified the difference between the two Luos, adducing evidence from the *Zhouli* and the *Yi Zhoushu*. Duan also noted that Cai Yong's 蔡邕 Xiping Stone Classics fragments for the *Documents* "Duo shi" chapter used the character Luo 雒, as did Zheng Xuan, in his commentary for *Zhouli* 周禮, when citing the *Documents* "Shao gao" 召誥 chapter. Duan concluded, "This shows that neither the Modern Script and Archaic Script *Documents* traditions used Luo 洛 for the capital in Henan, as neither Zheng nor Cai could have dared to emend the characters in the text." See Duan's commentary for Xu Shen, *Shuowen jiezi, pian* 11A.1, 529–30. [NB: Duan presupposes a clear distinction between *guwen* and *jinwen*.]

36. Liu, *Shangshu yuanliu jiqi zhuanben kao*, 134–36. The so-called "Kong Anguo Lesser Prefaces" treated three chapters—the "Kang gao," "Jiu gao" 酒誥, and "Zi cai" 梓才—in a single explanatory passage with the result that we have twenty-seven extant *xu* 序 (chapter postfaces/prefaces) for the chapters thought to belong to both the Han-era transmissions and the post-Han "Kong Anguo" version.

37. Duan, *Guwen Shangshu zhuanyi, juan* 32, p. 288. Duan stated that those commentators, according to the Tang *Documents zhengyi*, did not dare to emend the main text of the *Documents*, so they put the *Documents* Postface at the end of the main text of the Classic. In my view, many Qing scholars liked to call the *Documents* ascribed to Kong Anguo a "forgery," and their dismissive attitude carried over to the Postface/Preface that circulated with that version. For that reason, Duan Yucai urged us to follow Ma Rong and Zheng Xuan whenever possible. (The commentaries by Ma and Zheng now exist only in fragments, of course). Also, Duan Yucai preferred scholars to treat the Postface/Preface as a single, continuous chapter, rather than appending lines from it to the relevant chapters, as was done in the Ruan Yuan edition. In Qing, no matter whether a scholar was a proponent of Modern or Archaic Script texts, the Postface/Preface was appended as a single chapter or *juan* to the end of the *Documents* classic.

38. See Wang, *Shangshu Kongzhuan canzheng, juan* 22, 721.

39. Many scholars who have focused on the word "new" 新 contrast the new city not with the older Zhou capitals at Feng 豐 and Hao 鎬 in the Guanzhong basin, but with an older capital or at least an older city built by the Shang at Luo, in Henan, before the Zhou conquest, on the assumption that it was repaired or "renewed." In my opinion, such speculation, meant to resolve one difficulty, creates others. See below.

40. See note 36 immediately above. The "Shao gao" chapter says that Shaogong divined (*bu* 卜) the site before Zhougong's arrival to inspect it, and it gives time frames for various activities; it adds that Shaogong used the Yin men (old adherents of the conquered Shang) to build the site in the bend of the Luo River. The "bend of the Luo River" apparently refers to the north bank of the Luo River in Henan. Translator's note: the character *rui* 汭 has been interpreted in this way for a long time, but conceivably Rui refers to the name of a major river in the area.

41. The "Luo gao" narrates in great detail the coming of a "king" to the New City, who sacrificed at that site an ox (reddish brown) to Kings Wen and Wu, after which he had a copyist/annalist 作冊 Yi 逸 offer prayers and transcribe the events. By some traditions, the "king" in the "Luo gao" is not King Cheng, but Zhougong himself, speaking as king. These are but a few of the problems besetting interpretation of the "Luo gao."

42. Translator's note: one may cite Matsumoto Masaaki in this connection: "Shūkō ka to Shōkō ka."

43. Note that the characters for "Wu yi" are given in many combinations, and this chapter adopts the most common of the combinations. Note, too, that the order in which these chapters are given here differs from other early chapter orders, but agrees with today's post-Han chapter sequence, which may reflect activist editing from the fourth century CE or later. In today's order, the *Documents* chapters are listed as "Shao gao" (chap. 17), "Luo gao" (chap. 18), "Duo shi" (chap. 19), "Wu yi" (chap. 20).

44. *Shiji* 4.126.

45. Contrast *Shiji* 4.126, 133, 170 with *Hanshu* 28A.1555, 72.3055 (both of which date the transfer of the Nine Tripods to King Wu's reign). This phrasing does not appear in the *Shiji*, but it does appear in connection with the *Odes* and *Rites* traditions, and these Sima Qian seems to have had in mind.

46. See *Shiji* 4.129. Therefore, the "Basic Annals for the Zhou" says that King Wu announced, "Having settled our Western lands 定我西土, from the bend in the Luo river to the bend in the Yi River, you are to live at ease . . . by the Luo 雒 and Yi 伊 Rivers, not too far from the seat of Heaven [i.e., the court]," after which they proceeded to build Luo city.

47. In any case, such a major city (whether a ritual center or a secondary capital) ought to be a walled city, and indeed, the *Documents* commentators specifically describe building the city walls for it.

48. Translator's: readers may wish to note the following: in English, King Jing of Zhou may refer to either 周景王 (personal name Gui), the twenty-fourth king of the Zhou dynasty and the twelfth of Eastern Zhou (r. 544–520 BCE), or to 周敬王 (personal name Gai), the twenty-sixth king of the Zhou dynasty and the fourteenth of Eastern Zhou (r. 519–477 BCE).

49. A "King's City" (*wang cheng*) is mentioned for the year corresponding to 520 BCE in the *Annals*, also in the "Gongyang" *Tradition* for the year corresponding to 516 BCE.

50. *Shiji* 4.149: 東遷于雒邑, 辟戎寇 . . . 東徙雒邑 . . . 周東徙雒邑 . . .

51. *Shiji* 99.2716n4. NB: At the same time, as early as the *Shiji*, there is also talk of two cities, both identified with Henan, which are said to be Luo city 洛邑 and ChengZhou 成周. More on this below.

52. Properly speaking, *yi* 邑 can refer to a "settlement," and *need not be a capital or even a large city*. Ban Gu certainly, however, took Luoyi to be a secondary capital, replete with its own ancestral temple dedicated to the Western Zhou ancestors.

53. Michael Nylan, in her role as editor, notes how much weight was put upon content that originally circulated separately from the *Documents* classic itself.

54. Ruan Yuan, 15.218a: 成王在豐, 欲宅洛邑, 使召公先相宅, 作召誥。

55. Ruan Yuan, 15.224b. To recap: the "Luo gao" *xu* talks of ChengZhou as one site that is being built, whereas the "Shao gao" *xu* talks of building Luo city.

56. Ruan Yuan, 16.236a.

57. The chapter preface to the "Duo shi," says, "Once ChengZhou was completed, they moved the recalcitrant Yin there." The Kong Yingda subcommentary adds, "in order to complete the Zhou way" 以成周道, hence the name ChengZhou," making the connection still more explicit. See Ruan Yuan, "Duo shi."

58. It is important for readers to recall that dates are specified that seem impossibly short for the king to move with an entire entourage from inside to outside the Guanzhong Basin.

59. Translator's note: Michael Nylan would mention the potential for *three* major cities, because Prince Kang must have occupied a major site as well.

60. Ruan Yuan, 16.236a.

61. Ruan Yuan, *juan* 16, p. 236. Lee's word is *xia du* 下都.

62. Luoyang existed in Western Han, and also earlier, in late Chunqiu and Zhanguo times, as a major metropolis. The Kong Yingda subcommentary for the *Documents* could not be clearer on this point, as it says, "ChengZhou in Zhou was [the same as] Luoyang in Han. Luo city, as a king's capital, was called the (a?) 'lower [secondary] (*xia*) capital' 周之成周, 於漢為洛陽也; 洛邑為王都, 故謂此為下都, and it was to ChengZhou that the recalcitrant Yin population was moved." See Ruan Yuan, 16.236a.

63. I speak of the *Annals'* traditions, because apparently, the *Zuo Tradition* was one such inspiration (or perhaps the inspiration came from the early *Annals* traditions that eventually were compiled into something like today's *Zuo*).

64. Ban Gu then tells us about Wang Mang's renaming, but that is irrelevant here. Wang Mang called this site Yiyang 宜陽.

65. *Hanshu* 28A.1555.

66. *Hanshu* 28A.1555. Translator's note: the mention of Jiaru appears to be a commentarial tradition that may have been interpolated, and the same goes for the mention of the Nine Tripods (see ibid.). Identification of Jiaru goes back to three early sources, judging from the extant early commentaries to the *Shiji*: Du Yu's 杜預 (222–284) annotations for the *Zuozhuan*, *Kuodi zhi* 括地志, and Huangfu Mi's 皇甫謐 (215–282) *Diwang shiji* 帝王世紀. See *Shiji* 30.1700, 4.129.

67. Contrast *Shiji* 4.133.

68. *Hanshu* 27A.1555, when writing of Henan commandery, says: 周武王遷九鼎。周公致太平，營以為都。是為王城。至平王居之。

69. In the *Zuozhuan*, two places discuss the Tripods (Lord Huan, Year 2 and Lord Xuan, Year 3), but the relevant entries are not in complete agreement. The first entry in Lord Huan says that King Wu moved the Nine Tripods to Luo city 武王克商，遷九鼎於雒邑, whereas the Lord Xuan entry says that King Cheng moved the Nine Tripods to Jiaru. The commentary for the *Zuo* clearly identifies these as competing traditions. Translator's note: an additional explanation is found in Du Yu's commentary for Lord Yin, Year 8, but it is hardly clarificatory. See Du's *Chunqiu Zuozhuan zhushu*, 5.95a-b; 105b. In addition, while many of the stories that made their way into the *Zuozhuan* compilation plainly date to the pre-unification period, it is far from clear when the *Han Zuoshi chunqiu* was compiled, with suggested dates ranging from the fourth to the first century BCE. While few today still believe that the text was forged by Liu Xin (contra Kang Youwei), Ban Gu clearly presumed an impossibly early date of mid–sixth century BCE for the precursor *Zuoshi chunqiu* text.

70. *Hanshu* 28B.1641.

71. *Hanshu* 27B.2650. Note, however, *Hanshu* 28B.1650 leaves the question more open, as the early Western Zhou "capital" of Luo city "on all four sides" has nobles whose fiefs serve as buffer zones against invasion. This language leaves the precise location of ChengZhou less settled. *Hanshu* 28B.1651 describes the capital domain directly under Western Zhou rule as stretching from the Guanzhong Basin, where Hao (aka ZongZhou) is located, to Luoyang. The term "Xia du" or "lower capital" seems to refer to Luoyang in early Western Han times (see *Hanshu* 47.2119). But the term "lower capital" means something else in the *fu* of Sima Xiangru (see *Hanshu* 57B.2596).

72. Ruan, *Shijing zhushu*, 147b.

73. *Hanshu* 28B.1650. Zheng Xuan's idea of the "Shao gao" and "Luo gao" chapters is distinctly at odds with Sima Qian's version of the distant past. Whereas Sima Qian thought these two *Documents* chapters were written *after* Zhougong returned the rule to King Cheng, and so after Zhougong's seventh year, Zheng Xuan puts the composition of these chapters in Zhougong's fifth year of rule. More importantly for our purposes here, Zheng Xuan imagined two capitals, in the same manner as Ban Gu. As I note, Zheng Xuan's view informs the *Shangshu dazhuan* (*Great Commentary to the Documents*) reportedly "continued" by Zheng (and Zheng may, in fact, be responsible for the wording in that source). *Suishu* 27.431 (the bibliographic treatise) mentions Zheng Xuan's commentary to a *Shangshu dazhuan*, in three *juan*. Readers should know that there are several different chronologies supplied in early texts for the events in early Western Zhou, not just Zheng Xuan's schema.

74. *Shiji* 4.129n1.

75. Du comm., *Chunqiu Zuozhuan zhushu*, *juan* 53.932b (Lord Zhao, Year 32). Note that Du Yu thinks of the "King's City" as being built by King Cheng,

not by Zhougong, perhaps because after Wang Mang's close identification with Zhougong, many were anxious to say that the duke acted only at the king's behest.

76. *Hou Hanshu* 19.3398.

77. Li, *Shuijingzhu*, 250. Li states that "Henan" refers to "Henan county," the old land of Jiaru 郟鄏. Li further identifies Jia 郟 as the name of a mountain there, and Ru 鄏 as the name of a city there. He notes that the "King's City" (which he equates with the New City), at the SE corner of the city wall, had a Tripod Gate. Li surmises that the name reflects the historical event in early Western Zhou when the Nine Tripods passed through that gate. For that reason, he says, the area is also called "Tripod Center" (Ding zhong 鼎中).

78. Zhao's theories spun out from there to make a series of equations between sites and events: Zhao asserted, for example, that Zhougong built the "King's City" and the series of equations that made the unknown location Jiaru 郟鄏 the site of King Ping's city, in Henan (and the same site where the Nine Tripods were housed), while identifying Zhougong's ChengZhou with King Jing's capital at Cheng-Zhou/Luoyang. For Zhao, see *Gai yu cong kao* 陔餘叢考, juan 6, p. 155 (東西周 entry).

79. One may consult Liang's *Shiji zhiyi*, juan 3, p. 115. In my opinion, Zheng's belief in Luo city as some sort of an eastern metropolis still can be maintained, as can his identification of the capital (aka ZongZhou). But when Zheng identifies Luo as *the* eastern capital, that is anachronistic, for in early Western Zhou, such terms as eastern and western capitals were not yet in existence.

80. See *Shiji huizhu kaozheng*, juan 4, p. 79. For Takigawa, the relevant lines say: 周公又營下都, 以遷殷頑民, 是為成周, 則洛陽也。平王東遷, 定都於王城, 其時所謂西周者, 豐鎬也, 東周者, 王城也。及王子朝之亂, 敬王徙都成周.

81. Li, *Shuijingzhu*, 39.

82. One would do well to remember that no one seems to have known of or consulted the precursor to today's *Zuozhuan* before late Western Han.

83. Michael Nylan, as editor, notes that after Professor Lee wrote his chapter there was published one archaeological report that states definitively that the city walls date to a time no earlier than mid-Western Zhou, which means they must postdate Zhougong and King Cheng: Hou, "Lun Xi Zhou wanqi ChengZhou de weizhi ji yingjian beijing." Unfortunately, Li, *Social Memory and State Formation*, mixes sound archaeology and sheer fantasy, not to mention propaganda, perhaps because this archaeologist was not trained in anthropology or history.

84. Excavation work in Luoyang in the early days sought to avoid damage to the underground sites from the major construction sites, and so it mainly followed the lead of the National Academy of Sciences, Institute of Archaeology. Later, beginning in the 1980s, the second wave of Luoyang excavations began. Guided by a wide range of literary sources—the *Documents* or the *Chunqiu* (particularly today's *Zuozhuan*), or by Zhao Wanli's 趙萬里 rare edition of the *Yongle dadian* 永樂大典, with its numerous entries on Luoyang and maps of it—many Zhou sites have

been found in the Luoyang area. From the 1950s to the 1980s, such finds have generated a wealth of scholarly articles. But those at Hao still stand out, especially in the breadth of evidence. Worth consulting is the essay published in 1956 on the western suburbs of Luoyang co-authored by Guo, Ma, Zhang, and Zhou; also the more recent essay co-authored by Wei and Sun, "Han Wei Luoyang, Xi Zhou cheng, yu Xi Zhou Luoyi tansuo."

85. Luoyang Shifan xueyuan, He-Luo wenhua guoji yanjiu zhongxin, ed., *Luoyang kaogu jicheng*, 2.

86. This parenthetical material has been added by the translator.

87. Qian, *Shiji diming kao*, juan 7 ("On Dongxi Zhou" 東西周), 305–6, says, "ChengZhou is basically the general name for the 'King's City' (aka Luo city and Xiadu). But it came to be that Xiadu was used for ChengZhou in order to distinguish it from the 'King's City' (the palace?) proper" 成周本王城 (洛邑)、下都之總號, 至是以下都為成周, 以別於王城焉. NB: The proper noun Xiadu and the general name *xiadu* do not exist in the *Documents* classic.

88. See Peng, "Xin yi kao," 49–62. Wang Yuzhe 王玉哲 is a well-published Early China scholar whose work frequently appears in *Lishi yanjiu*. Shi Weile writes on many topics, including the "historical facts" (some of them factoids) about the legendary emperors.

89. See Kong Yingda, *Shangshu zhengyi*, 225a–b, 236a, 273b, for example.

90. He preferred to romanize his name as Chang-Chien Hwang.

91. The *Shangshu dazhuan* or *Great Commentary to the Documents*, which tradition ascribes to Fu Sheng 伏勝 in early Western Han, supplies the chronology mentioned just below. See Huang Zhangjian, "Shi Lingyi," 119.

92. That peace prevails is why the *Shangshu dazhuan* says that Zhougong began in the following year, Year 6, to regulate the rites and music, and in year 7, "returned the rule" (*fan zheng* 反政) to King Cheng, as the duke's extraordinary powers were no longer needed by the new king. Wang Guowei (d. 1927), in his *Yin Zhou zhidu lun*, identifies this as the "ruler's domain or hegemony" (*juntong* 君統), saying that the name ZongZhou marks the spot where the ancestral temples were located, as these conferred legitimacy upon the "ancestral line." Meanwhile, the *Shangshu dazhuan* chronology specifies this: in the first year Zhougong assumes the regency and rescues the dynasty from chaos; in the second year Zhougong [re]conquers Yin; in the third year he suppresses the rebels at Yan; in the fourth year he establishes a fief at Wei; in the fifth year he designed ChengZhou; in the sixth year he regulated the rites and music; and in the seventh year he returned the administration to King Cheng 周公攝政, 一年救亂, 二年克殷, 三年踐奄, 四年建侯衛, 五年營成周, 六年制禮作樂, 七年致政成王. See Zhong Qianjun, *Shangshu dazhuan*, in *Gujing jie huyuan*, juan 2, p. 34.

93. Huang Zhangjian translates the *mingbao* in the bronze inscriptional phrase "Zhougong zi mingbao" 周公子明保 as referring to Zhougong's younger son; see Huang's "Shi Lingyi," 119.

94. Only the fourth-century *xu* (preface) to the "Duo shi" chapter speaks of the *wan min* 頑民. The main contents of the "Luo gao" use a similar phrase: *Yin xian min* 殷獻民: "the offered Yin men."

95. Wang, *Yin Zhou zhidu lun*, 1–15.

96. Editors' note: Or perhaps King Cheng never went to Luo city at all (?).

97. Note that a single ode, in the "Cao Feng" 曹風 section, "Xia quan" 下泉, uses several expressions (周京、京周、京師) for the single place. Khayutina, "Western 'Capitals' of the Western Zhou Dynasty," concurs.

98. *Hanshu* 88.3589.

99. This verb could also mean he [King Wu] constructed the site.

100. *Shiji* 4.170 (Zhonghua shuju, ed.) mistakenly writes Luo city 洛邑, when it ought to be Luo city 雒邑.

101. *Shiji* 13.487: 太史公曰. 孔子因史文, 次春秋 . . . 至於序尚書, 則略無年月, 或頗有, 然多闕, 不可錄, 故疑則傳疑, 蓋其慎也。

102. This tradition was queried in He and Nylan, "On the Han-era Postface (*xu* 序)."

103. Li Xueqin (d. 2019) was one scholar who promoted the idea that the Tsinghua mss. showed the antique origins of the fourth-century Mei Ze version of the *Documents*.

104. See Asselin, *A Significant Season*, 79, where the goddess of the Luo, submerged beneath the waves, nonetheless propels the Luo waters on to the capital: "From its source in Xiong'er, / It draws together the Yi and Chan and Jian torrents / Carving a waterway from the source to the capital, / Drawing tribute from the wastelands and outskirts."

105. For the three layers of extrapolations that are important to archaeological reasoning, see Buccellati, *A Critique of Archaeological Reasoning*. Similarly, Giele's *Imperial Decision-making and Communication*, 78–79, acknowledges the severe limitations of the archaeological record.

106. Ruan Yuan 阮元, *Chunqiu Zuozhuan zhushu* 春秋左傳注疏 [hereafter Du Yu comm.], 50.871b.

107. Ibid.

108. Ibid.

109. Ibid.

110. *Chunqiu Gongyang zhuan zhushu*, also in Ruan Yuan [hereafter *Gongyang*], 24.304b.

111. *Gongyang* 24.304b.

112. He Xiu jiegu 何休解詁, *Chunqiu Gongyang zhuan jiegu* 春秋公羊傳解詁, in *Shisan jing guzhu* 十三經古注. Hangzhou: Yonghuaitang ben 永懷堂本, 1869; rpt. Taipei, 1976 [hereafter He Xiu], 24.167.

113. He Xiu, 24.167.

114. He Xiu, 23.165.

115. Du Yu comm., 25.351a.

116. Ruan Yuan, 50.879a.
117. Ruan Yuan, 50.879a.
118. *Gongyang* 24.304b.
119. Ruan Yuan, 52.900a.
120. Ruan Yuan, 25.931b.
121. Ruan Yuan, 25.931b.
122. Ruan Yuan, 53.932b.
123. Du comm., 53.932b–33b.
124. Ruan Yuan, 53.932a.
125. Ruan Yuan, 16.263a.
126. Ruan Yuan, 13.222b.
127. Ruan Yuan, 5.61a.
128. Ruan Yuan, 10.176b.
129. Ruan Yuan, 10.178a.

Asian-language Bibliography

Chen Shouqi 陳壽祺. *Zuohai jing bian* 左海經辨. n.p.: Sanshan Chen shi, 1840s.

Duan Yucai 段玉裁.

———. *Guwen Shangshu zhuanyi* 尚書古文撰異. n.p.: Qi ye yanxiang tang, 1821, accessed at HathiTrust Digital Library.

———. "Yi Luo guzi buzuo Luo kao" 伊雒古字不作洛考. In *Duan Yucai yishu* 段玉裁遺書 (Jingyun lou ji 經韵樓集). Taipei: Dahua shuju, 1986.

Du Yu 杜預, comm. *Chunqiu Zuozhuan zhushu*, in *Shisan jing zhushu* 十三經注疏, edited by Ruan Yuan (postface dated 1815). Reprint, Beijing: Zhonghua shuju, 1980.

Gong Zizhen 龔自珍. *Da shi da wen* 大誓答問. n.p.: Pang xi zhai, 1867.

Guo Baojun 郭寶鈞, Ma Dezhi 馬得志, Zhang Yunpeng 張雲鵬, and Zhou Yongzhen 周永珍. "Yijiuwusinian chun Luoyang xijiao fajue baogao" 一九五四年春洛阳西郊发掘报告, *Kaogu xuebao* 考古学报 1956.2: 1–14, 15–43.

Hanshu 漢書. Compiled by Ban Gu 班固, 32–92, Ban Zhao 班昭, 48–ca. 116, et al. 12 vols. All refs. to the punctuated edition. Edited by Yan Shigu 顏師古. Beijing: Zhonghua shuju, 1962.

Hong Gua 洪适. *Li shi* 隸釋. Taipei: Taiwan Shangwu yinshuguan, 1983.

Hou Hanshu 後漢書. Compiled by Fan Ye 范曄, 398–445/446. Commentary Li Zhao 劉昭 et al. 12 vols. All refs. to the punctuated edition. Edited by Li Xian 李賢 et al. Beijing: Zhonghua shuju, 1965.

Hou Weidong 侯卫东. "Lun Xi Zhou wanqi ChengZhou de weizhi ji yingjian beijing" 论西周晚期成周的位置及营建背景. *Kaogu* 6 (2016): 90–97.

Huang Zhangjian 黃彰健 [Chang-Chien Hwang]. "Shi Lingyi Zhougong Ziming bao zhi 'ming bao' ji Zhougong cizi Junchen, binglun Xi Zhou Cheng wang qinzheng Zhougong Shaogong fenzheng dongdu xidu" 釋令彝周公子明保之明保即周公次子君陳, 並論西周成王親政周公召公分正東都西都, in *Zhougong Kongzi yanjiu* 周公孔子研究, 107–29. Taipei: Zhongyang yanjiu yuan lishi yuyan yanjiusuo, 1997.

Kong Yingda 孔穎達.

———. *Shangshu zhengyi* 尚書正義 [*Correct Meaning of the Documents*]. See Ruan Yuan, *Shisan jing zhushu*.

———. *Chunqiu zhengyi* 春秋正義 [*Correct Meaning of the Annals*]. See Ruan Yuan, *Shisan jing zhushu*.

Li Daoyuan 酈道元. *Shuijing zhu* 水經注. Edited by Wang Xianqian 王先謙合校本. Beijing: Zhonghua shuju, 2009.

Liang Yusheng 梁玉繩. *Shiji zhiyi* 史記志疑. Beijing: Zhonghua shuju, 1981, reprint of 1888 ed.

Liu Qiyu 劉起釪. *Shangshu yuanliu jiqi zhuanben kao* 尚書源流及傳本考. Shenyang: Liaoning daxue chubanshe, 1997.

Luoyang Shifan xueyuan, He-Luo wenhua guoji yanjiu zhongxin, ed. *Luoyang kaogu jicheng: Xia, Shang, Zhou juan* 洛阳考古集成. 夏商周卷. Beijing: Beijing tushuguan, 2005.

Ma Heng 馬衡. *Han Shijing jicun* 漢石經集存. Beijing: Kexue chubanshe, 1957.

Ma Shiyuan 馬士遠. *Liang Han Shangshu xue yanjiu* 兩漢尚書學研究. Beijing: Zhongguo shehui kexue chubanshe, 2014.

Matsumoto Masaaki 松本雅明. "Shūkō ka to Shōkō ka: Shō kō hen no seiritsu o meguru shumondai" 周公家と召公家：召誥篇の成立をめぐる諸問題. *Tōyō gakuhō* 54, no. 2 (September 1971): 188–238.

———. *Shunjū Sengoku ni okeru Shōsho no tenkai: rekishi ishiki no hatten o chūshin ni* 春秋戰國における尚書の展開：歷史意識の發展を中心に. Tōkyō: Kazama Shobō, 1966.

Peng Yushang 彭裕商. "Xin yi kao" 新邑考. *Lishi yanjiu* 歷史研究 5 (2000): 49–62.

Pi Xirui 皮錫瑞. *Jinwen Shangshu kaozheng* 今文尚書考證. Beijing: Zhonghua shuju, 1989.

Qian Mu 錢穆. *Gu shi dili luncong* 古史地理論叢. Taipei: Dongda tushu gongsi, 1982; reprint, Beijing: Shenghuo dushu xinzhi Sanlian shudian, 2004.

———. *Shiji diming kao* 史記地名考. Hong Kong: Taiping shuju, 1962.

Ruan Yuan 阮元. *Shangshu zhushu* 尚書注疏, preface completed 1815, in *Shisan jing zhushu fu jiaokan ji* 十三經注疏附校勘記, at Academia Sinica Taiwan website http://hanchi.ihp.sinica.edu.tw/ihp/hanji.htm.

Shangshu dazhuan 尚書大傳: see Zhong Qianjun.

Shangshu zhengyi 尚書正義: see Ruan Yuan, in which this work is included.

Shiji 史記. Compiled by Sima Qian 司馬遷 et al. 12 vols. All refs. to the punctuated edition. Beijing: Zhonghua shuju, 1972.

Sima Zhen 司馬貞. *Shiji suoyin* 史記索隱, in *Shiji*.

Taiping yulan 太平御覽. Compiled by Li Fang 李昉 et al. Taipei: Taiwan Shangwu yinshu guan, 1968.
Takigawa Kametarō 瀧川龜太郎. *Shiji huizhu kaozheng* 史記會注考證. Taipei: Yiwen yinshuguan, 1975, reprint of 1932–34 ed.
Tan Qixiang 譚其驤. *Qingren wenji dili lei huibian* 清人文集地理類汇编. Hangzhou: Zhejiang Renmin chubanshe, 1987–.
———. *Zhongguo lishi ditu ji* 中國歷史地圖. Shanghai: ditu chubanshe, 1982–1987.
Wang Guowei 王國維. *Yin Zhou zhidu lun* 殷周制度論, in *Guantang jilin* 觀堂集林. Beijing: Zhonghua shuju, 1959, reprint of 1922 work.
Wang Niansun 王念孙. *Dushu zazhi* 讀書雜誌. Taipei: Shijie shuju, 1963.
Wang Xianqian 王先謙. *Shangshu Kongzhuan canzheng* 尚書孔傳參正. Beijing: Xu shou tang, 1904.
Wang Yinzhi 王引之. *Jingyi shuwen* 經義述聞. Taipei: Guangwen shuju, 1963, reprint of 1817 original ed.
Wei Chengmin 魏成敏 and Sun Bo 孫波. "Han Wei Luoyang, Xi Zhou cheng, yu Xi Zhou Luoyi tansuo" 漢魏洛陽西周城與西周洛邑探索. *Dongfang kaogu* 東方考古 9 (2012): 288–95.
Xu Shen 許慎. *Shuowen jiezi* 說文解字. Annot. Duan Yucai. Taipei: Shuming chubanshe, 1986.
Zhao Yiqing 趙一清. *Gai yu cong kao* 陔餘叢考. Taipei: Huashi chubanshe, 1975.
Zhong Qianjun 鍾謙鈞. *Gujing jie huiyuan* 古經解彙函, 64 vols. Yangzhou: Guangling shushe, 2012 [which includes *Shangshu dazhuan* in vols. 12–13].
Zhou Zhenhe 周振鶴. *Xi Han zhengqu dili* 西漢政區地理. Beijing: Renmin chubanshe, 1987.

Western-language Bibliography

Asselin, Mark. *A Significant Season: Cai Yong (ca. 133–192) and His Contemporaries*. New Haven: American Oriental Society, 2010.
Bielenstein, Hans. "Lo-yang in Later Han Times." *Bulletin of the Museum of Far Eastern Antiquities* 48 (1976): 1–142.
Buccellati, Giorgio. *A Critique of Archaeological Reasoning: Structural, Digital, and Philosophical Aspects of the Excavated Record*. Cambridge: Cambridge University Press, 2017.
Chavannes, Édouard, trans. *Les mémoires historiques de Se-ma Tsien*. Paris: E. Leroux, 1895–1905.
de Crespigny, Rafe. *Fire over Luoyang: A History of the Later Han Dynasty, 23–220 AD*. Leiden: Brill, 2017.
Giele, Enno. *Imperial Decision-making and Communication in Early China: A Study of Cai Yong's Duduan*. Wiesbaden: Harrassowitz, 2006.

He, Ruyue, and Michael Nylan. "On the Han-era Postface (*xu* 序) to the *Documents Classic*." *Harvard Journal of Asiatic Studies* 75, no. 2 (Summer 2015 [actual date: Summer 2016]): 377–427.

Holmgren, J. "Myth, Fantasy or Scholarship: Images of the Status of Women in Traditional China." *Australian Australian Journal of Chinese Affairs* 6 (July 1981): 147–70; also available at JSTOR.

Khayutina, Maria. "Western 'Capitals' of the Western Zhou Dynasty: Historical Reality and Its Reflections until the Time of Sima Qian." *Oriens Extremus* 47 (2008): 25–65.

Li Min. *Social Memory and State Formation in Early China*. Cambridge: Cambridge University Press, 2018.

Pauketat, Timothy. *Chiefdoms and Other Archaeological Delusions*. Plymouth: AltaMira Press, 2007.

4

Celestial Signs in Three Historical Treatises[1]

JESSE J. CHAPMAN 柴傑思

The topic of "celestial patterns" (*tianwen* 天文) is the subject of chapters in the earliest standard histories in the Chinese tradition, confirming its importance as a key technical discourse at the intersection of astronomy, classical learning, and the interpretation of omens. However, while the *Hanshu* 漢書 (*History of the Han*) "Treatise on the Celestial Patterns" (Tianwen zhi 天文志) borrows much of its prose from the *Shiji* 史記 (*Writings by the Senior Archivist*) "Treatise on the Celestial Offices" (Tianguan shu 天官書), it contains several new features that reflect the authority the Classics had gained in technical writing since the late Western Han. The similarity between the "Celestial Patterns" and the *Hanshu* "Wuxing zhi" 五行志 (Treatise on the *wuxing*) indicates that part of its function was shared, in this case, recording astronomical and meteorological phenomena that were also baleful signs. In particular, the chronicles at the conclusion of the "Celestial Patterns" treatise follow a chronological model like that of the signs interpreted in the "Wuxing zhi" rather than in the *Shiji* treatise. Another contrast is the presence of technical passages in the "Celestial Patterns" that integrate attributed citations from the *Changes*, the *Documents*, and the *Odes*. Comparison of the new passages with the borrowed passages casts into relief the increased value technical writers placed on classical authority in the late Western and Eastern Han dynasties.

From 100 BCE to 100 CE, the technical discourse on celestial signs came to rely on classical authority. Technical treatises in the *Shiji* and the *Hanshu* position celestial signs within a historical framework. Sima Qian's 司

馬遷 (ca. 145–ca. 86 BCE) "Celestial Offices" delimits technical information concerning stars, planets, inclement weather, and other celestial signs, framing these with a closing essay that draws attention to the critical role men such as he and his father, Sima Tan 談 (d. 110 BCE), had played at court since time immemorial. Sima Qian, as *Taishi ling* 太史令 (Senior Archivist), served as a reader, recorder, and interpreter of omens past and present.[2]

In the late Western Han, the authority of the *Annals* and the "Hongfan" 洪/鴻範 (Great Plan) chapter of the *Documents* became the basis for reading omens as exemplified in the "Wuxing zhi." Likely based in part on the late Western Han statesman Liu Xiang's 劉向 (79–8 BCE) "Hongfan wuxing zhuan lun" 洪範五行傳論 (Discourse on the Five Resources Tradition of the "Great Plan"),[3] the "Wuxing zhi" systematically reads historical signs in the Chunqiu 春秋 period (722–481 BCE), the Qin 秦 (221–207 BCE), and the Western Han 漢 (206 BCE–9 CE) to construct a sequel to the *Annals* classic. Just as the *Gongyang zhuan* 公羊傳 (*Gongyang Tradition*) for the *Annals* proceeded toward Kongzi's 孔子 (trad. 551–479 BCE) famous declaration that his way was at an end,[4] the final version of the "Wuxing zhi" developed teleologically toward the rise of Wang Mang 王莽 (r. 9 CE–23 CE) and the fall of the Western Han. The *Hanshu* "Celestial Patterns," compiled by Ban Zhao 班昭 (48–ca. 116 CE), Ma Rong 馬融 (79–166 CE), and Ma Xu 馬續 (ca. 79–after 141 CE)[5] perhaps a decade after Ban Gu's 班固 (32–92 CE) death, incorporated verbatim technical materials that made up much of Sima Qian's earlier treatise (see Appendix A to this chapter), but placed those materials in a framework that derived its authority from the Classics.[6]

I. Readers of Celestial Signs in History in the "Celestial Offices"

The concluding essay in Sima Qian's "Celestial Offices" couches technical information on celestial signs within a broader discourse on effective governance. It enumerates the constellations, describes the movements of the Five Planets and their omenological significance, explains what various sorts of halos and eclipses portend, outlines the significance of changes in color in particular stars, discusses the appearance of various sorts of *qi*, and details the practice of divining the harvest. Sima's concluding essay emphasizes the value of the observation of the heavens to courts ancient and contemporary:

The Senior Archivist said: Since the very birth of the people, who among the lords of the various ages has not calculated the movements of the sun, moon, planets, and stars? With the advent of the Five Houses[7] and the Three Dynasties, the practice continued and was further illuminated. In the inner lands, there were those who wore caps and belts, and in far-off places, there were the Yi and Di. The sages divided the central domains into twelve regions, and "they looked up to observe the signs in the heavens and looked down to emulate them upon the earth."[8] In Heaven, there are the sun and moon, and on Earth, *yin* and *yang*. In Heaven, there are the Five Planets, and on Earth, the Five Resources. In Heaven, there are the arrayed lodges, and on Earth, regions and borders. The Three Luminaries[9] are the essence of *yin* and *yang*. *Qi* originates in the Earth, and the sages unify it and put it in order.

太史公曰：自初生民以來，世主曷嘗不曆日月星辰？及至五家、三代，紹而明之，內冠帶，外夷狄，分中國為十有二州，仰則觀象於天，俯則法類於地。天則有日月，地則有陰陽。天有五星，地有五行。天則有列宿，地則有州域。三光者，陰陽之精，氣本在地，而聖人統理之。[10]

Sima Qian's essay opens by creating a powerful sense of the historical significance and sagely character of the practices detailed in the body of the treatise. For Sima, planetary observation and celestial observation in general are part of a much grander enterprise, one that had reached its present level of sophistication thanks to the acumen and perspicacity of the sage-kings of antiquity and the talented officials at their courts.

Sima Qian constructed a lineage of practitioners of the arts of celestial observation and astronomical calculation, a lineage to which he and his father, Sima Tan, who had served as *Taishi ling* from ca. 140–110 BCE,[11] belonged.[12] Beginning with its earliest practitioners in high antiquity, the lineage continues into the early Western Han: "As for those who have conducted the arts of celestial calculation since the beginning of the Han, Tang Du did so for the stars, Wang Shuo for *qi*, and Wei Xian for prognosticating the harvest" 夫自漢之為天數者，星則唐都，氣則王朔，占歲則魏鮮.[13] Tang, Wang, and Wei serve both as conservative transmitters of sagely knowledge and as innovators who kept that knowledge relevant and useful in their own world. While Sima Qian does not name himself as an heir to this lineage, in compiling the "Celestial Offices" he effectively summarizes and puts in order the sum total of knowledge needed to transmit the sagely enterprise. Sima's treatise not only encompasses the expertise of Tang, Wang, and Wei but also surpasses them, powerfully juxtaposing political history with the appearance of celestial signs:

> In the 242 years of the Chunqiu era, there were thirty-six lunar eclipses, three appearances of comets, and in the time of Duke Xiang of Song (r. 650–637 BCE), stars fell like rain.[14] The Son of Heaven had little leverage, and the local lords took power over the administration. The Five Hegemons[15] arose in turn, issuing their own commands in place of those of the ruler. From this time forth, the many did violence to the few, and the large absorbed the small. Qin, Chu, Wu, Yue—all Yi and Di became powerful hegemons. The Tian ministerial clan usurped power in Qi.[16] The Three Clans divided the domain of Jin.[17] All became warring states, struggling for glory and gain, repeatedly raising the weapons of war, again and again slaughtering [the populations of] entire cities, inaugurating [a period of] starvation, pestilence, burning, and bitterness. Lords and ministers alike feared calamities, and thus they considered the examination of fortune and the observation of planets and *qi* to be still more urgent.

> 蓋略以春秋二百四十二年之間，月蝕三十六，彗星三見，宋襄公時星隕如雨。天子微，諸侯力政，五伯代興，更為主命。自是之後，眾暴寡，大并小。秦、楚、吳、越，夷狄也，為彊伯。田氏篡齊，三家分晉，並為戰國。爭於攻取，兵革更起，城邑數屠，因以饑饉疾疫焦苦，臣主共憂患，其察禨祥候星氣尤急。[18]

Sima Qian portrays the observation of celestial signs as an art inextricably linked to the prognostication of political fortunes. The eclipses, appearances of comets, and meteors that "fell like rain" during the Chunqiu period accompanied the decline in power of the Zhou rulers and presaged the increasing tendency toward violence and fragmentation that characterized the Warring States (480–222 BCE) period.

Such signs continued to appear during moments of political crisis in the Qin and Han. Following the appearance of numerous comets during the time of Qin Shihuang 秦始皇 (r. as emperor 221–210 BCE), "with the rise of the Han, the Five Planets gathered in Eastern Well" 漢之興, 五星聚于東井, a most auspicious sign.¹⁹ In the generations to follow, however, ill omens appeared. The treatise associates a solar eclipse with the rebellion of members of the clan of Empress Lü 呂 (r. 187–180 BCE) and a comet with the Seven Kingdoms rebellion of 154.²⁰ Sima deploys vivid language to present the consequential relationship between the latter sign and the political events that follow it: "Heaven's Hound passed over the celestial field [corresponding to] Liang, and war broke out. Soon thereafter, blood gushed forth from beneath the fallen corpses" 天狗過梁野; 及兵, 遂伏尸流血其下.²¹

Sima Qian's essay, by discussing the appearance of celestial signs in conjunction with pivotal events in history, underscores the value of interpreting celestial signs. Moreover, by constructing a tradition extending back to the sage-kings, the essay declaims the essential role of expert interpreters. Sima concludes the essay with an assertion about knowledge of celestial prognostication that implicitly applies to his own treatise. "From the end back to the beginning, from antiquity up to the present, if we deeply examine the changes of the times, fathoming their finest points and crudest forms, then the Celestial Offices will be made complete" 終始古今, 深觀時變, 察其精粗, 則天官備矣.²² The "Celestial Offices" appeals to the hallowed authority of lineage and the strength of historical precedent. Despite its similarities to the later "Celestial Patterns," it does not, however, extensively employ frameworks drawn from the Classics or direct citations of classical texts. It belongs to a time prior to the intimate interweaving of classical and technical knowledge that would occur in the late Western and Eastern Han.

II. The "Wuxing zhi" and Its Classical Models

The "Wuxing zhi," in contrast to the "Celestial Offices," positions celestial signs and other omens within an interpretive framework largely derived from the Classics. The introductory lines in the treatise style it both as an extension of the "Great Plan" chapter of the *Documents* and as a Han expansion of the *Annals*. The "Wuxing zhi" provides many records of comets and eclipses for the Chunqiu and Western Han. It concerns the postmortem historiographical analysis of the fall of a dynasty—omenology writ large.²³

Its introduction analogizes the invention of the Eight Trigrams to the development of the "Great Plan." The treatise begins with a quotation from the "Tradition of the Appended Statements" (Xi ci zhuan 繫辭傳):

The *Changes* says, "Heaven suspended the starry signs (*xiang*) so as to make visible good and ill fortune, and the sages made images of them. The Yellow River gave forth the Chart. The Luo River gave forth the Writings. The sages used these as models."[24] Liu Xin 劉歆 (46 BCE–23 CE) took this to mean that Fu Xi as king was the successor to Heaven. He had received the River Chart and drawn up a chart modeled upon it; this was the Eight Trigrams. Yu had set in order the flooding waters and been given the Luo Writing, and produced a model emulating it; this was the "Great Plan." The sages followed [the same] Way and treasured its true qualities.

《易》曰：「天垂象，見吉凶，聖人象之；河出圖，雒出書，聖人則之。」劉歆以為虙羲氏繼天而王，受《河圖》，則而畫之，八卦是也；禹治洪水，賜《雒書》，法而陳之，《洪範》是也。聖人行其道而寶其真。[25]

These opening lines interweave the text of the "Great Plan" with the core unit of the *Changes*, the Eight Trigrams. "The River Chart and the Luo Writing relate to one another as warp and weft (*jingwei* 經緯), while the Eight Trigrams and the Nine Divisions become inner and outer for one another" 《河圖》、《雒書》相為經緯，八卦、九章相為表裏.[26] The Eight Trigrams and the Nine Divisions are both the products of sages who emulated sets of signs produced directly by Heaven, the River Chart and the Luo Writing. The core of the *Changes* and the "Great Plan," like the Classics and their commentaries, are bound together in turn with similar tightness, with each relying on the other for part of its meaning. The treatise next points to later sages, King Wen 文 (d. ca. 1046 BCE) and Kongzi, who produced extensions of the Eight Trigrams and the "Great Plan" during moments of historic decline. King Wen expanded the *Changes* by producing the sixty-four Hexagrams from the initial Eight Trigrams in the last generation of the Shang (ca. 1600–1046 BCE). Kongzi compiled the *Annals*, recording events that verified the "Great Plan," thereby demarcating the end of the Chunqiu period.[27] Both the creation of the sixty-four Hexagrams and the composition of the *Annals* constitute sagely responses to desperate times, profound and marvelous works whose origins may ultimately be traced to Heaven itself. King Wen and Kongzi lived in times when the will of Heaven seemed to have been ignored or forgotten, yet they produced texts that could be used to recover and reenact that will.

The *Hanshu*, compiled nearly a century after the collapse of the Western Han, places the "Wuxing zhi" within the lineage of the *Annals*, and thus within the lineage of the "Great Plan" and the Luo Writing. Western Han omen readers from Dong Zhongshu 董仲舒 (ca. 195–ca. 104 BCE) to Gu Yong 谷永 (fl. 36–9 BCE) had explained contemporary signs by comparing them with classical precedents. The compiler concludes: "I have gathered together ... the explanations of various court affairs up until the time of Wang Mang, for a total of twelve generations, to serve as a supplement to the *Annals*, and compiled this chapter" 是以攬 … 所陳行事, 訖於王莽, 舉十二世, 以傅《春秋》, 著於篇,[28] thereby designating the treatise as the culmination of Western Han omenology and an extension of the Classics.

The "Wuxing zhi" is at once an expansive set of commentaries to the "Great Plan" and an annotated *Annals* for Han times. It employs and distinguishes among the interpretations of omens, during both the Chunqiu period and the Han, from the most prominent Han omenologists and experts on the *Annals*, the *Documents*, and the *Changes*. The macrostructure of the "Wuxing zhi" is derived from three of the Nine Divisions in the "Great Plan": the Five Resources (*wu xing* 五行), the Five Modes of Engagement (*wu shi* 五事), and the Sovereign's Standard (*huang ji* 皇極). The materials included in these major sections might be read as commentary on the "Great Plan" itself. The first major subsections place omens in categories associated with each of the Five Resources: Wood (*mu* 木); Fire (*huo* 火); Earth (*tu* 土); Metal (*jin* 金); and Water (*shui* 水). The second major section categorizes omens under each of the Five Modes: Demeanor (*mao* 貌), Speech (*yan* 言), Sight (*shi* 視), Hearing (*ting* 聽), and Thinking (*si* 思). The final major section includes omens related to the Sovereign's Standard. Each category is further subdivided into specific kinds of disasters and anomalies that issue from imbalances in the Five Resources, problems related to the Five Modes, or imperfect adherence to the Sovereign's Standard. Some omens fit into their categories according to a transparent logic, while in other cases the relationship between a specific kind of phenomenon and the category into which it falls is opaque. The Mode of Speech category, for instance, includes such likely suspects as "speech that is not yielding" (*yan bu cong* 言不從) and "rhymed warnings" (*shi yao* 詩妖), but also includes strange occurrences that bear no obvious relationship to speech, such as "constant sunshine" (*heng yang* 恆陽), "signs involving dogs" (*quan huo* 犬禍), and "white omens" (*bai xiang* 白祥).[29]

The treatise uses the motley assortment of signs associated with the Five Resources, the Five Modes, and the Sovereign's Standard as the basis for its organization. The treatise introduces each subsection with a quotation from a certain tradition (*zhuan* 傳), or set of traditions, of uncertain provenance,[30] detailing the sort of problems associated with the resource or mode under discussion. Like the *Annals*, the treatise organizes individual types of signs chronologically. First, the "Wuxing zhi" lists entries in the *Annals* that record appearances of a given type of sign and Western Han interpretations. Second, the treatise lists instances of appearances of the same kinds of signs in the Qin and Western Han, likewise with Western Han interpretations. The "Wuxing zhi" formally mirrors the *Annals*.

Celestial signs, along with certain human signs, such as people growing horns or spontaneously changing sexes, occur almost exclusively in the final major division of the "Wuxing zhi," the section on the Sovereign's Standard:

> The tradition says, "When the sovereign does not attain the standard, this is called failing to establish it. Its sign of blame is the loss of vision, its penalty constant *yin*, and its extreme manifestation is weakness. At times there are archery warnings, at times dragon and snake signs (*nie* 孽), at times signs (*huo* 禍) involving horses, at times afflictions [leading] the people below to attack those above them, and at times the sun and moon stray from their courses, or the planets move in retrograde against the stars.
>
> 傳曰：「皇之不極，是謂不建，厥咎眊，厥罰恆陰，厥極弱。時則有射妖，時則有龍蛇之孽，時則有馬禍，時則有下人伐上之痾，時則有日月亂行，星辰逆行。」[31]

The lengthy explanatory section that follows makes the appearance of celestial signs, and other signs included in the same section the product of lapses in all of the Five Modes. Taken as a group, these lapses constitute a failure to attain the standard. Cloudy *qi* rises from the mountains, covering the heavens, so that the "*qi* of Heaven is disordered" 天氣亂. The ruler is like the arrogant, high-flying dragon in the Line statements to the first hexagram in the *Changes*, Qian 乾 (Pure Yang).[32] Because, as it says in the *Changes*, "clouds follow dragons" 雲從龍,[33] dragon and snake signs are born. Signs involving horses occur because the hexagram Qian corresponds to both rulers and horses. The explanatory section proclaims that the tradition, like the *Annals*, in fact conceals a powerful critique of the ruler:

When the Way of the ruler comes to harm, sickening the *qi* of Heaven, and the text does not say that the Five Resources have damaged Heaven, but instead states that "the sun and moon stray from their courses," or "the planets move in retrograde against the stars," as if those below did not dare to damage Heaven, this is comparable to the *Annals*' statement that "the King's forces suffered defeat at Maorong."[34] It does not say who defeated him, but uses language suggesting that he was defeated of his own accord. The meaning of doing so is to exalt those who are worthy of exaltation.

凡君道傷者病天氣, 不言五行沴天, 而曰「日月亂行, 星辰逆行」者, 為若下不敢沴天, 猶《春秋》曰「王師敗績于貿戎」, 不言敗之者, 以自敗為文, 尊尊之意也。[35]

The "Wuxing zhi" places responsibility for the appearance of celestial signs squarely on the head of the ruler. While it claims that the tradition avoided, out of ritual respect, any indication of the ruler as the source of disorder in the cosmos, it suggests that disorder in the heavens issues from disorder at court. The idea that celestial signs are rooted in ongoing processes in the human world colors the interpretation of such signs both in the *Annals* and in the omenology of the Western Han.

III. The "Wuxing zhi" Chronicle of Comets

The "Wuxing zhi" repeatedly tells two stories. One is the story of the Chunqiu period, the slow descent toward the chaos of the Warring States period. The second has a different arc, beginning with Liu Bang's 劉邦 (d. 195 BCE) triumph over Qin Ershi 秦二世 (r. 209–207 BCE) and over his rival Xiang Yu 項羽 (232–202 BCE). Yet the second story, which always ends with the rise of the usurper Wang Mang, is likewise a narrative of decline. Almost any chronicle of almost any category of inauspicious sign can serve in the "Wuxing zhi" as a chronicle of the decline and fall of the Western Han. The set of records pertaining to comets reflects broader formal features of the chronicles, presenting a typical, if elaborate, example of the narrative arc in which the "Wuxing zhi" examines various signs in the Chunqiu, Qin, and Western Han periods. Within these chronicles, the "Wuxing zhi" proves to be a true supplement to the *Annals*, an *Annals* for Han times.

The *Annals* for the Chunqiu period recorded only three comets. The Han chronicle of comets includes detailed readings of these records attributed to Dong Zhongshu, Liu Xiang, and Liu Xin, as well as relevant passages from the *Zuozhuan* 左傳 (Zuo Tradition). Liu Xiang's commentary also cites the *Xingzhuan* 星傳 (Tradition of Stars and Planets)[36] several times. In the *Annals* classic, as it is read in the "Wuxing zhi," comets appear at crucial moments. A comet sweeps through the Northern Dipper in 613 BCE, presaging the murders of the lords of Qi 齊, Song 宋, Lu 魯, Ju 莒, and Jin 晉 over the next twenty-eight years. A comet near the Great Star 大辰 (Antares) prompts one reader of celestial signs, whose words are reproduced in today's *Zuozhuan*, to make a memorable and often repeated claim: "Broom stars are the means by which the old is swept away[37] and the new is spread out" 彗所以除舊布新也.[38] Many once-glorious houses were indeed swept away following the final comet in the *Annals* (482 BCE). The "Wuxing zhi" records, but does not differentiate between, the interpretations of Dong Zhongshu and Liu Xiang: "Dong and Liu thought as follows: The reason it does not say the name of the lodge is because it was not imposed on a lodge. When a celestial body rises following the sun, disorderly *qi* occludes the brightness of the lord. The next year, the events in the *Annals* came to a conclusion" 董仲舒、劉向以為不言宿名者，不加宿也。以辰乘日而出，亂氣蔽君明也。明年，《春秋》事終.[39] The final comet appeared to herald the end of the Chunqiu period.

The chronicle of comets during the Western Han presents an arc that might be heuristically divided into five stages: the founding of the dynasty; early rebellions; military conquests; strife at court; and the rise of Wang Mang. The first stage, which includes a single comet appearing in 204 BCE, announces the rise of Liu Bang, the first Han emperor:

> In the seventh month of the third year of the High Emperor of Han (204 BCE), a comet appeared near Great Horn (Arcturus), and only after more than ten days did it set. Liu Xiang thought [as follows:] At the time of these events, Xiang Yu was the King of Chu, and was hegemon over[40] the local lords, but the Han had already pacified the three districts of Qin,[41] and was distant from Xiang Yu, in the city of Xingyang. But the hearts of the whole world gravitated to the Han, and Chu was about to be destroyed. Thus, the broom-star cleared the throne. One source says: "Xiang Yu massacred [surrendered] Qin soldiers,[42] burned

palaces, murdered Yidi (r. 208–205), and brought disorder to the throne. Thus, the broom-star was imposed upon him."

高帝三年七月，有星孛于大角，旬餘乃入。劉向以為是時項羽為楚王，伯諸侯，而漢已定三秦，與羽相距滎陽，天下歸心於漢，楚將滅，故彗除王位也。一曰，項羽阬秦卒，燒宮室，弒義帝，亂王位，故彗加之也。[43]

This comet proves a baleful omen, but not for Liu Bang. A series of bad omens can be auspicious indeed for one who is bent on toppling a sitting ruler. The "Wuxing zhi" gives two possible interpretations, both inauspicious for Xiang Yu. Liu Xiang's interpretation identifies Xiang Yu as a king. After briefly occupying the throne, Xiang Yu is swept away. In the second interpretation, from an unknown Western Han source, Xiang Yu brings disaster on himself through a series of atrocities culminating in the murder of the previous King of Chu, Yidi (the Dutiful Emperor), whom Xiang Yu himself had enthroned only a few years before.

Comets in the second stage mark rebellions, respectively occurring in 157 BCE and 135 BCE, only a few years prior to the Seven Kingdoms rebellion and the rebellion of Liu An 劉安 (d. 122 BCE), the King of Huainan. Liu Xiang's interpretation of the 157 BCE comet, which appeared soon after the death of Emperor Wen 文 (r. 179–157), points to both the coming rebellion of the Seven Kingdoms and the ultimate victory of Emperor Jing 景 (r. 157–141) three years later.[44] The second comet in this stage seems to blend intrigue at court with rebellion. It points to Liu An's visit to court in 134 BCE and the "haughty and self-indulgent" 驕恣 behavior of Empress Chen 陳皇后 (fl. 130 BCE).[45] Liu An's rebellion was suppressed, and Empress Chen was deposed.[46] Comets announce rebellion and, in the case of Empress Chen, potentially destructive influences within the palace walls. The second stage emphasizes the capacity of early Western Han rulers to respond effectively to challenges.

The first comet of the third stage, identified as "The Banner of Chiyou" 蚩尤旗,[47] occurs only two months after the last comet of the previous stage. Stretching across the sky, the comet elicited an anonymous prognostication announcing "the King will march in conquest on the Four Directions" 王者征伐四方. It appeared to herald the expansionist policies of Emperor Wu 武 (r. 141–87 BCE). An interpretation of the second comet in this stage, also without an identified source, points to the intensification of expansionist policies: "In the fourth month of the fourth year of the Yuanshou reign period (119 BCE), a long star again emerged in the northwest. At this time, the campaigns against the Hu grew especially extreme" 元狩四年四月, 長星又出西北, 是時伐胡尤甚.[48] Whereas the prognostication accompanying the first comet supported an expansionist campaign, the interpretation of the second points to the excessive and potentially dangerous point the expansionist policies had reached. The second comment is not a prognostication, but a retrospective, critical judgment representing a particular historical perspective. It works through a simple act of juxtaposition, artlessly bonding the trajectory of the history of the Western Han to the appearance of celestial signs.

The fourth stage marks in earnest the decline of the Western Han, but that decline is punctuated by a restoration under Emperor Xuan 宣 (r. 74–48). The three comets in this stage all accompany strife at court. The first occurs in 110 BCE, presaging the disastrous fighting near the end of Emperor Wu's reign, during which the Crown Prince would be executed and Empress Wei 衛 (d. 90 BCE) condemned for witchcraft.[49] The second appeared during the last year of the life of Huo Guang 霍光 (d. 68 BCE), the de facto ruler of the empire since Emperor Wu's death. It anticipates Huo's demise and the destruction of his family two years later.[50] Huo's rule marks a period in which power did not rest in the hands of a Liu clan emperor. Like Wang Mang, Huo could not be a legitimate ruler, although, unlike Wang Mang, Huo never declared himself emperor. With Huo's death, legitimate rulers returned to power. Emperor Xuan's reign ushered in a mid-dynastic renaissance. For three generations, the line of succession would be uninterrupted. But after the coronation of his grandson, Emperor Cheng 成 (r. 33–7 BCE), the process of decline began again with the appearance of a comet in the first month of 32 BCE. It would be read, by the end of the new emperor's reign, as a sign that no son would survive to succeed him. Both Liu Xiang and Gu Yong retrospectively interpreted the comet as a sign that a pregnant woman would be murdered, an act that would cut off the line of succession. Emperor Cheng would, in the end, nominate a nephew as his heir.[51] And while Emperor Ai 哀 (r. 7–1 BCE) proved no friend to Wang Mang, his early death without issue set the stage for Wang to become regent and, ultimately, to take the throne.

The final stage in the chronicle consists of a single, lengthy entry narrating an appearance of Halley's Comet near the end of Emperor Cheng's reign (12 BCE).[52] The entry consists of three main components: a description of the spectacular trajectory of the celestial sign, the views of Gu Yong and

Liu Xiang, and, finally, a description of the events portended by the comet up to the rise of Wang Mang. Gu identifies its appearance with treachery in the rear palace and widespread revolt, while Liu compares it with signs that accompanied the fall of the previous dynasty.[53] The treatise summarizes the events between the appearance of the comet and the fall of the dynasty. After three successive emperors who died without sons, Wang Mang would finally seize power.[54] The chronicle of comets ends, as do the chronicles of many types of signs in the "Wuxing zhi," with the fall of the Western Han.

IV. Etiology, Chronology, and Classicism in the "Celestial Patterns"

Ban Zhao and the Ma brothers synthesized aspects of the "Celestial Offices" with aspects of the "Wuxing zhi" in their "Treatise on Celestial Patterns." They appropriated much of Sima Qian's treatise wholesale. Large parts of the "Celestial Patterns" correspond very closely to the "Celestial Offices," with sections on the Five Planets and the sun and moon incorporating the bulk of new technical material. However, the "Celestial Patterns" distinguished itself from Sima's treatise in three important ways. First, the introduction to the "Celestial Patterns" replaces Sima's closing essay. Where Sima's comments framed technical materials as the patrimony of a lineage of experts, the opening of the "Celestial Patterns" emphasizes the etiology of celestial signs, declaring them to be direct reflections of events in the human realm. Second, the "Celestial Patterns" draws directly and explicitly upon classical authority, including several technical passages that cite the Classics as proof texts. The classicism of the "Celestial Patterns," moreover, plays a vital role in the network of taxonomic correspondences within which it positions planetary signs. In contrast to Sima's treatise, it ties each of the Five Planets to one of the Five Modes in the "Great Plan" and to the "Monthly Ordinances" of the *Record of Rites* (Liji 禮記). Finally, the "Celestial Patterns" closes with a chronicle of celestial signs that tells the story of the rise and fall of the Western Han.

Where the *Shiji* treatise speaks of the pragmatic value of celestial observation to the court, linking it to the traditions of high antiquity, the "Celestial Patterns" interprets celestial aberrations as signs that the ruler has endangered himself and his court on account of his ritual failures. Before launching into a description of the constellations of the Five Palaces nearly identical to the first section of the "Celestial Offices," the "Celestial Patterns" introduces the treatise, claiming a pronounced resonance between such celestial signs and the court:

> [Portents] are what originate in the earth and erupt above into the heavens. When governance fails (down) here, then aberrations appear (up) there, just as shadows are cast in the images of their forms and echoes are responses to sounds. This is why the clear-sighted ruler sees them and awakens, putting himself in order and rectifying his affairs. Should he set his mind to repenting his crimes, then calamity can be avoided and blessing brought forth. This is the tally that comes about on its own.
>
> 其本在地，而上發于天者也。政失於此則變見於彼，猶景之象形，鄉之應聲。是以明君覩之而寤，飭身正事，思其咎謝，則禍除而福至，自然之符也。[55]

While the "Celestial Patterns" contains much of the same information as the "Celestial Offices," it frames that information differently. Sima Qian's treatise moves directly into the technical details of which constellations are called what and located where. The "Celestial Patterns" contains the same information, but prefaces it by claiming that the ruler and his court are responsible for the appearance of celestial signs. Inauspicious signs result from faults in governance or in ritual. Actions can be taken to avoid the calamitous outcomes associated with signs that have already appeared; prognosis is not to be confused with fate. The same technical information, presented differently, might seem inevitable. The "Celestial Patterns" emphasizes that bad fortune can be turned into good. Auspicious signs signal that the court has earned the favor of the heavens.

The "Celestial Patterns" tells stories similar to those in the "Celestial Offices"; the passages it includes on Chunqiu-era portents and the political events that accompany them correspond closely to those in the *Shiji*, and much the same is true when it recounts the tales of Qin and early Western Han. The appearance of celestial signs as evidence that Heaven has vested authority in the ruler is, however, articulated far more explicitly in the "Celestial Patterns" than in the "Celestial Offices." Where the "Celestial Offices" reports the conjunction of the Five Planets that accompanied the "rise of the Han,"[56] the "Celestial Patterns" alone refers to this sign as the "Tally that the Founding Emperor had received the charge" 高皇帝受命之

符.⁵⁷ The "Celestial Offices" argued for the pragmatic value of the sagely enterprise of reading celestial signs to the court. The "Celestial Patterns" claimed that celestial signs could indicate that a particular individual had received the right to rule from the heavens, but, more often, celestial signs indicated ritual or administrative failure.⁵⁸

V. Linking Signs to the Classics in the "Celestial Patterns"

While the "Celestial Offices" no doubt drew from a variety of textual materials, it tended to efface rather than highlight its relationship to its textual antecedents. Despite the high level of correspondence between the main body of the "Celestial Patterns" and the "Celestial Offices," the "Celestial Patterns" contains far more numerous references to the Classics than does its predecessor. It used the Classics to underscore the authority of its technical calculations. Moreover, as is the case in the "Wuxing zhi," traditions linked to the "Great Plan" shape the manner in which planetary omens are interpreted, associating anomalies with specific types of ritual or political failures.

Whereas Sima Qian claimed a sagely lineage for the practice of reading celestial signs, the "Celestial Patterns" invests the reading of celestial signs with the authority of the Classics. One technical passage in the "Celestial Patterns," but not the "Celestial Offices," integrates the movements of the moon into a spatial schema built on the *Changes* and bolstered by citations from the *Odes* and *Documents* classics:

> The stars in Basket are wind. They are the stars of the northeast. Northeast serves Earth, and it is Heaven's Position. Thus, the *Changes* says: "In the northeast a friend is lost."⁵⁹ The trigram *Yielding* lies in the southeast and is wind. Wind is *yin* within *yang* and is the counterpart of the Great Minister. Its stars are those in Axletree. When the moon deviates from the middle path [i.e., the ecliptic], it moves northeast and enters Basket, or if it moves southeast and enters Axletree, then there is much wind. The west is rain. Rain is the position of lesser *yin*. When the moon deviates from the middle path and moves west to enter Net, then there is much rain. Thus, when the *Odes* says, "The moon is caught in Net! How it makes the waters run,"⁶⁰ it speaks of there being much rain. And when the *Tradition of the Stars and Planets* says, "When the moon enters Net, there will be those

among the generals and great ministers who commit crimes on account of their households," it speaks of the overabundance of *yin*. When the *Documents* says, "In the stars there is fine wind; in the stars there is fine rain . . . when the moon follows the stars, there is wind and rain,"⁶¹ it speaks of [the moon's] deviation from the middle path as it moves from east to west. Thus, the *Tradition of the Stars and Planets* says, "When the moon drifts south and enters the Ox Herder, the south is warned that there will be a plague among the people; when the moon drifts north, enters Grand Tenuity, and emerges to the north of the throne, as if it is transgressing against the throne, then the men below will plot against those above."

箕星為風，東北之星也。東北地事，天位也，故《易》曰「東北喪朋」。及《巽》在東南，為風；風，陽中之陰，大臣之象也，其星，軫也。月去中道，移而東北入箕，若東南入軫，則多風。西方為雨；雨，少陰之位也。月去中道，移而西入畢，則多雨。故《詩》云「月離于畢，俾滂沱矣」，言多雨也。《星傳》曰「月入畢則將相有以家犯罪者」，言陰盛也。《書》曰「星有好風，星有好雨，月之從星，則以風雨」，言失中道而東西也。故《星傳》曰「月南入牽牛南戒，民間疾疫；月北入太微，出坐北，若犯坐，則下人謀上。」⁶²

By citing the *Changes*, the *Documents*, and the *Odes*, the "Celestial Patterns" reinterprets the Classics themselves. Fragmentary though these citations may be, their use in this context implies that the Classics speak directly to such issues. The discourse of celestial signs, cobbled together from sundry technical texts, the "Celestial Offices," and only a few other named traditions, thus obtains classical authority.⁶³

The "Celestial Offices" and the "Celestial Patterns" contrast sharply in their treatments of the planets, and this contrast issues in part from the classicism of the *Hanshu* treatise. The "Celestial Patterns" introduces several planetary correspondences linked specifically to the Classics that do not appear in the "Celestial Offices." In addition to the familiar correspondences with direction, season, and resource, the "Celestial Patterns" ties each of the Five Planets to one of the Five Constants (*wu chang* 五常) and to one of the Five Modes.⁶⁴ The emperor's failure in these domains, we see in the "Wuxing zhi," is liable to produce all manner of baleful signs. The "Celestial Patterns" identifies specific actions that the ruler and his court should take by associating planetary aberrations with failure to adhere to the ritual, economic, and military prescriptions of the "Monthly Ordinances" (Yue ling 月令) of the *Record of Rites*.⁶⁵ The case of the Dazzling Deluder, Mars, presents a typical example:

> The Dazzling Deluder is said [to correspond to] "the south, summer, and fire."⁶⁶ It is the rites, and the (Mode of) Sight. When the rites diminish and there are failures in the (Mode of) Sight, or when the Summer Ordinances are violated, this damages the fire *qi*, and the penalty appears in the Dazzling Deluder.
>
> 熒惑曰南方夏火, 禮也, 視也。禮虧視失, 逆夏令, 傷火氣, 罰見熒惑。⁶⁷

As is the case with its treatment of the other visible planets, the "Celestial Patterns" here evokes the cosmology of the "Wuxing zhi" and the Classics from which it derives its framework. It places the aberrations in the motion or color of the Dazzling Deluder into a constellation of signs already established in the "Wuxing zhi." Retrograde Mars becomes readable against and amplified by the broader network of signs it is woven into through its association with the ruler's failure to properly engage in the ritual Mode of Sight. Moreover, while Mars had long been associated with summer, the "Celestial Patterns" specifies that the Dazzling Deluder produces meaningful signs when the "Summer Ordinances" are violated.⁶⁸ Declaring that the Dazzling Deluder strays from its path because of a fault in the Mode of Sight or a failure to implement the "Summer Ordinances," Ban and the Ma brothers relocate the meaning of the sign into the Classics.

VI. The "Celestial Patterns" as a Dynastic History

The "Celestial Offices," like the *Shiji* as a whole, covers a period from antiquity until the adulthood of its primary author, Sima Qian. The "Celestial Patterns" is part of a dynastic history; it covers only the Western Han dynasty. The "Wuxing zhi" and the "Celestial Patterns" tell the story of the rise and fall of that dynasty. The chronicle of celestial signs at the end of the "Celestial Patterns" contains some four dozen entries that tell of the triumph of the dynastic founder Liu Bang, troubling attempts to usurp power early in the dynasty, the glories and eventual excesses of Emperor Wu, and the rising power of the Wang clan following the death of Emperor Yuan 元 (r. 48–33 BCE).

The chronicle effectively begins with the Chunqiu period, reproducing in somewhat different language events that appear in the "Celestial Offices" through the rise of the Han. The "Celestial Patterns" chronicle then goes on to provide a detailed account of celestial signs through the end of the Western Han. A typical entry reads: "In the sixth year of Emperor Wu (135 BCE), the Dazzling Deluder guarded Cart Ghost.[69] The prognostication said: 'It is an aberration of fire. There will be a funeral.' This year, in the Exalted Garden there was a fire. Dowager Empress Dou died" 六年，熒惑守輿鬼。占曰：「為火變，有喪。」是歲高園有火災，竇太后崩。[70] Four distinct units may be identified in this brief passage: 1) the time of occurrence; 2) the features of the celestial sign; 3) a citation of a prognostication text; and 4) events that transpired in the terrestrial or human realm.

Toward the end of the Western Han dynasty, increasingly inauspicious signs occur that, read conveniently in retrospect, presage its fall. The final mention of Saturn in the chronicle, in 27 BCE, tells of the Dazzling Deluder following the Quelling Star (*zhen xing* 填星) westward across the sky during the last ten days (*xun* 旬) of the tenth month. In the first week of the eleventh month, Mars and Jupiter pass Saturn, all of them in retrograde motion. The prognostication predicts "warfare and mourning at court and abroad" 外內有兵與喪 and that "there will be newly established kings and dukes" 改立王公. Major political upheavals, the chronicle explains, take place over the next two years: the execution of King Xin 歆 of Yelang 夜郎 for treason, the burning of public temples, the theft of imperial seals, and, finally, the death of King He 賀 of Liang.[71]

The "Celestial Patterns" records the dramatic appearance of a "flowing star" that glowed red and white and stretched across the sky more than 100 degrees in length during the fourth month of 12 BCE. Smaller bright objects, perhaps meteors issuing from the comet's tail, fell behind the flowing star. The chronicle interprets this sign with respect to both a similar sign recorded in the *Annals* classic and the coming fall of the Western Han:

> In the commanderies and kingdoms it was always said that they were falling stars. In the *Annals* "stars falling like rain" is a portent of a king losing power and the local lords raising up a

hegemon. After this occurred, Wang Mang took the reins of the kingdom.[72] The Wang clan began to flourish during the reign of Emperor Cheng, and on account of this, the sign of the falling stars occurred. Later still, Wang Mang would usurp the empire.

郡國皆言星隕。《春秋》星隕如雨為王者失勢諸侯起伯之異也。其後王莽遂顓國柄。王氏之興萌於成帝[時]，是以有星隕之變。後莽遂篡國。[73]

The penultimate entry cites the appearance of a bright, long-lasting comet near Ox Herder. It seemed to confirm the arguments of Xia Heliang 夏賀良 (d. ca. 5 BCE) that the mandate of the Western Han needed to be renewed, citing the age-old axiom that broom stars sweep away the old and usher in the new.[74] The court instituted a number of reforms aimed at renewing the mandate, establishing a new reign period and imperial title and changing the number of divisions in the clepsydra from 100 to 120. It is the abolition of Xia's reforms, however, that most directly presages the rise of Wang Mang:

On the *dingsi* day of the eighth month, all of these reforms were abolished. Xia Heliang and his clique in all cases were executed or exiled from the capital. Following this, the calamity in which Wang Mang seized the empire finally occurred.

八月丁巳，悉復蠲除之，賀良及黨與皆伏誅流放。其後卒有王莽篡國之禍。[75]

Xia and his allies must have appeared remarkably prescient in the Eastern Han. The Western Han did fall, and its mandate did indeed need to be renewed. This did not occur, however, until the rise of the Emperor Guangwu 光武 (r. 25–57 CE), a distant relative in the Liu clan, but a Liu nonetheless.

Both the treatise and the dynasty effectively end with the final entry in the chronicle:

In the eleventh month of the first year of the Yuanshou reign period (2–1 BCE), the Year Star (Jupiter) entered Grand Tenuity, and moved retrograde toward the Right Enforcer of the Law.[76]

The prognostication said: "When the great ministers are in a superior position, the Enforcers of the Law are punished, as if they themselves were culpable." On the *wuyin* day of the tenth month of the second year, the Marquis of Gao'an, Dong Xian 董賢 (d. 1 BCE), resigned his post as Marshal of State, returned to his domicile, and committed suicide.

元壽元年十一月，歲星入太微，逆行干右執法。占曰：「大臣有憂，執法者誅，若有罪。」二年十月戊寅，高安侯董賢免大司馬位，歸第自殺。[77]

Wang Mang is not mentioned explicitly in the chronicle's final entry, but Dong Xian's death nonetheless marks the end of Liu clan power in the Western Han. Dong's fall came with Emperor Ai's death, when he was accused of refusing to provide the young emperor with medical aid.[78] The next Marshal of State, de facto ruler for the remainder of the Western Han, and founder of the next dynasty, the short-lived Xin 新 (9–23 CE), was Wang Mang himself.

VII. Conclusion

The *Hanshu* "Celestial Patterns" marked out the history of a dynasty through the appearance of celestial signs, and it was this treatise, rather than its predecessor in the *Shiji*, that established the template for later "Tianwen" treatises.[79] It shared the division of labor for recording astronomical and meteorological phenomena with the various "Wuxing" treatises in most subsequent dynastic histories. The two types of treatises recorded the ongoing observations of the heavens by court astronomers and diviners throughout the imperial period.[80] These treatises not only set a precedent for assiduous observations of the heavens in imperial China, but also created a repository of records of global value that scientists continue to employ for research on environmental history, historical novae, and historical solar activity.[81]

The chronicle at the conclusion of the "Celestial Patterns" is emblematic of the shift in the social significance of celestial signs between the *Shiji* and the *Hanshu*. The "Celestial Patterns" at a glance appears to be a near facsimile of Sima Qian's "Celestial Offices"; a great deal of technical material is shared between the two treatises. However, for Sima, that material is the patrimony of men like him and his father, readers of celestial signs who had served sages, or been ignored by mediocrities, since high antiquity. Narratives recounting appearances of celestial signs gained their authority through their association with authoritative figures more than their inclusion in authoritative texts. The "Wuxing zhi" drew not only authority, but its very structure from the *Annals* and the "Great Plan." All manner of signs were organized into categories drawn from the "Great Plan"—the Five Resources, the Five Modes, and the Sovereign's Standard. Each type of sign presented examples drawn from the *Annals*, in chronological sequence, followed by examples in the Qin and Western Han. While Liu Xiang's "Discourse on the Five Resources Tradition of the 'Great Plan'" could not have included the rise of Wang Mang, the *Hanshu* "Wuxing zhi" did, making each chronicle of each type of sign point toward the ultimate fall of the Western Han. Ban Zhao and the Ma brothers effectively synthesized the technical materials of the "Celestial Offices" with the teleology and classicism of the "Wuxing zhi." In the "Celestial Patterns," heavenly signs directly reflected the failure of the ruler to fulfill the Five Modes; technical knowledge became dependent on classical authority, and records of the appearance of celestial signs became a chronicle of the downfall of a dynasty.

Appendix A: Organization of the *Shiji* "Celestial Offices" and *Hanshu* "Celestial Patterns"

Shiji "Celestial Offices"	*Hanshu* "Celestial Patterns"	Comments
	Introduction	No correspondence with *Shiji*
The Stars		
Circumpolar region Eastern Palace Southern Palace Western Palace Northern Palace		Close correspondence throughout

continued on next page

Shiji "Celestial Offices"	Hanshu "Celestial Patterns"	Comments
The Five Planets		
Jupiter	Jupiter	The Hanshu adds new correspondences, including the Five Constants and the Five Modes of the "Great Plan," and reorders the planets so that Saturn, associated with the Center and with Virtue (de), appears in the final position. Correspondence between treatises is limited.
Jupiter cycle		
Mars	Mars	
Saturn	Venus	
Planetary interactions	Mercury	
Venus	Saturn	
Mercury	Planetary interactions	
	Regional field allocations	Closely corresponds to section in Shiji
	Sixty-day cycle, regional correspondences	Partial correspondence with Shiji, ten-day cycle, regional correspondences
	Planets and regions	
	Jupiter cycle	Cites Shi shi 石氏 and Gan shi 甘氏
	Retrograde motion and eclipses	Cites Classics to argue against claims that eclipses and retrograde motion can be considered regular behavior
Regional field allocations		Shiji section contains one additional line
Sun and moon		Little correspondence with cognate section in Hanshu, significantly shorter
Ten-day cycle, regional correspondences		Most of this section is also found in Hanshu, sixty-day cycle, regional correspondences

Celestial Signs in Three Historical Treatises | 203

Shiji "Celestial Offices"	*Hanshu* "Celestial Patterns"	Comments
Emergent "Stars" (primarily designations applied to comets and novae)		Close correspondence throughout
	Sun and moon	More developed than cognate *Shiji* section, with several citations of Classics
Clouds and *qi*		Close correspondence throughout
Tetrasyllabic poem		Poem describes various baleful signs, including earthquakes, collapsing mountains, crop failures, and floods. Close correspondence throughout
Divining the harvest		Methods include examining weather, wind direction, tone emitted from populace, and the location of *Taisui* (a hypothetical planet with a position opposite Jupiter) at the beginning of the year. Close correspondence throughout
Sima Qian's Postface		Parts of Sima Qian's Postface are incorporated into the *Hanshu* treatise, especially its concluding chronicle
	Chronicle through Aidi (7–1 BCE)	

Appendix B: Comparison of Planetary Correspondences[82]

Planetary Correspondences in the "Celestial Offices"

Planet	Direction	Resource	Season	Days	Punishment for loss of…
Jupiter	East	Wood	Spring	*jia* and *yi*	Duty 義
Mars	South	Fire	Summer	*bing* and *ding*	Ritual 禮
Saturn	Center	Earth	Late Summer	*wu* and *ji*	Duty, Ritual, Reduction, and Punishment
Venus	West	Metal	Autumn	*geng* and *xin*	Reduction 殺
Mercury	North	Water	Winter	*ren* and *gui*	Punishment 刑

Planetary Correspondences in the "Celestial Patterns"

Planet	Direction	Season	Resource	Five Constants 五常	Five Modes 五事	Seasonal Ordinance	Qi
Jupiter	East	Spring	Wood	Humaneness 仁	Demeanor 貌	Spring	Wood
Mars	South	Summer	Fire	Ritual 禮	Sight 視	Summer	Fire
Venus	West	Autumn	Metal	Duty 義	Speech 言	Autumn	Metal
Mercury	North	Winter	Water	Wisdom 知	Hearing 聽	Winter	Water
Saturn	Center	Late Summer	Earth	Promise-keeping 信	Thinking 思心		Earth[83]

Notes

1. This chapter benefited greatly from the advice given by Nathan Sivin, Luke Habberstad, Robert Ashmore, and Daniel P. Morgan, all of whom read the work at various stages. Any remaining errors or infelicities are mine alone.

2. The *Shiji* was a private work, initiated by Sima Qian's father, that included much more than omens: 1) "Benji" 本記 (Basic Annals) that chronologically narrate major events from high antiquity through the reign of the sitting emperor; 2) "Biao" 表 (Tables); 3) "Shu" 書 (Treatises) on subjects ranging from rituals to canals; 4) "Shijia" 世家 (Hereditary Houses) chapters that told the stories of powerful families; 5) "Lie zhuan" 列傳 (Arrayed Biographies) that related the lives of prominent people and in many cases preserved their writings. Though Sima Qian's duties as *Taishi ling* certainly did not entail the composition of such a history, it could not have been composed without access to imperial archives. For historical background, see Gu Jiegang, *Handai xueshu shilue*.

3. The medieval historian Liu Zhiji 劉知幾 (661–721) claimed that the "'Wuxing zhi' was derived from Liu Xiang's work on the 'Great Plan' chapter of the *Documents*" 五行出劉向洪範. See his *Shitong* 3.8. Liu Xiang's *Hanshu* biography suggests that his work had much in common with the "Wuxing zhi," describing it as "a record of auspicious signs, disasters, and anomalies from high antiquity through the Chunqiu, Warring States, Qin, and [Western] Han" 上古以來歷春秋六國至秦漢符瑞災異之記 that Liu organized into categories derived from the "Great Plan." The title is not listed in the bibliographic treatise to the *Hanshu*. However, a work titled "Wuxing zhuan ji" 五行傳記 (Records Concerning the Traditions on the Five Resources) in eleven *juan* 卷 (fascicles) rather than eleven *pian* 篇 (chapters) is credited to Liu Xiang therein. See *Hanshu* 36.1950 and 30.1705.

4. Kongzi, better known in the West as Confucius, realized that his sagely method of governance would not be instituted in his own lifetime and declared "My way is at an end" 吾道窮矣 when the mythic unicorn (*lin* 麟) was captured by a firewood collector in 481 BCE. The legendary event, recounted in the final passages of the *Annals*, marks the conclusion of the Chunqiu period. See He, *Chunqiu Gongyang zhuan zhushu*, 28.624.

5. For the compilation of the "Tianwen zhi," see *Hou Hanshu* 84.2784–85.

6. Most probably because it appears in the same historiographical project as the "Wuxing zhi," much material that otherwise fits into the "Celestial Patterns" is left out; the two *Hanshu* treatises are complementary.

7. Two interpretations of the term *Wujia* 五家 obtain: First, according to Sima Zhen's 司馬貞 (early eighth century CE) *Shiji suoyin* 史記索隱 (*Commentary on the* Writings by the Senior Archivist), these are experts in the Five Cycles (*wuji* 五紀) outlined in the "Great Plan" chapter of the *Documents*: the cycles of Jupiter (*sui* 歲), the moon (*yue* 月), the sun (*ri* 日), the planets and the constellations (*xing chen* 星辰), and calendrical calculations (*lishu* 曆數). Zhang Shoujie's 張守節 *Shiji zhengyi* 史記正義 (*The Correct Meanings of the* Writings by the Senior Archivist; preface dated 737 CE) identifies five sage-kings as the *wujia*: The Yellow God 黃帝, Gaoyang 高陽, Gaoxin 高辛, Yao of Tang and Shun of Yu 唐虞堯舜. See *Shiji*, 27.1343n1–2. Rendering the phrase as Five Houses, I follow Zhang Shoujie's interpretation.

8. These two lines are a quotation from the "Tradition of the Appended Statements" (Xici zhuan 繫辭傳) to the *Changes*. *Zhouyi zhengyi* 8.74b in Ruan, *Shisan jing zhushu*.

9. The Three Luminaries are the sun, moon, and stars.

10. *Shiji* 27.1342. For a complete annotated translation of this treatise, see Pankenier, *Astrology and Cosmology*, 444–511.

11. Loewe, *Biographical Dictionary*, 486.

12. Michael Nylan has argued that Sima Qian sought to commemorate his own family line by composing the *Shiji*. See her "Sima Qian," 245.

13. *Shiji* 27.1349.

14. This event occurred in the sixteenth year of Duke Xi of Lu (644 BCE) (*Shiji* 27.1345n4).

15. Zhao Qi 趙岐 (d. 201) identifies the *Wubo* as the Five Hegemons: Duke Huan of Qi 齊桓 (r. 685–643 BCE), Duke Wen of Jin 晉文 (r. 636–628 BCE), Duke Mu of Qin 秦穆 (r. 659–621 BCE), Duke Xiang of Song 宋襄 (r. 650–637 BCE), and King Zhuang of Chu 楚莊 (r. 613–591 BCE) (See *Shiji* 27.1347n6).

16. This occurred in the twenty-third year of King An 安 of Zhou (379 BCE) (*Shiji* 27.1345n8).

17. In the twenty-sixth year of King An of Zhou (376 BCE), Jin was divided into three domains: Wei 魏, Han 韓, and Zhao 趙 (*Shiji* 27.1345n9).

18. *Shiji* 27.1345.

19. *Shiji* 27.1348.

20. Ibid.

21. Ibid.

22. *Shiji* 27.1351.

23. While the "Wuxing zhi" may be built to a considerable extent on the work of Liu Xiang, the critical teleological aspect of the treatise could not have issued from Liu Xiang, given his death nearly two decades prior to the rise of Wang Mang. Entries after 8 BCE may have been added by Liu Xin.

24. "Xici zhuan" in *Zhouyi zhengyi* 7.70b.

25. *Hanshu* 27A.1315.

26. *Hanshu* 27A.1316. Literally, the [Yellow] River Chart (Hetu 河圖) and Luo [River] Writing (Luoshu 雒書).

27. Ibid.

28. *Hanshu* 27A.1317.

29. A more detailed discussion of the terminology and organizational features in the "Wuxing zhi" may be found in Michael Nylan's chapter in this volume.

30. This tradition is sometimes identified as Fu Sheng's 伏生 (d. after 156 BC) *Shangshu dazhuan* 尚書大傳 (*Great Tradition of the* Documents), though this identification is by no means certain. See Michael Nylan's chapter in this volume.

31. *Hanshu* 27D.1458.

32. Ibid. The treatise cites the phrase "for the arrogant dragon there is cause for regret" 亢龍有悔 from the Line statement to the upper nine in the Qian hexagram. The *Hanshu* passage goes on to cite the Master's comment on that phrase in the "Xici zhuan" (see *Zhouyi zhengyi* 1.2a and 7.67c–68a).

33. This phrase is found in the *Wenyan* 文言 ("Commentary on the Words") comment to the nine in the fifth place in the Qian hexagram (*Zhouyi zhengyi* 1.4b).

34. This event occurred in the first year of Duke Cheng (590 BCE).

35. *Hanshu* 27D.1458–59. For Maorong, see *Hanshu* 36.1941n25.

36. The *Tradition of Stars and Planets* is cited a total of seven times in the *Hanshu*. Six citations occur in the "Celestial Patterns" and one in the "Wuxing zhi." No text under the title *Xingzhuan* appears in the *Hanshu* bibliographic treatise.

37. Here, *chu* 除 might also be rendered as "swept away." The image of sweeping is implied by the word for the comet—i.e., "broom star." See Yang Bojun's note in his *Chunqiu Zuozhuan*, 1390.

38. *Hanshu* 27E.1515.

39. *Hanshu* 27E.1515–16.

40. Translation follows Yan Shigu's suggestion that *bo* 伯 be read *ba* 霸 (*Hanshu* 27E.1516n1).

41. *San Qin*, or the three districts of Qin, is the collective name given to Yong 雍, Di 翟, and Sai 塞, areas roughly corresponding to modern western, northern, and eastern Shaanxi, respectively, following the destruction of Qin in 206 BCE. See *Shiji* 6.275.

42. For Xiang Yu's massacre of surrendered Qin troops, said to number more than 200,000, see *Shiji* 7.310; 8.376; 92.2612; *Hanshu* 1A.44.

43. *Hanshu* 27E.1516. Liu Bang cited Yidi's murder in a speech calling for a campaign against Xiang Yu (*Hanshu* 1A.34).

44. See *Hanshu* 5.142–43.

45. Empress Chen was demoted in status in 130 BCE because she had not borne a son, and owing to accusations that she had committed imprecations against Wei Zifu 衛子夫, who was to become Emperor Wu's second empress. See Loewe, *Biographical Dictionary*, 31–32.

46. *Hanshu* 27E.1517.

47. For a discussion of the Banner of Chiyou in both excavated and received texts, see Loewe, "The Han View of Comets."

48. This refers to campaigns initiated by Wei Qing 衛青 (d. 106 BCE) and Huo Qubing 霍去病 (ca. 140–ca. 117 BCE) against the Xiongnu. See Tomiya and Yoshikawa, *Kanjo Gogyôshi*, 346n2.

49. The Western Han capital was located near the area that had once been the political center of the Qin domain.

50. See Loewe, *Crisis and Conflict*, 113–53; Liu Pakyuen (Liao Boyuan), *Les institutions politiques*, 36–170.

51. For the nomination of Emperor Cheng's heir, see Vankeerberghen, "Pining for the West," 359–60.

52. For the identification of this entry as an appearance of Halley's Comet, see Wen Shion Tsu, "The Observations of Halley's Comet."

53. We must, however, exercise caution in reading the comment as having been intended by Liu Xiang to address this particular comet. Whereas the treatise introduces Gu Yong's statement with *dui yue* 對曰 (said in response), a phrase that suggests Gu Yong was responding to the appearance of this particular comet, Liu Xiang's statement is introduced with *yi yue* 亦曰 (also said). This leaves open the possibility that Ban Gu, if he is indeed the treatise's final compiler, is quoting a statement in Liu Xiang's work that was never specifically directed at this comet, at least not explicitly. To be certain, it would have been a dangerous, potentially treasonous act to compare Emperor Cheng with the rulers of the short-lived Qin dynasty, or with Xiang Yu. Ban Gu, however, living in the Eastern Han, could use Liu Xiang's writings about celestial signs that appeared under ill-fated earlier rulers as a powerful explanation of historical developments leading to Wang Mang's rise.

54. *Hanshu* 27E.1518.

55. *Hanshu* 26.1273.

56. *Shiji* 27.1348.

57. *Hanshu* 26.1301. Cf. Ban Biao's 班彪 (3–54 CE) "Wang ming lun" 王命論 (Discussion of the Charge of Kings) (*Hanshu* 100A.4211–12).

58. For omens prompting criticism, see Bielenstein, "An Interpretation of the Portents." For auspicious signs celebrated in hymns of praise, see Kern, "Religious Anxiety."

59. This phrase is from the hexagram statement to *Kun* 坤 (*Zhouyi zhengyi* 1.5c).

60. Mao 232. Translation modified from Waley, *The Book of Songs*, 221.

61. This is an elliptical citation of the "Great Plan."

62. *Hanshu* 26.1295–96. Basket, Axletree, and Net, respectively, correspond to stars in Sagittarius, Corvus, and Taurus. Ox Herder corresponds to Altair (α Aquilae). The sprawling Grand Tenuity corresponds to parts of Virgo, Coma Berenices, and Leo.

63. Another new technical passage responds to advances in mathematical astronomy that had assimilated planetary retrograde motion and eclipses into newer almanacs. Appealing to the Classics and related traditions, it argues that such phenomena are meaningful signs ("minor aberrations" or *xiao bian* 小變) rather than mere "regular motion" (*zheng xing* 正行) (*Hanshu* 26.1290–91). For a global discussion of the development of mathematical astronomy in Han times, see Cullen, *Heavenly Numbers*.

64. For the taxonomic correspondences drawn between the Five Planets and other features of the cosmos in the "Celestial Offices" and "Celestial Patterns," see Appendix B to this chapter.

65. The *Liji* does not appear to have taken its present form before the White Tiger Hall debates of 79 CE. See Riegel, "Li chi," 294.

66. This appears to be a quotation, but it does not correspond exactly to any received text (*Hanshu* 26.1281). The association between Mars and the ritual Mode of Sight is not found in any earlier astronomic treatise or in the "Seasonal Ordinances" chapters pertaining to summer in the *Record of Rites*.

67. *Hanshu* 26.1281. Aberrations in the movements or color of the planet manifest the penalty for lapses in the rites associated with the Duty of Sight or Summer Ordinances.

68. The section on the last month of summer in the "Summer Ordinances" chapter of the *Record of Rites* contains many sets of correspondences that had been linked to Mars since at least the early Western Han. The last month of summer is tied to Zhurong, to fire, and to the days *bing* and *ding*. However, the "Monthly Ordinances" chapters also contain a series of policy and ritual recommendations. The Son of Heaven is to reside in the right chamber of the Mingtang 明堂 (Hall of Illumination), ride in a red carriage, order fishermen to seek out sharks and turtles, and command his people to labor to provide sacrifices for various deities. Should the emperor employ the wrong ordinances for the seasons, dire consequences occur. Employing the fall ordinances during the summer, for instance, produces unseasonable rains, failed crops, and disasters falling upon women in his court, or perhaps in the empire as a whole. See *Liji zhengyi*, 16.142c–43c.

69. Cart Ghost corresponds to stars located in Cancer.

70. *Hanshu* 26.1305.

71. *Hanshu* 26.1310.

72. In 8 BCE, four years after this occurrence, Wang Mang became Marshal of State; Loewe, *Biographical Dictionary*, 537.

73. *Hanshu* 26.1311.

74. *Hanshu* 26.1312.

75. Ibid.

76. Translation of *zhifa* 執法 after Pankenier, *Astrology and Cosmology*, 464. Pankenier suggests that there is some uncertainty concerning the identification of the Left and Right Enforcers of the Law, but he regards η and β Virginis as most plausible.

77. *Hanshu* 26.1312.

78. Loewe, *Biographical Dictionary*, 68.

79. Further research is required to examine how that template changed over time and whether or not later dynastic histories continued to tie the appearance of astronomical and meteorological phenomena to the teleological decline of dynastic fortunes. A recent volume on technical treatises in the *Sui shu* might be taken as a model for additional work on the remaining Standard Histories. See Morgan and Chaussende, *Monographs in Tang Official Historiography*.

80. Several of the Standard Histories do not include technical treatises. These include the general histories of the Northern and Southern Dynasties (the *Bei shi* 北史 and *Nan shi* 南史), as well as histories of the Three Kingdoms period (221–280), and the Liang (502–556), Chen (557–589), Northern Qi (550–577), and Northern Zhou (557–581) dynasties. Of the remaining dynastic histories, only the history of the Khitan-ruled Liao dynasty (907–1125) includes neither an astronomical nor an omenological treatise. The *Xin Wudai shi* 新五代史 (*New History of the Five Dynasties* [907–960]) includes an astronomical treatise but not an omenological one. See Wilkinson, *Chinese History*, 512.

81. For a concise discussion of the value of historical astronomical records to modern science, see Neuhäuser, Neuhäuser, and Posch, "Terra-Astronomy," 145–47.

82. Tables are derived from *Shiji* 27.1312, 1317, 1319–20, 1322, 1327 and *Hanshu* 26.1280–85.

83. The phrase *tu qi* never occurs in the "Celestial Patterns" section on Saturn; however, *tu* occurs four times and is clearly associated with the planet.

Asian-language Bibliography

Gu Jiegang 顧頡剛. *Handai xueshu shilue* 漢代學術史略. Taipei: Qiye, 1975 reprint of Shanghai, 1949 edition of the same work under a different title.

Hanshu 漢書. Compiled by Ban Gu 班固 (32–92). Annotated by Yan Shigu 顏師古 (581–645). 12 vols. Beijing: Zhonghua shuju, 1962. Reprint, 1996.

He Xiu 何休 (129–182 CE) comm. Xu Yan 徐彥 (Tang) subcomm. *Chunqiu Gongyang zhuan zhushu* 春秋公羊傳注疏 [*Commentary and Subcommentary to the Gongyang Tradition of the Annals*]. Beijing: Beijing daxue, 1999.

Liji zhengyi 禮記正義 [*Correct Meanings of the* Record of Rites]. See Ruan Yuan, *Shisan jing zhushu*.

Liu Zhiji 劉知幾 (661–721). *Shitong* 史通 [*A Complete Understanding of History*]. In *Siku quanshu* 四庫全書 [*Complete Books of the Four Chambers*]. Wenyuange edition. Intranet version, 2007.

Ruan Yuan 阮元 (1764–1849). *Shisan jing zhushu fujiao kanji ben* 十三經注疏附校勘記本 [*Commentary and Subcommentary to the* Thirteen Classics, edited] (postface dated 1815). Beijing: Zhonghua, 1980.

Shiji 史記 [*Writings by the Senior Archivist,* aka *Historical Records*]. Compiled by Sima Qian 司馬遷 (?145–?86 BCE). 10 vols. Beijing: Zhonghua shuju, 1959. Reprint, 1982.

Shitong. See Liu Zhiji.

Tomiya Itaru 冨谷至 and Yoshikawa Tadao 吉川忠夫, eds. *Kanjo Gogyôshi* 漢書五行志 [The History of the Han "Treatise on the Five Resources"]. Tokyo: Heibonsha, 1986.

Yang Bojun 楊伯峻. *Chunqiu Zuozhuan jizhu* 春秋左傳集注 [*Collected Commentaries to* the Zuo Tradition of the Annals], vol. 4. Beijing: Zhonghua, 1981.

Zhouyi zhengyi 周易正義 [*Correct Meanings of* the Changes of Zhou]. See Ruan Yuan, *Shisan jing zhushu.*

Western-language Bibliography

Bielenstein, Hans. "An Interpretation of the Portents in the *Ts'ien-Han-Shu.*" *Bulletin of the Museum of Far Eastern Antiquities* 22 (1950): 127–43.

Cullen, Christopher. *Heavenly Numbers: Astronomy and Authority in Early Imperial China*. Oxford: Oxford University Press, 2017.

Kern, Martin. "Religious Anxiety and Political Interest in Western Han Omen Interpretation: The Case of the Han Wudi Period (141–87 BC)." *Studies in Chinese History* 10 (2000): 1–31.

Liu Pakyuen (Liao Boyuan) 廖伯源. *Les institutions politiques et la lutte pour le pouvoir au milieu de la dynastie des Han antérieurs* [Political Institutions and the Struggle for Power in the Western Han Dynasty]. Paris: Collège de France, 1983.

Loewe, Michael. *A Biographical Dictionary of the Qin, Former Han, and Xin Periods: 221 BC–24 AD*. Leiden: Brill, 2000.

———. *Crisis and Conflict in Han China, 104 BC to AD 9*. London: George Allen & Unwin Ltd., 1974.

———. "The Han View of Comets." In *Divination, Mythology, and Monarchy in Han China*, 61–84. New York: Cambridge University Press, 1994.

Morgan, Daniel P., and Damien Chaussende, eds. With the collaboration of Karine Chemla. *Monographs in Tang Official Historiography: Perspectives from the Technical Treatises of the* History of the Sui (Sui shu). Cham, Switzerland: Springer, 2019.

Neuhäuser, Ralph, Dagmar L. Neuhäuser, and Thomas Posch. "Terra-Astronomy— Understanding Historical Observations to Study Transient Phenomena." *Proceedings of the International Astronomical Union* 14, no. A30 (2018): 145–47. doi:10.1017/S1743921319003934.

Nylan, Michael. "Sima Qian: A True Historian?" *Early China* 23/24 (1998–99): 203–46.

Pankenier, David W. *Astrology and Cosmology in Early China: Conforming Earth to Heaven*. Cambridge: Cambridge University Press, 2013.

Riegel, Jeffrey. "Li chi." In *Early Chinese Texts: A Bibliographical Guide*, edited by Michael Loewe, 293–97. Berkeley: University of California, 1993.

Vankeerberghen, Griet. "Pining for the West: Chang'an in the Life of Kings and Their Families during Chengdi's Reign." In *Chang'an 26 BCE: An Augustan Age in China*, edited by Michael Nylan and Griet Vankeerberghen, 347–66. Seattle: University of Washington Press, 2015.

Waley, Arthur, trans. *The Book of Songs: The Ancient Chinese Classic of Poetry*, edited by Joseph R. Allen. New York: Grove Press, 1996.

Wen Shion Tsu. "The Observations of Halley's Comet in Chinese History." *Popular Astronomy* 42 (1934): 191–201.

Wilkinson, Endymion. *Chinese History: A Manual*. Revised and enlarged edition. Cambridge: Harvard University Asia Center, 2000.

5

On *Hanshu* "Wuxing zhi" 五行志 and Ban Gu's project

MICHAEL NYLAN 戴梅可

> 災異天事，非人力所為
>
> Disasters and anomalies are Heaven's affairs,
> not products of human efforts.[1]

The *Hanshu* "Treatise on the Wuxing" has elicited due attention from many fine scholars: Hans Bielenstein, Wolfram Eberhard, Martin Kern, and Yang Shao-yun, among others.[2] Though each piece evinces greater sophistication than the last, all previous scholarship has operated on the plausible assumption that the treatise comments on ruler-subject relations in Ban Gu's own Eastern Han times, with a jaundiced eye. In this chapter, I adopt a different approach, premised on the belief that finely crafted rhetoric never maps neatly onto the current sociopolitical realities, as Ban Gu, who styled himself first and foremost as a poet and eulogist, knew all too well when compiling his history.[3] Thus, it is incumbent on historians to try to situate the chapter within the context of the larger *Hanshu* project, and, more specifically, the *Hanshu*'s claims about the role of court omen experts espousing rival theories about good governance.[4] To that end, this chapter first discusses related chapters in the *Hanshu*, in particular *Hanshu* 75, a collective biography of omen experts, as well as other *Hanshu* chapters in passing, including the biographies of Dong Zhongshu and Liu Xiang. As I will argue, the timing for the court's decision to offer regular court appointments to omen experts (not just consult them on an ad hoc basis, as crises loomed) may be one key to a greater understanding of the treatise under investigation, but other motives are equally important: the necessity for officials to work in concert and for the ruler to consult widely in policy matters. In any event, only after establishing the likely rhetorical context for the *Hanshu* omen treatise does this chapter turn to an analysis of the omen chapter's contents, building on recent insights culled from the work of two Chinese scholars, Su Dechang 蘇德昌 and Chen Kanli 陳侃理, and several Japanese experts.

My chapter consists of six parts. Part I travels outside the omen treatise to ascertain the *Hanshu*'s general approach to omen experts consulted at court. Part II reviews the main structure and arguments of the *Hanshu* "Wuxing zhi" to aid the many readers who may be unfamiliar with it, highlighting Ban's interest in the *Changes* classic. Part III discusses probable authorship and dating of the treatise.[5] All good scholars cannot but wonder whether Ban Gu imported another's views wholesale (as he seems to have done for much of his geographic treatise), revised the earlier sources he used beyond recognition, or wrote this treatise more or less from scratch. After all, the treatise has no ready counterpart in either the Classics or the *Shiji*. Part IV focuses on the theories attributed in the *Hanshu* essay to Dong Zhongshu and Liu Xiang to consider what the treatise prefers to downplay or occlude. Here, as with some other technical arts discussed in this volume (see Chemla's chapter, for example), the pervasive influence of the *Annals* classic, with its "subtle teachings" conveyed often by concealment, spurs further reflection. Part V mulls over the continual repetition of the phrase "Heaven's warning seemed to say" (*Tian jie ruo yue* 天戒若曰), an adaption of a formula from the *Documents* classic. Interpretations associated with no fewer than three of the Five Classics are, then, foregrounded in Ban Gu's omen treatise (the *Changes*, the *Annals*, and the *Documents*), with the content of the two remaining Classics, the *Odes* and the *Rites*, figuring occasionally as subjects under investigation.[6] Part VI details a few final conundrums before offering its preliminary conclusions.

I. *Hanshu* on Omen Experts

The collective biography of late Western Han omen experts (*Hanshu* 75) is the obvious starting place if we hope to situate the lengthy omen treatise of five *juan* 卷 (normally, "chapters") within Ban Gu's larger rhetorical project. For, by Ban Gu's own account, *Hanshu* 75 treats nearly all the most renowned omen experts of the second century of Western Han, whose omen pronouncements then play their parts in Ban's omen treatise.[7] Only a few omen experts (e.g., Dong Zhongshu, Liu Xiang, and Yang Xiong) have separate *Hanshu* biographies elsewhere, either because their stories belonged to the first century of Western Han rule (Wudi's reign and before) or because their enormous talents were too varied to share a collective biography with others.[8] That said, it is striking how pervasive an activity omen reading by high-ranking officials had become by the second century of Western Han, with a majority of biographies in that period attesting the practice. (In this connection, it is probably not coincidental that this second century is precisely the time when the auspicious reign names [*nianhao* 年號] were adopted to signify the blessings that the reigning emperor had received, with these *nianhao* themselves soon figuring as omenological interventions.)

Curiously, then, *Hanshu* 75 unambiguously states that its subjects have a great deal to answer for, as becomes evident when *Hanshu* 75 is contrasted with *Hanshu* 72, most of whose subjects shared a strong interest, professional and personal, in omen interpretation.[9] Let us begin with a review, then, of the concluding Appraisals for the two *Hanshu* chapters, both of them exceptional, in that Ban Gu is seldom this forthright in expressing his views, as he generally deems indirection the wiser course:

> Regarding the Mystery and clarity, the invisible and visible, the gods of heaven and earth:[10] Nothing is clearer than the *Changes* and *Annals* classics when it comes to communicating and harmonizing the Way of heaven-and-human. Yet Zigong still remarked, "Now the master's writings and ritual performances we can be apprised of, but we have been unable to hear anything of our master's theories about human nature in relation to heaven's Way."[11] When the Han arose, there were those who inferred yin/yang *qi* theories when speaking of disasters and anomalies. During the reign of the filial emperor Wudi, there was Dong Zhongshu and Xiahou Shichang; during the reigns of Zhaodi and Xuandi, Sui Meng and Xiahou Sheng; during the reigns of Yuandi and Chengdi, there was Jing Fang, Yi Feng, Liu Xiang, and Gu Yong; and during Pingdi's reign, there was Li Xun and Tian Zhongshu.[12] These were the most celebrated figures of the time who gave advice to the rulers of their day.[13]
>
> If we investigate their theories, they *seem* to have had a single starting point 一端, the Dao itself.[14] They made use of the

Classics to set up what they deemed right, trusting to images and reasoning by categories. Did they not, however, sometimes simply "hit the target by dumb luck after a million tries"?[15] Dong Zhongshu was sent down for trial. Xiahou was imprisoned and manhandled. Sui Meng was slaughtered. Li Xun was exiled. May these examples constitute a great warning to students [of the predictive arts]. Jing Fang acted on the smallest pretext, without weighing major versus minor.[16] Dangerous words, sharp ridicule ensnared the powerful officials in resentment, until they were charged with crime upon crime. That their indiscretions cost them their lives, how sad it is!

As if to underscore the thrust of the foregoing passage, the verse summary for *Hanshu* 75 offered by Ban Gu reiterates the inherent dangers of using the predictive arts at court to further politicize policy discussions:

By the ancient methods of divination one can predict the future. The mysteries and the clarity [of the *Changes* lines], the gods of heaven-and-earth [are in the *Changes*].[17] But should it not be the right person [who performs a divination], the Dao will not be transmitted, as easily as if wafted on air. Learning will then be tenuous; and techniques, benighted. Sometimes they may have a *semblance* of truth, such as may seem to vanquish doubts and dangers.[18] But when the readings offend the crowd and affront the age, even superficial mistakes become errors and regrets; and deeper ones, unalloyed harms. And thus I narrate the chapter. . . .[19]

Were a reader to get this far and no further, she might well conclude that Ban Gu was a "secular humanist," in the Chen Lai 陈来 mold,[20] opposed to the superstitions of omen reading. Nothing could be further from the truth, however, which becomes plain as soon as the reader immerses herself in Ban's dynastic history. Agreeing that numerous prodigies are beyond human ken ("being not the product of human effort" 非人力所為),[21] Ban never condemns an interest in foretelling the future or diagnosing the present per se. Twice Ban heaps praise on a professional diviner, Yan Junping 嚴君平 (aka Zhuang Zun 莊遵), once in the introductory remarks for this *Hanshu* chapter 72 and once in his geographic treatise.[22] And let us not forget that, for better or worse, Dong Zhongshu 董仲舒 is one of Ban Gu's heroes.[23] Unlike the experts in *Hanshu* 75, Dong supposedly *did* root his omen theories in the unitary Dao 一端, a clear sign from Ban that readers should regard Master Dong as a sage worthy of emulation.²⁴ That Ban Gu was proud of his own relative, Ban Bo 伯, who had written a lengthy critique of Xu Shang's 許商 *Wuxing zhuanji* 五行傳記 (no longer extant), is probably also a factor here.

In his concluding Appraisal for *Hanshu* 72, a collective biography of late Western Han remonstrating officials, Ban Gu provides a positive characterization of his chapter's subjects, many of whom offered omen readings to the court—a positive characterization that provides a sharp contrast to Ban Gu's negative portrayal of their peers depicted in *Hanshu* 75:

> The *Changes*, when discussing the Way of noble men, speaks of some men in service, and some out; and of some maintaining silence and some speaking up. It means that each type of livelihood has its own distinctive form of principled conduct. One may compare it to vegetation, where there is a huge variety of species. Therefore they say, Once the men of the mountain forests leave, they cannot find it within themselves to return; and once men enter court service, they find they cannot bring themselves to leave. The two ways of life (reclusion and court service) each have their own shortcomings. From the Chunqiu era, with its ministers and chancellors, down to the Han, with its famous officials, many of them *harbored ambitions for salary and favors, and so duly lost their lives.* For this reason, the men in service who had purer principles deserve to be honored.
>
> Still, a great proportion of them can govern themselves but they cannot govern others. To have the talent and capacities of Wang Ji and Gong Yu is better than being a Gong Sheng or a Bao Xuan, for Wang and Gong, at least, maintained themselves until death while doing good. Gong Sheng was treading that path [when he refused to serve Wang Mang]. Xue Fang nearly approached the ideal of being "firm in his principles but not credulous" 貞而不諒. . . .²⁵

This concluding Appraisal for *Hanshu* 72 is clear enough, in its contrasts between those in court service and out. Yet judging from the lengthy introductory remarks for *Hanshu* 72, even to divine the future for a fee in a local marketplace could be respectable not only in theory, but also in practice, under these conditions: the man of honor must be above mere profit-making; he must promote the ethical "constants" in human relations that conduce to sociopolitical order, and he must never interfere with local or imperial administration. Simply put, profit-making, as a form of self-interest, versus officeholding, with its commitment to serve the public interest,[26] represented two separate spheres that must never meet, even while no clear demarcation separated the realm of the unseen from the realm of the living. (That understanding, at least, Ban Gu and others credited to Yan Junping.)[27]

So what, precisely, is it that Ban Gu finds wrong with the celebrated omen experts in *Hanshu* 75, when all is said and done? To investigate Ban Gu's complaints, I begin with a speech made by Yi Feng 翼奉, one of the omen experts whose biography appears in *Hanshu* 75, which surely met the criteria for Ban Gu's warm approval:

> I heard my teacher say, Heaven and Earth took their places, and sun and moon were suspended in the sky; the starry constellations were arrayed in the sky, and yin *qi* separated from yang. Then the four seasons were fixed, and the Five Phases, so these could be seen by sages, who named them [rightly] "the [cosmic] Dao, or Way." Once the sages caught sight of the Dao, they understood the signs for kingly rule, and so they drew the demarcations for provinces [as sagely Yu once did], they established the ruler-subject relations, they invented the pitch pipe standards and calendar, they laid out criteria for perfection versus defeat. To show these to the worthies, they named these [rightly] the constants, and once the worthy men caught sight of these constants, they understood, at long last, the work that goes into the human Way, and then came the *Odes*, the *Documents*, the *Changes*, the *Annals*, and the rites and music. The *Changes* has its theories of yin and yang; the *Odes*, its references to the Five Boundaries;[28] the *Annals* has its disasters and anomalies.[29] Each of these arranges the cycles; deduces gains and losses, examines the heart and intentions of heaven, in order to speak of the stable and dangerous ways to rule as king. . . . [Here follows the standard Eastern Han critique of oppressive Qin rule, whose swift downfall followed massive unrest.]
>
> Now your majesty is a sage-ruler of great perspicacity who has taken to heart the essential way . . . who saves those in dire distress and poverty, supplying them with healing physics, and coffins and cash [to bury their dead]. Your grace and favor

have been particularly generous. Moreover, you have taken up straight-talking admonitions and sought to learn of your faults and failings. Splendid is your virtue, complete and pure. The realm is fortunate in the extreme.[30]

Yi Feng's remonstrance, delivered in the most tactful way, urges his emperor to become his best self, an enlightened sage, by focusing on the *jing* 經 (at once the "constants" of the sociopolitical and cosmic orders and the Classics) and the Way (again resonant on the two levels of the cosmic and sociopolitical), in what are clearly perilous times. Perhaps we can discern in Yi Feng's polite style of speech his long training from Hou Cang 后蒼, the most esteemed master of the rites. In any event, Yi Feng's successive omen readings urged his emperor, Yuandi, to consider undertaking much needed reforms, in particular to reduce expenditures and weaken the influence of the Wang clan, all measures that were designed to reestablish the ruling house on a firmer footing, with reduced responsibilities, restricted outlays, and a stronger strategic position. Like his fellow omen experts, Yi Feng did not shy away from visionary proposals (including a proposal to move the capital back to a new location styled ChengZhou, near Luoyang?).[31] And although nearly none of Yi Feng's proposals became policy, some doubtless planted a seed, because we find echoes of those proposals in Gu Yong's 谷永 sweeping recommendations during Chengdi's reign.[32] We are not surprised, then, when Yi Feng receives the court's favors and lives to a ripe old age, dying in office as an Advisory Counsellor.

By contrast, Sui Hong (aka Sui Meng), a second omen expert early in the second century of Western Han, correctly divined that a spate of omens signaled the rise of a commoner to the throne, but instead of honors, Sui was forced to undergo the excruciating and humiliating death of a convicted traitor. The substance of Sui's remarks, that the young Zhaodi, Xuandi's predecessor, should abdicate in favor of a yet-to-be-named commoner, could hardly have failed to rile Zhaodi's regent Huo Guang 霍光, so Sui Meng was remanded to trial and swiftly executed on charges of gross immorality (i.e., treason).[33] (By way of reparations, after Xuandi ascended the throne, Sui Meng's son—after Huo Guang's death, presumably—was belatedly appointed to the coveted post of palace courtier.) Five years later, the future Xuandi, then living as a commoner, was indeed installed as Son of Heaven, and, possibly because of this confirmation of Sui's powers, Sui's claim that the Han ruling line had descended from Yao was met with ever increasing approval by the Ban family members, most certainly, and by others of their acquaintance. With the benefit of hindsight, Ban Gu could accept the wisdom of Sui Meng's theories, insofar as descent from Yao and Xuandi's divine election were concerned, while still harboring grave doubts about the wisdom of extrapolating Sui Meng's larger rhetorical message. For after all, Sui Meng had reminded everyone in his hearing that Sui's own master Dong Zhongshu had prophesied that a legitimate successor to the Han throne might be dislodged by a commoner at any time, a potentially traitorous notion that Ban Gu, like his father before him, sought to condemn.[34] It gets worse. . . .

Xiahou Sheng joined Huo Guang and Zhang Anshi 張安世 in their efforts to depose Liu He 劉賀, later Changyi wang 昌邑王 (aka "Haihun hou" 海昏侯 of archaeological fame), with Xiahou Sheng's omen readings providing a particular help when it came to persuading the dowager empress. (Zhang Anshi, nominally the person in charge, did almost nothing to help, presumably because he-preferred to play it safe.)[35] As a *Documents* expert from a long line of authoritative *Documents* experts, Xiahou Sheng immediately became a well-respected and powerful proponent of the idea that the oldest Classics function as repositories of correct omen theories.[36] But very soon afterward, Xiahou Sheng jeopardized all the power and prestige he had accrued for helping to elevate Xuandi to the throne, when he went out of his way "alone" and "on his own authority" (*du* 獨) to denounce the plans of the newly installed Xuandi. Shortly after coming to the throne, Xuandi had announced his decision to confer upon Wudi, his great-grandfather, a lofty posthumous title and splendid ancestral temple, it being his "fondest wish" to express his filial devotion to his forebear. In response to the edict, Xiahou Sheng promptly proceeded to enumerate Wudi's many crimes and stupidities, concluding with the assertion that Wudi had very nearly cost the Han ruling house its throne. The outcome of his caustic dissent at court was predictable: as Xiahou Sheng had acted on his own, he was accused of treason, and the ancestral temple for Wudi was built according to Xuandi's plans. And when Huang Ba 黃霸, a very junior official, had the temerity to try to defend Xiahou Sheng, he was promptly thrown into the same prison cell. Awaiting execution, so they believed, Xiahou Sheng instructed Huang Ba in the Classics. But when later a series of earthquakes brought great loss of life, the court ordered their release (a sign that some members of the court deemed their punishment unjust).[37] Xiahou Sheng was soon reinstated to high office, and eventually Xuandi named him Senior Tutor for his heir apparent. Not coincidentally, Xiahou's omen readings, rooted in the *Documents* traditions, were soon awarded honors as well. And when Xiahou

Sheng died at an advanced age, he was buried with extraordinary honors. Significantly, however, Xuandi did not order Xiahou Sheng's portrait to be painted among his "loyal ministers" for a state visit in 51 BCE, despite the vital role that Xiahou Sheng had played in Xuandi's elevation.

Ban Gu concurred with Xiahou Sheng's negative assessment of Wudi's reign. Not only did he quote Sheng's protest memorial at length; he also registered his agreement with its main message numerous times. In fact, no fewer than fourteen chapters, not counting the omen treatise, present Xiahou Sheng as an authoritative figure, while the omen treatise praises him four times.[38] In addition, Ban Gu's decision to list so many in Xiahou Sheng's line who later performed well in office signals Ban's belief that Xiahou Sheng deserved commemoration for his strong family ties and a strong classical legacy. So what was Ban thinking in *Hanshu* 75? Evidently, Ban condemned the combination of bluntness and "acting alone" as counterproductive and offensive. For Ban consistently argued that indirect or allusive admonition was far more likely to effect real improvements in the throne's deliberations, especially when a good ruler was in place.[39] "Knowing when to stop" was a virtue for Ban Gu, as for many others of his time; not to know this and not to act in concert with others smacked of self-preoccupation and self-aggrandizement.

One final biography included in *Hanshu* 75, that of Jing Fang, clinches the case I am constructing here, insofar as it showcases Ban Gu's profound distaste for any court omen experts who used their theories (true or not) to try to get their own way. Clearly Ban Gu thought highly of Jing Fang's acumen, as Jing Fang's theories are cited in the omen treatises more often—some seventy times, according to Su Dechang's count—than those of any other single expert, given that the theories of Dong Zhongshu and Liu Xiang are often merged, quite unaccountably.[40] Like Jing Fang, Ban Gu himself thought the *Changes* readings represented the most important strand in the Five Classics corpus for the science of omen prediction, and Jing's teacher Jiao Yanshou 焦延壽 had proven himself an able local administrator capable of using his predictive powers rooted in the knowledge of the hexagrams to forestall bandits and corruption. Jiao's care for the locals, moreover, propagated a civilizing influence throughout his jurisdiction, as we are told.[41] Ban Gu would have approved of Jing Fang's warning Yuandi about factions at court who were gaining too much power over decision making; Jing Fang was an implacable enemy of Shi Xian 石顯,[42] Yuandi's favorite eunuch and the wily head of the Palace Writers, whom Ban Gu portrayed in a unfailingly bad light. Nevertheless, Jing Fang and his ally

Zhang Bo were both publicly executed in the marketplace for their troubles in 37 BCE, when Jing Fang was only forty-one *sui*, or forty years old. The problem was twofold: not only had Jing Fang risked open confrontations at court; he had also fabricated or misused ominous "coverings of *qi*"[43] to try to advance his own interests at another's expense (see his sealed memorials dated ca. 37 BCE).[44] So on the grounds that Jing Fang pressed his points too far, ignoring the proper forms of polite engagement during court discussions, and that, worse, he was motivated by selfish aims, Ban Gu assigns him to the collective biography in *Hanshu* 75,[45] with its very negative concluding Appraisal. There is no suggestion that his political instincts or his opposition to Shi Xian merited blame.

Hanshu 75, it is easy to forget, captures a time when the Western Han was undergoing a mid-dynastic restoration of a sort during the four reigns (Zhaodi through Chengdi, impressively with Xuandi) that reestablished the dynasty on a sound footing after Han Wudi's disastrous foreign wars of aggression and other excesses had bankrupted the court and fostered rampant corruption.[46] Ban Gu, living under the best emperor that Eastern Han was to see, surveyed that mid-dynastic restoration from his own vantage point, that of the miraculous mid-dynastic restoration to which Ban Gu and his father Ban Biao had devoted their lives.[47] He was intent upon using history as a mirror to warn the rulers and officials of his day that they must not overstep their spheres of authority. To again quote Ban Gu's summary of the omen experts, "But when the readings *offend the crowd and affront the age*, even superficial mistakes become errors and regrets; and deeper ones, unalloyed harms."[48] Long reading in *Hanshu* returns readers often to this same theme: Officials are not to seek their own personal advantage, lest they arrogate powers and privileges to themselves that are the ruler's (*shan* 擅). The ruler, for his part, is not to transgress the boundaries of ritual propriety, either by competing with the commoners in profit-making or by monopolizing decision-making power at court, running roughshod over his officials (both actions characterized by *zhuan* 專, arrogant and unlawful monopolizing).[49]

The foregoing provides a brief summary of the context for the *Hanshu* "Treatise on the *wuxing*." What I have tried to establish above is this: that Ban Gu's rhetorical aims are invariably complex, insofar as he intends the *Hanshu* history to serve as a "mirror" for his own Eastern Han. Ban asserts his twin beliefs that delicate, complex, and protracted negotiations must govern relations at court and that due attention must be given to long-standing precedents and unexpected signs if the court composed of ruler and oligarchy is to succeed in keeping the stable hierarchical orders

Ban terms the prized "constants." This sketch of Ban Gu's thinking will be useful when we come to consider the rhetorical aims that led him to break precedent by supplying an omen treatise.

II. Structure and Contents of the *Hanshu* "Wuxing zhi"

a. Preliminaries

Several years ago, not long after mulling over Liu Zhiji's 劉知幾 (661–721) monumental *Comprehensive Analysis of Historiography*, which takes nearly an entire chapter to excoriate Ban Gu for including his omen treatise in what Liu deems an otherwise nearly perfect work, I began research on the "Tables" in the *Shiji* and *Hanshu* and on the *Suishu* omen treatise. That research immediately raised new questions.[50] So, naturally, when confronting the *Hanshu* "Treatise on the *wuxing*"—recall that I do not title it "Treatise on Five Phases" for reasons explained below—I began with the historian's obvious initial questions about authorship and probable context for the compilation to ascertain whether and how it fits with Ban Gu's larger historical project.[51] Such ostensibly simple questions have been far from easy to answer, for the *Hanshu* omen treatise has no clearly identified author or date of compilation, just the traditional attribution to Ban Gu, which hampers movement beyond speculation when trying to assess its contents. Careful research by previous scholars into the *Hanshu* treatises plainly shows that the final compiler of other *Hanshu* treatises drew liberally from disparate sources, some of much earlier date.[52] Students of Han-era manuscript culture must be prepared to accept the indisputable fact, which further complicates matters, that one part of the piece of writing may derive from one source or one set of sources, while other parts bespeak different origins and inspirations.[53] Confronted with such complexities, we must often turn to structural matters, searching for clues for possible breaks within the text. We need not lose heart, for what we do know about the omen treatise compels our interest: the treatise's insertion into the history represents a major innovation (with Ban Gu a man generally averse to innovations, at least in the crude modern sense), rather than a continuation of any part of the *Shiji*. Meanwhile, as good readers of the *Hanshu*, we must consider not only the explicit message of the treatise but also its silences on certain matters.[54] That a great deal of time and energy went into this compilation is probable, given its stupendous length.[55]

224 | *Michael Nylan*

Since, at this remove in time, certainty about the "Treatise on the *wuxing*" is elusive, this section of my chapter seeks mainly to offer a fair, if necessarily preliminary, introduction to the *Hanshu* omen treatise, to spark greater interest in the text. The omen treatise (still untranslated into any Western language, so far as I know) has seldom figured in general treatments of Han history or even in more specialized accounts of Han divination practices, doubtless because of its complexity and length, but also, perhaps, because many feel that a lengthy work of superstition is hardly worth in-depth analysis.[56] Scholars of other antique civilizations do not feel the same. And it is time, past time really, to remedy the gap, because a few scholars have marshaled so much evidence for us in superb treatments devoted to the subject.[57] As an omen treatise, *Hanshu* 27 fits quite nicely within the customary rubric of divination or even "historical divination" (Mansvelt Beck's term).[58] Its introductory remarks cast it as a revelation text, a type of classicizing text whose significance in the pre-Daoist and pre-Buddhist world is ripe for further exploration.[59] Ideally the current effort will help readers to refine their own understanding of some of the distinctive features of Han world-making beyond the impoverished "correlative cosmology" concept, whose meaning is ill defined, if not downright incoherent.[60] (All pre-print cultures rely heavily on numbered lists as mnemonic aids, early China being no exception in this.)

The conclusions tentatively reached here are grounded in the secondary scholarship devoted to this *Hanshu* omen treatise. As early as 1929, Miu Fenglin identified seven types of content in Ban Gu's omen treatise, and Miu's work still serves as a good starting point:[61]

1. citations of the Classic, i.e., the "Great Plan" chapter of the *Documents*;

2. *zhuan yue* 傳曰 (the commentary or tradition says);

3. *shuo yue* 說曰 (the explication or theory says);

4. *yi yue/huo yue* 一曰/或曰 (alternative views);[62]

5. passages pointing out where Liu Xin's views differ from those of Xiahou Shichang (with no notification given when they are in agreement);

6. views expanding on the *shuo* 說 readings from disparate thinkers down to Liu Xin, which are analogized to subcommentaries; and

On *Hanshu* "Wuxing zhi" and Ban Gu's Project | 225

7. value judgments regarding the fundamental disagreements among views.

Some readers may recall that nearly all the portent specialists in early Western Han are said to be *Gongyang* or *Changes* specialists, with Dong Zhongshu prominent among them, but by the second century of Western Han, beginning with Xiahou Sheng, readings keyed to the *Documents* classics come into ever-increasing prominence in court discussions. On the relative values accorded rival theories, Miu had decided views. He felt sure that the compiler was disinclined to rate Liu Xin's theories more highly than those of Dong Zhongshu and Liu Xiang, because Liu Xin's name seldom appears first in the entries specifying several omen readings. Conceivably, however, placement within the omen treatise reflects a decision on the final compiler's part to follow chronological order, and it is hard not to attach great significance to the fact that the entire omen treatise begins by citing Liu Xin as omen authority (see below). That said, the combined views of Dong and Liu Xiang generally meet with approval, whereas twice Liu Xin's views are dismissed out of hand, with the curt "this is wrong" (此說非是 or 非是).[63] In the main, the *Hanshu* is inclined to praise Liu Xiang for elucidating the resonant connections between heaven and human 天人之應, and Liu Xiang mostly followed Xiahou Shichang's omen readings, the omen treatise says.

b. Organization of the Hanshu Omen Treatise

At first glance, the structure of the *Hanshu* omen treatise clearly derives from the "Great Plan" ("Hongfan" 洪範) chapter of the *Documents* classic, four sections of which (1, 2, 5, and 8) the omen treatise explicitly and repeatedly cites.[64] Closer readings show, however, that the treatise diverges from the famous *Documents* chapter, even as it expands upon it. The nine-part structure of the *Documents* "Plan" is as follows:

section 1: Five Material Resources (*wuxing* 五行), Wood; Fire; Earth; Metal; and Water;

section 2: Five Modes of Engagement (*wushi* 五事); aka Five Conducts, Demeanor; Speech; Sight; Hearing; and Thinking;

section 3: Eight Policy Concerns (*ba zheng* 八政)

section 4: Five Cycles (*wu ji* 五紀)

section 5: *huangji* 皇極 (Sovereign's Highest Standard), understood to confer the ultimate sovereign powers

section 6: Three Treatments (*san de* 三德)

section 7: Seeing to Doubts (*ji yi* 稽疑)

section 8: Many Proofs (*shu zheng* 庶徵)

section 9, in two parts: Five Blessings + Six Ultimate Afflictions (*wu fu, liu ji* 五福, 六極)

How closely the *Hanshu* treatise cleaves to the nine-section "Hongfan" *Documents* chapter can be seen below from its highly complex structure, which reconfigures the Plan, even as it draws heavily from sections 1, 2, 5, and 8 of the Plan:

From section 1 of the "Great Plan," the *Hanshu* treatise borrows the term *wuxing* for its first five parts, utilizing the term largely, if not wholly, in its original meaning as Five Resources;[65]

From section 2 of the "Plan," the *Hanshu* treatise extracts the concept of the Five Modes of Engagement, which greatly facilitates theorizing the relation between bad behavior and ill omens;

From section 5 of the "Plan," in its part 11, the *Hanshu* treatise finally takes up *huang* 皇 (the "sovereign" or "august"), concluding with the ruler's role in influencing his subjects, before turning indisputably to the Five Phases theories.

In a final section, the *Hanshu* omen treatise includes entries on celestial events that do not appear at all in the original "Great Plan" chapter of the *Documents.*

Structural features of the omen treatise stand out, even upon first reading. First, as the *Hanshu* omen treatise is five *juan* (originally silk scrolls, chapters) in length, it is the single longest chapter in the *Hanshu* by far. Second, the "Wuxing zhi" devotes a disproportionate amount of its attention to the Five Modes of Engagement, discussion of which occupies roughly half the entire treatise. It reassigns the ominous weather signs (constant rain, constant drought, etc.) that appear in section 8 in the original "Plan" to these Five Modes of Engagement, taking these signs to be the clear result of specific acts of misconduct on the part of the ruler, and thus reaffirmations of the underlying message transmitted through the Five Modes of Engagement section. Third, the *Hanshu* omen treatise then adds, in its fifth and final *juan*, many entries devoted to signs in the skies that do not correspond precisely to any section in the "Hongfan," despite the brief mention in section 4 of the "Plan" of the heavenly bodies as markers of time. Fourth, the *Hanshu* treatise meanwhile treats *in extenso* brand-new theories that are utterly foreign to the "Great Plan," in particular those theories articulating *wuxing xiangli* 五行相沴 or *liu li zuo jian* 六沴作見, which represent types of *qi* destroying each other.[66] Fifth, if length is a prime consideration (which it may not be), the treatise slights section 5 devoted to *huangji* 皇極, which is indisputably the heart and longest section in the "Great Plan" chapter, in that the omen treatise assigns no exemplary cases, auspicious or inauspicious, to the category of *huangji* as epitome and standard of good governance. In short, the *Hanshu* omen treatise adopts vocabulary and some structural elements from the "Great Plan," but the treatise does not, in the end, closely adhere either to the overall organization of the *Documents* chapter or to its main arguments. Mansvelt Beck nearly came to a similar conclusion when he wrote, "the system becomes less tightly organized, and in the end all connection with the Five Phases is lost."[67] My sole complaint is that Mansvelt Beck presumes that all appearances of *wuxing* describe the Five Phases, an understanding distinctly at odds with the "Great Plan" chapter in the *Documents*, where the *wuxing* plainly represent five material resources on earth.[68]

If we tarry for the moment on the question of possible motivations for slighting *huangji* in the omen treatise, four come to mind. Perhaps as an ideal, *huangji* conduct cannot elicit warning signs from Heaven. Or perhaps kingly legitimacy is above theorizing, to all intents and purposes. Perhaps discussing this issue was simply too dangerous to undertake. Or perhaps no ironclad instances of it can be adduced in history since Eastern Zhou.[69] Quite unaccountably, the treatise redefines the central concept of *huangji* 皇極 as a function of one term that precedes it in the original "Great Plan" and one term that follows it, ignoring the obvious fact that those two terms are listed in separate sections and categories in the original "Plan."[70]

Speculation aside, delving more deeply into the structure of the *Hanshu* omen treatise proves rewarding. The chapter opens with a summary of Liu Xin's elaborate theory concerning two revelation texts known as the [Yellow] River Chart (Hetu 河圖) and Luo [River] Writing (Luoshu 雒書),[71] mythic guides for the antique rulers that supposedly took the form of magic squares with attached texts.[72] Liu Xin's theory must have been controversial, for it traces the Hetu and Luoshu not only back to the primeval Fuxi 伏羲 (tradit. 2852 BCE), inventor of the "core text" of the Eight Trigrams in the *Changes* classic, but also to the mythical flood-queller Yu the Great (tradit. r. 2205–2197 BCE), who figures prominently in the first three Han-era *Documents* chapters, the "Canon of Yao" 堯典, "Counsels of Gao Yao" 皋陶謨, and most famously, the "Tribute of Yu" 禹貢. All three chapters, it need hardly be said, justify and sanctify unified empire, but they do far more: they construct successful empire as the product of ruler and minister working in concert, "shared rule" being both "natural" and "preordained."[73] (Note, meanwhile, that Fuxi is absent in the *Documents* classic, which begins with the trio of mythical antique sage-rulers by the names of Yao, Shun, and Yu.) Liu Xin's theory, as reported in *Hanshu*, makes the River Chart somehow the divinely revealed precursor for the Luo Writing,[74] even as it styles Jizi 箕子, advisor to both the bad last Shang ruler and Western Zhou founder, as the inheritor and disseminator of Yu the Great's teachings.[75] As a result, the focus never strays far from Luoyang, perhaps intentionally. While revelation texts fulfill one role in the Bible (to signal rupture), this particular revelation text in early China serves instead to confirm the ties between the cosmic and sociopolitical orders, renewing age-old commitments to constructive activities. But in the early literature that survives, there is no evidence before Liu Xin for the claim that the River Chart was ever associated with Yu the Great.[76] In the first century of Eastern Han, to foreground Liu Xin's relatively new theory, when Liu Xin was so controversial a figure, given his strong affiliation with the usurper Wang Mang—this alone attests the profound respect the final compiler(s) had for Liu Xin's extraordinary erudition and love of systematicity, a systematicity that the omen treatise, curiously, does not itself evince and possibly does not even aim for, if my working hypothesis is correct.[77]

After these initial framing remarks, the omen treatise organizes the ominous events recorded from hoary antiquity[78] down to Wang Mang's 王莽 Xin 新 interregnum (9–23 CE), according to the typologies loosely spun from the *Documents* "Great Plan" chapter, as noted above: the Five Resources (*wuxing*) appear in the first *juan* of the treatise; the Five Modes of Engagement (*wu shi*), in the second, third, and first half of the fourth *juan*; the Five Phases (also *wuxing* 五行) and marginally *huangji*, in the second half of the fourth *juan*.[79] The fifth and final *juan* of the omen treatise consists of reports on comets and equally fearsome omens in the sky, portending the ruler's good or ill conduct—signs never mentioned in the "Great Plan" at all! By contrast, the three preceding *juan* consign all notices of unusual weather to entries on the Modes of Engagement, evidently in the belief that their significance does not merit separate treatment.

c. Discrepancies between the Great Plan and the Hanshu "Wuxing zhi"

The importance of one innovation in the *Hanshu* omen treatise must be heavily underscored, if only because so many to date have ignored it: the Five Phases of cosmic *qi* do not figure at all in the original "Great Plan" chapter of the *Documents*, because that concept and its correlations represent a late Western Han construction and a substantial modification of the original meaning of *wuxing*, which originally refers to the Five Material Resources on earth.[80] As if to register that fact, the omen treatise assigns ominous events traced to the operations of the Phases to a discrete section, placed *after* the entries it assigns to the Five Resources and Five Modes but preceding the passage devoted to *huangji*.[81] Thus it inserts the Phases in the middle of much older ideas, perhaps because omens categorized according to this fairly recent cosmological concept could not be so easily squared with entries culled from older portent writings.[82] In consequence, the treatise title and contents employ the term *wuxing* in two opposing ways: initially in the sense of Five Material Resources and later in the opposing sense of "phased cosmic *qi* operations," with contrastive formulae deployed for each of the two separate senses.[83] Notably, the omen treatise does not dwell at length upon Five Phases rhetoric, it being far more preoccupied with familiar descriptions of the good or bad ruler's modes of engagement that date from as early as Zhanguo times. At the same time, the outright opposition between dueling Five Phases units (usually operating in pairs) seems to threaten the ultimate in chaos to those residing on earth.

Other departures from the "Great Plan" are equally noteworthy. Both the Resources and the Phases are listed in the sequence of wood, fire, earth, metal, and water, that is, the Five Phases "Generative Order" instead of the earlier "Conquest Order" derived from or read into the *Documents*

"Great Plan" chapter.[84] The Generative Order directs the ruler to place the highest priority on suasive rule, charismatic grace, favors, and rewards (all *de* 德) in preference to punitive measures, on an analogy with beneficent heaven, which begins the calendar year with the genial months of spring and summer before introducing the harsher regimes of autumn and winter. (The choice of the Generative Order tallies with the *Hanshu*'s celebration of Dong Zhongshu—a celebration distinctly at odds with Dong's shabby treatment as a ridiculous omen theorist in the *Shiji*.)[85]

Meanwhile, the *Hanshu* treatise stipulates a complex process whereby Heaven issues repeated warnings to the unrepentant ruler in successive *stages* so as to alert him to his misconduct. These stages supposedly unfold with ever-increasing harshness and frequency.[86] (NB: The "Great Plan" chapter, for its part, is silent on the punishments for the unwary ruler, intent as it is to portray the wise ruler fully alive and attentive to his duties who thus attains the ultimate in good governance.) Within the *Hanshu* treatise, these stages are defined in language that is far more precise than that found outside the treatise (or indeed, in some of its own entries) and apparently innovative:[87]

1. *yao* 妖 (nascent warnings),

2. *nie* 孽 (signs in plants, but later extended to other signs),

3. *huo* 禍 (signs in domestic animals),

4. *e* 痾 (afflictions to the body, on a small, then larger scale),[88]

5. *sheng xiang* 眚祥, defined often as "the birth of anomalous beings" 異物生, a subclass of which, if efficacious changes are made, become the *xiang* 祥, and

6. the culminating occurrences of *wuxing xiangli* 五行相沴 (i.e., types of *qi* destroying each other).

Worth noting is this: within each of first five stages there are finer gradations in the warning process: for example, it is far worse if the afflictions of the body called *e* attack the vital organs, instead of the skin, tongue, eyes, and ears.[89]

In the main, the *Hanshu* omen treatise takes the form of "If X, then Y, and then Z" where X is nearly always one or more acts of misconduct on the part of those in power (usually but not invariably the ruler, as we will see), Y is the response to misconduct, and Z the explanation attached to the sign. Only very rarely does a ruler's good conduct register or elicit a favorable sign. Thus, in stages 1–5, by the *Hanshu* theory, Heaven issues repeated warnings to the ruler and the members of his inner circle to reform. If no appropriate response to Heaven's warnings then occurs, prompting a set of major reforms in behavior and policy, the onset of stage 6 ensues, when neither the ruler nor the members of his court have the luxury to see the error of their ways, to amend their behavior, or to alter their policies to ward off the looming disaster to the ruler's line, his realm, and his person. The crisis in stage 6 (*always* preceded by earlier omens), in effect, signals a point of "no return." Either an anthropomorphic Heaven has become so incensed at repeated wrongdoing that it will rain down punishment on the recalcitrant powers-that-be, or the totality of the cosmic powers signified by the shorthand term *tian* have become so imbalanced and discordant (*bu he* 不和) that wide-ranging disturbances inevitably result, as ripple effects accumulate and intensify.[90]

Although the precise theology remains unclear, the emphatic gradualism of the omen treatise should strike us, especially when compared with the omen analyses offered by other Eastern Han histories,[91] for according to the omen treatise (following a *zhuan*, a tradition), the effects of "the warnings only gradually increase" 言寖深也, before portending the worst.[92] No less striking is the usual form the final-stage warnings take: they typically involve pairs, destroyer and destroyed, most often in Wood-Metal, Fire-Water, and Earth-*huangji* combinations. Theoretically, any power in the Five Phases can destroy Earth, signifying the center, the ruler, and, once at least, the superiority of the Central States civilization.[93] At the same time, it was evidently unthinkable to posit the destruction of Heaven or the heavens, as the source of goodness, by the Five Phases *qi* modalities, singly or in combination. Nor could one or more Five Phases modalities entirely destroy the ultimate *principle* of good rule (*huangji*), which persists somehow, through long centuries of misrule, thanks to multiple sages' belief and trust in the potential for Great Peace under a more promising ruler.[94]

When all is said and done, the simple insertion of Five Phases theory into older omen readings makes for a different sort of omen interpretation. The principal reconfiguration takes place when it comes to the Five Phases' insertions, which counter the priority placed on beneficence elsewhere in the omen treatise. And whereas earlier omen readings had cast *yin-yang* largely in terms of above-below relations (most often pertaining to ruler-subject or ruler-minister relations), the Five Phases introduced a host of new factors to be analyzed, with location of the ominous event and its color assuming new prominence alongside graphs and sounds as special indicators of ominous import. Perhaps this helps to explain why the final compiler of the *Hanshu* omen treatise cautions readers that the omen treatise can offer the court only imperfect guidance. As Espesset astutely notes, the compiler (Ban Gu for him) "warned . . . of the *limited accuracy* of the explanatory process: some signs might simply never appear, and some may manifest only *in the wake of the events they supposedly forbode*" (italics mine).[95] So while warnings as a general rule should precede (*xian* 先) calamities, and calamities follow or accompany (*sui* 隨) human actions, sometimes a calamity might occur *before* any sign announcing it was observed, conflating sign, cause (practical and logical), and effect, in a manner known in omenology right 'round the antique world.[96]

What seems not particularly novel in the *Hanshu* treatise, judging from the extant sources, is some standard vocabulary (e.g., "rough analysis" or *gaishuo* 概說; "blame" or *jiu* 咎; "penalty" or *fa* 罰; and the phrase *shi ze* 時則)[97] and the omen treatise's attention to the conduct and demeanor of the powerful, which inevitably meant, in turn, somewhat less attention to natural disasters, as "signs" of more resonances and regularities in the cosmos. By long-standing traditions, omens were cast as the "outer signs of the king's doings" (王事之表), and these traditions are duly upheld in the *Hanshu* omen treatise.[98] Standard, too, is the limited attention paid in the omen treatise to auspicious events or forms of behavior, although some omens naturally betoken mixed fortunes, good and bad, for the parties involved;[99] after all, the conquest of one kingdom represents good fortune for the conqueror and bad for the conquered. But the treatise aims to forestall the conclusion that Heaven or the cosmic powers grant victory to any winner who succeeds by luck or good timing rather than by moral cultivation and effective suasive rule; the omen treatise more than once states unambiguously that a temporary victory may pave the way for final defeat, so good governance sooner or later wins out. This is what we would expect if the final version were carefully edited by Ban Gu or by Ban Zhao, whose father, Ban Biao 班彪 (3–54), was insistent upon this very point, in support of the Eastern Han rulers' claims to legitimacy.[100]

That said, one of the curious features of the *Hanshu* "Wuxing zhi" is its scope: its comments on events in Western Han history are far outnumbered by entries devoted to pre-Han. More curious still, a mere fraction of the potential categories outlined in the *zhuan* 傳 ("traditions," usually uncritically ascribed to Fu Sheng) and *shuo* 說 ("explications" of unknown origin) appear to be verified by corroborating signs in the omen treatise.[101] To give readers some idea of this feature of the treatise, I cite one fairly typical passage from the earliest part of the treatise that specifies the bad behavior (mainly the ruler's, as we will see) that gives rise to dislocations in society and cosmos, in relation to Wood. The "Hongfan" *zhuan* 傳 (attached "tradition"?) forbids:

> Hunting out of season,
> 田獵不宿，
> Food offered without due decorum,
> 飲食不享，
> Immoderation in comings and goings,
> 出入不節，
> Stealing the people's time for farming,
> 奪民農時，
> And letting treasonous plots arise.[102]
> 及有姦謀。

In other words, at court audiences or on journeys out from the palaces, and when hunting or feasting, the ruler was to exercise moderation, paying particular attention to the corvée labor demands, lest the burdened farmers lack the energies needed to farm the land and secure their livelihoods during the spring, summer, and early autumn months. This would respect seasonality and timing.

After stipulating these lengthy preliminaries about the category, the *Hanshu* omen treatise then surprisingly supplies only a single case under the heading of Wood, and a case drawn from the Chunqiu period at that, long centuries before Western or Eastern Han. Dated precisely to the sixteenth year of Lord Cheng of Lu (575 BCE), the case concerns one Shusun Xuanbo 叔孫宣伯 (aka Qiaoru 喬如), an officer who came to foment a rebellion against the Lu ruling house after having illicit relations with Lord Huan's consort Mu Jiang 穆姜.[103] Even more surprisingly, the single case adduced in evidence has no obvious connection either with the prohibitions listed in the relevant "tradition" (*zhuan* 傳) or the incitements to good behavior found in the "explication" (*shuo* 說) that is summarized or quoted.[104]

Admittedly, very occasionally the connections between the "Hongfan" chapter, its traditions and explications, and the ominous occurrences assigned to its categories seem more coherent and a better fit. For instance, talk of Fire seems more focused on misbehavior by the palace women, and Metal, on occasions of warfare.[105] Yet this seeming "coherence" may simply reflect the fact that the omen treatise repeats so many complaints about interference by palace women in court politics and courts' unwise decisions to launch wars. Palace ladies interfering with the orderly succession by legitimate heirs certainly seems germane to late Western Han politics, and it was a prominent theme in many *Hanshu* chapters. Thus, the steady repetition of such themes provides a semblance of coherence when there may, in fact, be little or none. At least, if modern readers understand that the "Wuxing zhi" was retrofitting an older tradition based on Five Resources to newer theories positing Five Phases *qi*—a task that can never have been all that easy[106]—they will be less likely to find themselves outraged, puzzled, or defeated by the disjointed rhetoric found in so many entries in the *Hanshu* omen treatise.

Turning to its treatments relating past history to seminal events in Western Han, the *Hanshu* omen treatise mainly criticizes outright rebellions and virtual coups d'état, casting the coups as part and parcel of the numerous instances of interference in the legitimate succession. For Western Han, the omen treatise is wonderfully even-handed in its "praise and blame" in that every single emperor, regent, or ruler after the Western Han founder is blamed for his or her misdoings, and none escapes harsh censure. Dowager Empress Lü encouraged her family members to usurp Liu clan power and mistreated Liu Bang's favorite after his death;[107] Wendi failed to heed his good advisors;[108] it was Jingdi's aggressiveness that led to the Seven Kingdoms Rebellion (in 154 BC); and Wudi's own son and heir had cause to rebel against his erratic father. Under Zhaodi, the Huo family went a long way to try to usurp the throne.[109] Xuandi's trust was misplaced in his favorites, which calls into question his powers of discernment. Yuandi relied on his corrupt and vicious favorite, Shi Xian 石顯.[110] Chengdi paid incognito visits to the suburbs of the capital,[111] after which he elevated two sisters of low origins to the two highest-ranking positions in the back palace, those of empress and Brilliant Companion (Zhaoyi 昭儀).[112] After Chengdi came the Wang family's gradual ascent to supreme power, which had begun as early as Xuandi's or Yuandi's reign, and this ended in Wang Mang's illegitimate takeover.[113] Barbarian incursions on the crown territories also merit mention in the treatise whenever they posed serious threats to capital and court, as did what we would call "natural disasters."[114]

A sample entry devoted to Western Han omens reads like this:

> In the second year of Heping, the first month [27 BC], in Chengdi's reign, the foundry for the Iron Office in Pei commandery had a case where the iron would not pour down. There was a huge boom with a thunderous sound; also the sound of a drum. Thirteen ironworkers fled the scene in terror. When the sound stopped, they went back to have a look at the site. The earth had caved in to a depth of several feet; the furnace had split into ten pieces. In each the metal had splattered like a meteor shower, flying up and away. (This is the same image as for the second year Zhenghe [91 BC].) That summer, five maternal uncles of the emperor were enfeoffed; they were called the Five Marquises. The eldest uncle, Wang Feng 鳳, became General-in-Chief with the concurrent title of Marshal of State, and it was he who grasped the reins of government. Two years later, a rift developed between Chancellor Wang Shang 商 and Wang Feng, so Wang Feng slandered Shang. Removed from office, Wang Shang committed suicide.
>
> The next year Wang Zhang 章, Governor of the Capital, pled Wang Shang's case, saying that he was loyal to the throne and a straight-talker (i.e., able remonstrant). Zhang also alleged that Wang Feng monopolized power, so Wang Feng maligned Wang Zhang, charging him with the crime of *dani* 大逆 (treasonous activity). Wang Zhang was remanded to prison, where he died, and his wife and children were removed to Hepu 合浦 [the hinterlands, in modern Guangxi province]. Later [Chengdi's] Empress Xu 許 was tried and deposed for *wugu* 巫蠱 (concocting poisons by witchcraft). Then Zhao Feiyan 趙飛燕 became empress, and her younger sister, a Brilliant Companion. The sisters harmed the imperial sons, so Chengdi had no heirs. The Zhao sisters, empress and Brilliant Companion, both admitted their crimes [sometime later].
>
> One theory says, "Iron flying belongs to the omen category 'Metal does not conform to the mold' 金不從革 [instead of the category to which it was assigned, 'Fire does not flame up' 火不炎上]."[115]

The entry gives the impression that the final compiler was determined not only to retain alternate theories regarding the same omen but also to have the single explosion refer to complex and protracted relations among the various members of the Wang clan. Dowager empress Wang Zhengjun and her kinsmen interact with various empresses in chaotic fashion that mimics the scattershot of the metal that misperforms. It is hard to know where to assign blame, with so much blame to go around, and murders, exiles, executions, and egregious slanders. With multiple entries like this one, it seems inadequate to observe, as Mansvelt Beck does, that the treatise "gives natural catastrophes . . . political significance,"[116] since the focus so seldom strays from the interactions among the political actors. So perhaps a contrast with Fan Ye's *Hou Hanshu*, with its narrow focus on the collusion between palace women and palace eunuchs, is more illuminating: despite the dizzying complexity of its accounts, the *Hanshu* omens generally assign final culpability to the reigning Han ruler for all misconduct that occurs at the court he nominally heads, imparting a moralizing "praise and blame" tone to the treatise distinctly at odds with the complexities of court politics it portrays.[117]

This reader confesses to being taken aback by the wide spectrum of omen experts, each equipped with rival interpretations, whose views the omen treatise cites, also the willingness of the omen treatise to contradict itself openly ("One theory says . . ."; "Someone says. . . ."). And yet, an omen expert as famous as Jing Fang (d. 37 BCE), mentioned some seventy times in the omen treatise, is never explicitly identified as authoritative, unless no rival readings exist.[118] (See section IV below.).[119] Considered as a whole, then, the contents of the *Hanshu* omen treatise foreground fierce, ongoing debates among Han omen experts who sought to "read" pre-Han and Han events.[120] At first, this author hypothesized that the treatise's refusal to promote one set of ideas over another simply revealed a disinclination on the part of the Bans to risk incurring the displeasure of the early Eastern Han emperors they served, partly because of their own family histories at court: their esteemed great-aunt had narrowly escaped execution on false charges; Ban Gu in his youth had been thrown into jail on suspicion that his writings about the Han rulers might be treasonous, and even Ban Zhao had witnessed her palace patron driven to consider suicide.[121]

Upon further consideration, it seems equally plausible that this notable feature of the *Hanshu* omen treatise may reflect trends shared by members of the capital oligarchy, regardless of the specific identity of the final compiler of the omen treatise. For instance, the *Hanshu*'s exhaustive treatment of the

Chunqiu period (770–475 BCE) likely points not only to the importance of Confucius to the revised Eastern Han notions of legitimate power but also to the political processes that were undermining centralized empire and leading to more decentralized forms of power sharing,[122] like those known in pre-unification days. I hazard this speculation in large part because the court-sponsored *Bohu tong* 白虎通, compiled in Ban Gu's time, certainly reflects a heightened awareness of the complex relations pertaining between local lords and magnates and their kinfolk vis-á-vis the central court, in making more people, male and female, and not just the reigning emperor himself, bear the ultimate responsibility for dynastic order.[123] Accordingly, in this treatise the omen experts exercising their prerogatives assume as much pre-eminence as the titled political agents whose steps and missteps the treatise depicts.[124] This sort of thinking undergirds my current working hypothesis, as outlined below.

A sample entry drawing upon pre-unification tales may prove instructive in this regard. In an entry dated to the fourteenth year of Lord Huan of Lu (corresponding to 698 BCE), lightning struck the granary of the ancestral temple of Lu, setting it ablaze.[125] To interpret the omen, the treatise cites no fewer than three omen experts, Dong Zhongshu, proponent of the *Gongyang Tradition* for the *Annals*; Liu Xiang, who preferred the *Guliang Tradition*; and his son Liu Xin, who promoted ideas of the *Zuo Tradition*, although we cannot know, in the vast majority of cases, whether the views ascribed to these thinkers represent long-standing filiations or their personal innovations.[126] (NB: none of the experts cited were famed for their special expertise in the *Documents*, which by late Western Han had become so influential at court.)[127] Dong thought the fire in the ancestral temple was caused by the general laxity of the officials of the Lu court, who did not try to "protect the temple during their lives"; Liu Xiang blamed the consort of the Lord of Lu for her incestuous behavior, which ultimately ended in an act of regicide a full four years later; Liu Xin thought the ominous events should be traced to some unclearly specified breach in the social norms and institutions.[128] The final compiler of the treatise (presumably Ban Gu) does not openly adjudicate between these incommensurate views, which treat ominous occurrences simultaneously as generated by past human events and as harbingers of events to come.[129] This entry is representative of the *Hanshu* omen treatise, which tends to offer multiple views for nearly all of the occurrences discussed, except for those rare cases where it offers no interpretation whatsoever.[130]

Moderns cannot hope to discern at this remove why the final compiler of the *Hanshu* omen treatise chose to ignore a slew of ominous events in late

Western Han, many of them detailed elsewhere in the *Hanshu*, with these omissions a source of bitter complaints by Liu Zhiji.[131] Did the compiler intend to highlight those occurrences most easily tied to dramatic changes in the situations of the powerful at the late Western Han courts? This is plausible, but we simply do not know. Confronting uncertainties, I have one thought: while the final compiler's contributions to the treatise may seem minimal—perhaps confined in the main content to six overt comments marked by the formula *huo yue* 或曰 ("some say"), that is, omen readings for which no authority whatsoever is cited, and/or the still-rarer comments addressed to readers on the correctness of a given reading[132]—two decisions by the final compiler—the decision to foreground Liu Xin's theory about the River Chart and Luo Writing and the decision to merge the often disparate theories of Dong Zhongshu and Liu Xiang—had monumental consequences for the way readers would interpret the treatise, and the final compiler must surely have foreseen that effect.

III. Probable Authorship, Dating, and Motivation: Speculations

It strikes me as distinctly odd that none of the senior experts on Han omenology writing in Western languages seem to have appreciated the problem of the treatise's authorship. Hans Bielenstein, Wolfram Eberhard, and Martin Kern, to name but three of the reputable experts who weighed in, all assumed that the *Hanshu* omen treatise transparently represents Ban Gu's era and his own individual point of view, as does the Chinese expert Shi Ding 施定.[133] (Espesset seems more hesitant in this regard, as does Yang Shao-yun.)[134] But surely it makes a difference whether Ban only applied some finishing touches to previously compiled materials or authored the treatise more or less *de novo*, with both potentials attested during the manuscript culture of the time. After all, that culture put a premium on splicing older passages with newer texts to produce an elegant piece of persuasive writing carrying an admirable whiff of the antique through a compilation process known as *zhuwen* 屬文. It should matter equally to moderns what the *Hanshu* compiler thought of the discipline of omen reading as laid out in at least three authoritative works that shared more or less the same title, *Traditions on the Great Plan's wuxing* (*Hongfan wuxing zhuan* 洪範五行傳), written by experts a century and more before Ban Gu's time: Xiahou Shichang, Liu Xiang, and Liu Xiang's son, Liu Xin.[135] (Modern scholars who trace this line of thinking further back to Fu Sheng 伏生 in early Western Han are simply wrong.)¹³⁶ It is not inconceivable that a single finished piece of writing, the *Traditions*, came to incorporate all three experts' views, for in early manuscript cultures such additive processes were common.

As inheritor of Xiahou Shichang's teachings, Xiahou Sheng is further mentioned in this connection, as is his student Xu Shang 許商 (d. after 8 CE), compiler of a *Record of wuxing Traditions* (*Wuxing zhuanji* 五行傳記),¹³⁷ and Xu's student Li Xun 李尋.¹³⁸ Others to whom an omen treatise modeled loosely on the "Great Plan" is credited include at least two members of the Ban family in previous generations. Ban Bo 伯 had written a lengthy analysis of Xu Shang's *Record*, and Ban Bo's brother, You 斿, had once worked in the imperial palace libraries under the direction of Liu Xiang, the omen specialist.¹³⁹ We tend to forget the degree to which erudition remained largely a "family business" in early manuscript culture. Still, for the Bans of Eastern Han to celebrate these forebears required nimble rhetoric, for the Ban family, like most members of the governing elite in the late Western Han capital, had allied itself with the very Wang Mang who had destroyed the Western Han dynasty¹⁴⁰ and who was condemned, naturally enough in Eastern Han, as "usurper."

Let us start with the basics: in all likelihood, the phrase *zhuan yue* 傳曰 ("the tradition says") signifies the teachings of Xiahou Shichang, transmitted to Xiahou Sheng, and then to other *Documents* experts, for Liu Xiang's readings purportedly tallied with those of the *Documents* experts Xiahou Shichang, Xiahou Sheng, and Xu Shang. As a result, "only Liu Xin's teachings were at odds" (*wei Liu Xin zhuan du yi* 唯劉歆傳獨異) with the dominant *Documents* interpretations of ominous events.¹⁴¹ For this reason, the Qing experts on the *Hanshu* omen treatise were inclined to believe that the writings (now lost) of Liu Xiang represented the "core" materials around which the contents of the *Hanshu* omen treatise coalesced. For instance, Yao Zhenzong 姚振宗 (1843–1906), in his *Hanshu Yiwen zhi tiaoli* 漢書藝文志條理,¹⁴² identifies Liu Xiang's lost work as the principal source for the *Hanshu* omen treatise.¹⁴³ And Wang Mo 王謨 (*jinshi* 1778), another Qing expert, concurred in the preface to his collection of fragments from Liu Xiang's lost corpus.¹⁴⁴ The modern scholar Chen Kanli 陳侃理 has continued this line of argumentation, deciding that, based on the *Suishu* "Wuxing zhi," the "explications" found in the *Hanshu* omen treatise basically reflect Liu Xiang's *Hongfan wuxing zhuan* (now lost).¹⁴⁵ But given the number of times that Liu Xiang and Liu Xin are mentioned by name in connection with specific omen readings in the *Hanshu* omen treatise, it is at least as likely that neither of them is the principal (i.e., most substantial) contributor to the *Hanshu* compilation. For it is hard to imagine why the final compiler, presumably Ban Gu, would choose not to ascribe one Liu as author, given the multigenerational ties binding the Liu and Ban clans through intermarriage and the willingness of *Hanshu* to credit Liu Xin with compiling major works, including the *Classic of the Ages* (*Shijing* 世經).[146]

As Liu Xiang is said to follow Xiahou Shichang, both Su Dechang in Taiwan (currently, Taipei, Neihu Gaozhong 台北內湖高中) and He Ruyue 何如月 in the People's Republic of China (Shaanxi Shifan daxue) have concluded instead that the base text and main source of the *Hanshu* omen treatise probably derives from Xiahou Shichang or his immediate disciples, to which Ban Gu made final additions and emendations. Their reasoning is similar: that the *Hanshu* omen treatise typically names the principal experts it draws upon, but the *Hanshu* treatise explicitly refers by name to Xiahou Shichang but twice in its narratives.[147] Perhaps Xiahou Shichang *is* the principal compiler who contributed most substantially to the treatise, but I myself doubt that the Five Phases language inserted into the omen treatise was already current during Wudi's era, when Xiahou Shichang taught, because we fail to find such Five Phases language being used even as late as Jing Fang's time (d. 37 BCE).[148] Similarly, it seems no more likely that the Hexagram-*qi* numerological explications predates Sui Meng's *Changes* teachings during the reign of Zhaodi.[149]

To state the obvious, many works based loosely on the title "Great Plan" or *wuxing* seem to have been in circulation in the second century of Western Han and the first of Eastern Han, and conceivably Ban Gu simply imported these writings into the omen treatise, just as we know he imported large chunks of his geographical treatise from palace documents drawn up nearly a century before his time.[150] However, He Ruyue argues that the compiler(s) of the *Hanshu* "Wuxing treatise" had a specific agenda: to remove important powers—powers that had been famously abused in the past—from the hands of the professional court diviners by generating a systematic theory that could be understood and, crucially, *applied* by non-specialists at court, as needed.[151] The probable historical context for the treatise's final compilation lends a measure of credence to Professor He's theory, for the early Eastern Han courts had witnessed angry debates over the reliability of the apocryphal traditions attached to the Classics, with Ban Biao, Ban Gu's father, deploring the Eastern Han rulers' frequent resort to the apocrypha.[152] Certainly, her hypothesis would not contravene any of the introductory comments to the chapter, which imagines that each of the great dynasties of Xia, Shang, and Zhou received a multipart revelation text, thanks to a single wise advisor to the throne, a revelation whose receipt appreciably extended the life of the dynasty and peace in the realm, and thus could do so for the "restored" Eastern Han.[153]

Utilizing a different tactic, Su Dechang speculates that Ban's goal was not to advance his own ideas regarding such matters, but rather to demonstrate how far successive Western Han theorists had diverged from earlier, especially pre-Qin readings of the Classics.[154] Su's theory certainly addresses the question "Why did Ban Gu, in compiling a dynastic history, include so many pre-Han omens in his treatise?" And it acknowledges the implied parallel drawn between the twelve Han reigns down to Wang Mang with the twelve reigns of Lu, supposedly recorded in a fashion to issue "praise and blame" by Kongzi/Confucius in his *Annals* (*Chunqiu*) classic. Unfortunately, Su's logic is ultimately unpersuasive for two main reasons: First, the compiler (definitely Ban Gu for Su) seems to relish many of the latest Han theories, including those of Liu Xin, whose remarks receive open approbation at the beginning of the treatise, and, even more importantly, the compiler seldom intervenes to identify the single best omen reading when there are competing interpretations, so as to enlighten the reader. Second, several of the other *Hanshu* treatises consider Western Han events against the backdrop of older histories and theoretical writings, not just this one omen treatise.[155] (Notwithstanding, we are all deeply indebted to Su for the most comprehensive and enlightening study yet produced on the *Hanshu* omen treatise.)

Keeping the insights of He Ruyue and Su Dechang in mind, I entertain an alternative working hypothesis, mindful as I am that seemingly small tweaks to a manuscript may have major implications, as we know from earlier careful work in Early China studies.[156] By my current working thesis, when assembling the "Wuxing zhi" 五行志, the final compiler (presumably Ban Gu) borrowed many ideas from Liu Xiang (as the text clearly indicates) but equally those of Liu Xin (although Xin is mentioned less often).[157] Curiously, despite the apparent tendency of the final compiler of the *Hanshu* omen treatise to slightly favor the categories put forward by Liu Xin for the all-important Five Modes of Engagement and failures thereof, the final compiler seldom notes how much the omen theories set forth by Liu Xiang and Liu Xin are at variance, though one such admission figures in the text.[158] So while the final compiler(s) must have shared the same general motivation as Liu Xiang—to warn the ruler and the dynasty about court policy decisions that were liable to endanger the ruling Liu clan—the treatise equally endeavored to advance the son's (Liu Xin's) theories, as Ban Gu does elsewhere in the *Hanshu*. (In writing this, I dispute Miu Fenglin's 1929 views, needless to say.) So, by this current working hypothesis, the final compiler (Ban Gu?) quietly announced his preference for some of Liu Xin's more daring theories, which were supposedly corroborated by the *Santong* 三通 (Triple Concordance) calendrical system of Liu Xin's devising, the calendar accepted by Eastern Han.[159] At any rate, study of the *Hanshu* "Wuxing zhi" testifies to the pervasive influence of Liu Xin in Eastern Han, a topic worth further exploration.[160] (Obviously, the foregoing need not mean that Ban Gu's own readings always had to match those of Liu Xin.)[161]

A brief word on systematicity and accuracy is in order: we should not overlook the fact that the final compiler of the *Hanshu* omen treatise, like the final compiler of the *Bohu tong* in Ban Gu's day, did not hesitate to give alternate readings and interpretations to supplement the main theoretical stances put forward in the compilation. While moderns are wont to aim for absolute accuracy and total control (at their peril), no one in Han times would have been so benighted. Instead, we discern an epistemological modesty in the Han writings devoted to the cosmic Mystery in relation to sociopolitical interactions. After all, the Zhangjiashan strips award the highest honors to a diviner whose predictions are fulfilled 70 percent of the time; similar figures are given for the best healers in several Han texts.[162] Because it was the two Han courts' announced intention to capture a full range of opinions bearing upon policy matters "as in a net," it should not shock us if the *Hanshu* omen treatise proffers multiple readings without expressing the need to settle upon one.[163] Indeed, eliciting and then weighing multiple proposals was precisely what the important institution of Han court conferences was designed to do. Elsewhere, the *Hanshu* evinces a very strong interest in showing that the best policy decisions are arrived at through these elaborate court conferences, where different interest groups (high- and low-ranking court officials and technical experts, and often the emperor as well) judged competing policy proposals.[164] In this connection, the surprising frequency with which the omen treatise indicates by one of two formulae ("one says" or "someone says") the existence of unnamed experts contributing their views gains significance.[165] For at least we can say that *if* Ban Gu is the final compiler of the omen treatise, it would be entirely consistent with his preoccupations to have compiled an omen treatise as a repository of credible readings by reputable experts,[166] whose inherent applicability and authority would be determined by future events as they unfolded and modified by further debates.[167] Compare Ban Gu's remark in the bibliographic treatise (or Liu Xiang's or Liu Xin's that he borrowed) on anecdote collections:

> As Kongzi said, "Even the byways offer sights that are worth contemplating, but he who would travel far must fear being detained, and the noble man therefore ignores them." *Yet neither does he destroy them.* The insights that villagers have attained based on some minor form of knowledge should also be written down and kept, in case they contain one observation that is worth adopting."[168]

So even if we suspect, based on evidence from the other *Hanshu* treatises, that Ban Gu may have done little more than put a few finishing touches on an earlier omen treatise whose component parts had been compiled before him, perhaps by a Xiahou or a Liu, the introductory remarks that appear in the omen treatise strongly direct the reader to think along certain lines and deter the reader from asking certain tricky questions. Remember that there was no "general reader" in Han-era manuscript culture; also that small tweaks to a manuscript might have major implications, as we now know.[169] Treatises were written by and for members of the court about matters that frequently came before the court to adjudicate. Indeed, high cultural literacy was a minimum requirement, in that the *Hanshu* omen treatise necessitates prior knowledge of the *Annals, Documents,* and *Changes* classics, the vast "Monthly Ordinances" (Yue ling 月令) literature, at least one of the three *Rites* classics, plus the late Western Han cosmological and alchemical theories in the apocryphal traditions, not to mention the Daybooks designed "to dispel doubts."

Regardless of the identity of the final or principal compiler(s) of the omen treatise, the *Hanshu* generally and the omen treatise in particular supply a ready explanation for the impulse to compile such a treatise when it "cites" Wang Yin 王音, d. 15 BCE, as a "good" and loyal member of the usurping Wang family, who held high office for more than a decade. Cognizant of his emperor's misbehavior, especially his propensity to heed favorites who had slandered good officials, Wang Yin risked a great deal when he informed the emperor about one bad portent that had taken the form of pheasants flocking into the palaces and squawking loudly. After this, Wang Yin discussed with others his fears that the emperor would not bother to take the warning to heart. "But if Heaven itself [by issuing portents] still cannot move the emperor [to mend his ways], what hope have I, a mere official?" asked Wang rhetorically.¹⁷⁰ The only alternative that remained to him, as Wang saw it, was to speak boldly and calmly "await execution."

Clearly, this anecdote testifies to the belief that a record of ominous events—far more effectively than "empty theorizing"—might in some cases, if not all, persuade rulers to redouble their efforts to do their best. Put another way, an omen treatise might help prepare the reigning Eastern Han rulers and high-ranking ministers to fend off challenges from a wily conniver in the mold of Wang Mang.¹⁷¹ The omen treatise offers remarkably few instances where those in power duly heed heaven's warnings.¹⁷² Jesse Chapman has written in his chapter that all the Han omens included in the "Celestial Patterns" treatise (*Hanshu* 26) seem "implicitly or explicitly" to comment on events leading up to the downfall of the Western Han dynasty and the rise of the "usurper" Wang Mang. True enough, but the final compiler of *Hanshu* 27 intends more, in my view: to emphasize how long that downfall took and how many powerful people connived at that outcome, by their silences and their self-interested or disinterested actions and attitudes.¹⁷³ This was Ban Gu's message in the rest of the *Hanshu*, and Ban Gu had elsewhere expressed intense interest in omens and revelations, for instance, in the final lines of his "Two Capital *fu*." We leave it at that.

Two fundamental distinctions thread through this section on dating and authorship. The first pertains to the difference between an "author" (someone who writes more or less *de novo*) and a "compiler," a person who assembles other people's writings, often adding paratextual comments or bridging paragraphs, to better frame earlier writings in order to provide a sense that they are coherent, even when they are a composite drawn from disparate sources. A second distinction would disambiguate the initial or principal compiler(s) from the final compiler(s), given that these categories in manuscript culture may not coincide in the same person¹⁷⁴ and traditional attributions can be little more than fond fictions. If we are to continue to make progress in the Early China field, we are going to have to do our basic homework when it comes to rethinking the traditional attributions for the received, excavated, and "found" texts we study.

IV. On Dong Zhongshu and Liu Xiang as Omen Experts

For Early China scholars today, the *Hanshu* omen treatise represents one of the chief repositories of Dong Zhongshu's 董仲舒 (d. ca. 107 BCE) thinking, as both the topics of legal judgments and omens (as Heaven's judgments) were said to be dear to the heart of that Confucian master.[175] And yet, as scholars before me have noticed, nearly none of the pronouncements ascribed to Dong in the *Hanshu* omen treatise match the remarks ascribed to Dong in the *Elegant Crown for the Annals* (*Chunqiu fanlu* 春秋繁露).[176] While one can easily "remove" this anomaly by dating the entire *Chunqiu fanlu* compilation, not just the substantial number of Five Phases chapters in it, to the post–Han period,[177] as some are wont to do, modern readers, not unreasonably, still expect to find traces of Dong's omen theories lingering in the writings ascribed to Dong before the Six Dynasties period. How should a modern historian grapple with this apparent problem?

The *Hanshu* omen treatise pairs and contrasts Dong's theories rather consistently with those of Liu Xiang 劉向 (79/8–7 BCE) through a variety of precise formulae:

(a) the two thinkers' ideas are cast as parallel 董仲舒以為 . . . 劉向以為 (two times);

(b) the two thinkers' ideas are conflated 董仲舒、劉向以為 . . . (thirty-one times); and

(c) the two thinkers' ideas are "in rough agreement," even if Liu Xiang's theories dominate the discourse 劉向以為 . . . 董仲舒指略同/指略如 (twelve times).

Of those thirty-one cases of type b where Liu Xiang's theories are said to concur with those of Dong Zhongshu, seventeen concern solar eclipses, and four are fires.[178] (The preoccupation with these events is traditional, judging from the *Annals* and today's *Zuozhuan*.)[179] Notably, only with type c does there appear the momentous phrase 天戒若曰 ("Heaven's warning seemed to say . . ."). Equally notably, this phrasing occurs *only* in the *Hanshu* in the omen treatise, and it appears nowhere else in Western or Eastern Han literature until the very end of Eastern Han, when two authors, Xun Yue and Ying Shao, both well-versed in the *Hanshu*, adopt it for their own omen theories. Probing further, alert readers will have noted that whereas Dong's cautious theories tie the neglect of "rites and music" to instances of political decline during the Chunqiu period, many centuries before Dong's own era, Liu Xiang boldly extrapolates from the records of the antique omens to derive correctives for the political failures in his own dynasty of Western Han—sometimes at the very courts he has loyally served. Possibly under the influence of the omen expert Jing Fang, Liu Xiang was more targeted and less abstract in his readings, whereas Dong often vaguely talked of lapses in "rites and music" that upset "those below" or "those above."[180] In addition, Liu Xiang places particular emphasis on the "end of days" character of the last Lu reigns of Dukes Ding and Ai, presumably to conjure an equally serious threat to contemporary malefactors, in light of omen readings predicting the imminent collapse of Western Han rule.[181] Certainly we learn from Liu Xiang's *Hanshu* biography that Liu Xiang felt it his duty as a senior member of the Han imperial house to use the "Great Plan" *Documents* chapter as well as the *Annals* to offer criticism about the dangers the Wang family posed to the Han imperial house.[182] So even in the instances where the compiler deems the theories of Dong Zhongshu and Liu Xiang to be virtually identical, not-so-subtle distinctions color their styles of omen interpretation.

Hirasawa Ayumu shows a few instances where Dong Zhongshu and Liu Xiang disagreed in their readings, and the omen treatise glosses over their disagreements.[183] Certainly Dong and Liu Xiang agreed in one key respect: all of their omen readings considered *yin-yang*, not Five Phases theories. Additionally, Liu Xiang, given his personal history, was far more inclined to trace ominous events to the consort clans or male favorites than was Dong, whose *Annals* training would have made him more preoccupied with the misappropriation of imperial powers and prerogatives by the local lords and high-ranking ministers. As noted above, Liu Xiang's explication of *The Great Plan's wuxing traditions* (*Hongfan wuxing zhuan lun* 洪範五行傳論) in eleven *pian*[184] clearly borrowed from the Xiahou family's *Documents* interpretations, which treated the *Documents* classic as a repository of discrete revelation texts. By such treatments, all talk of omens was expected to reveal a cosmic or Heavenly charge to the Han rulers to govern well.[185]

In this context we may consider the set formula "Heaven's warning appeared to say" (Tian jie ruo yue 天戒若曰), which occurs no fewer than thirty-one times in the *Hanshu* omen treatise, but without unambiguous attribution to a particular thinker.[186] The question of Heaven's identity remains amorphous as well. In the formula, is Heaven conceived of as single anthropomorphic god reigning over all, or does Heaven simply stand in for the totality of sky powers, including the royal and imperial ancestors residing in the afterlife in the sky plus the complete pantheon of gods? Or is it rather a shorthand way of alluding to the complex operations of the Five Phases? My suspicion: the theology underlying this concept, a variation on the *Documents'* formula *wang ruo yue* 王若曰,[187] is left purposely vague, the better to circumvent the age-old problem of evil and justice, for why would a beneficent deity or group of divinities allow misconduct on the part of those in charge to plague the lives of the innocent in their charge "below"?[188] The formula only implies that the cosmic workings are more than mechanistic (i.e., products of direct relations of cause and effect), being more reliant on sympathies and antipathies, with the harmonies or improprieties at court inclined to induce ripple effects for good or ill throughout all-under-Heaven.

While Dong Zhongshu was known to admonish the ruler about Heaven's warnings,[189] in a tradition that ultimately goes back at least to the *Mencius*,[190] the specific formula *Tian jie ruo yue* seems to have been Liu Xiang's innovation, because the vast majority of the entries employing this formula appear in the summaries of Liu Xiang's theories, and not in connection with Dong Zhongshu.[191] Nor is this formula found elsewhere in Dong's attested writings that have survived (e.g., the three memorials in Dong's *Hanshu* biography).[192] The implication—that Liu Xiang was the more systematic or admired thinker—tallies with testimony recorded elsewhere in the *Hanshu* itself, also with the reliance on ideas and vocabulary that the omen treatise ties to Liu Xiang.[193] Certainly, in one case where the two thinkers' opinions were said to concur, we know that Liu Xiang improved upon the ideas of Dong Zhongshu,[194] with the result that he has the last word. And it was Liu Xiang who was the more aware—or the more outspoken—of the two, it seems, when it came to saying that only Heaven, in the guise of natural disasters or weird occurrences or the historical record of omens and anomalies, has sufficient power to deter a ruler's wrongdoing in many cases. Ergo the stories found in Liu Xiang's major compilations, the *Shuoyuan* 說苑, *Xinxu* 新序, and *Lienü zhuan* 列女傳. This prompts the question of why the omen treatise maintains silence when it comes to certain of Liu Xiang's pet views.[195] Most likely, the Eastern Han compiler made the decision not to underscore Wudi's excesses, because thinkers at the late Western Han courts had already, in the *haogu* 好古 ("passion for antiquity") movement, advertised his reckless disregard for good governance.[196] After all, at points the *Hanshu* lionizes Wudi, chiefly for his patronage of the classicists—contradicting the historical judgments rendered of Han Wudi by the late Western Han reformers who were themselves most steeped in classical learning.[197]

So where does that leave us when considering the role Ban Gu may have played in the compilation of the *Hanshu* omen treatise? Let us backtrack a bit: Bielenstein was basically right, if overly simplifying, when he characterizes the *Hanshu* omen treatise as a powerful tool in the hands of those who would trim imperial power. (I would complicate his assessment, noting that the treatise is a powerful tool in the hands of anyone who would clip the power of any power holder, given how many kings and nobles and elite females elicit bad omens in the *Hanshu* treatise.) By their careful selection and presentation of omens, Bielenstein argued, they could deliver blame to the emperor without incurring inordinate personal risk for their remarks. That so many bad omens are catalogued for the Han argues for a late Western Han date of initial compilation, insofar as officials in that period tended to be most forthright in their criticisms of the reigning emperors.[198] Countering the dangers is Heaven's continuing favor to Han, shown in the steady issuance of extraordinary signs. *If* Ban Gu was the final compiler, he would have needed to do very little to an earlier omen treatise to make this collection of omens provide cosmic witness to the repeated signs of Heaven's favor for the ruling Liu clan, in the manner of his father, Ban Biao.

V. Final Conundrums and Conclusions

The more time one invests in the *Hanshu* "Wuxing zhi," the more apparent its complexities and puzzles become. First, we have no idea why certain events mentioned in the *Annals* and other histories were selected for review in the *Hanshu* omen treatise when comparable events were not,[199] nor do we know why the narrative sometimes reverses the chronological order of events.[200] With so many authoritative omen experts weighing in, it is hard to ascertain which expert's views should be adopted as "more persuasive" when explaining any given occurrence, let alone whether the principal or final compiler of the omen treatise fully intends to convey an overarching message to readers. (For this reason, my tentative hypothesis makes a virtue of this, rather than assuming it to be an inadvertent failing.) Further complicating any analysis, a single portent can represent a response to more than one contributing cause, as when a fire is traced to the building of too many ancestral temples, contra the sumptuary regulations, but likewise to the failure of Lord Ai of Lu to make full use of Kongzi's services.[201]

Portent theories, which gain increasing dominance over the course of Western Han, only come to be fully elaborated in Eastern Han, by nearly every classicist whose work we still have. One need only compare the theories of Dong Zhongshu or Liu Xiang with those of He Xiu 何休 and Ying Shao 應劭 in late Eastern Han to demonstrate the point.[202] The Han classical scholars, whom some today prefer to imagine as steeped in secularism, were continually dabbling in mysteries in the belief (which seems wise to me) that a large chunk of human experience cannot be captured, pace Horatio and big data, by any rational theories. This is the "big picture" not only for the two Han dynasties but also for the immediate post-Han period. In addition, as soon as a thinker posits the existence of a benevolent Heaven, he has to some degree embraced the fundamental part of *Tian ren zhi ying* 天人之應 (exchanges/responses between Heaven and human),[203] by which theory benevolent Heaven issues warnings to those in power and intervenes in sociopolitical affairs to "save" the innocent from massive harm. Simply put, portent theories, which strike us as so profoundly "illogical," have a logic of their own, and long millennia of thinkers in China found them highly compelling (as did and do many of their Western counterparts, of course).

The foregoing exploration of certain technical matters acknowledges multiple questions and puzzles that probably cannot be resolved now, at least by me, chief among them the following:

(1) In the treatise, is there a hierarchy of techniques, with turtle and milfoil divination placed higher than watching the sky patterns (Tianwen) or divining from dreams? Did there exist different hierarchies among the different populations of expert practitioners in real life?

(2) Yan Shigu's 顏師古 (581–645) commentary for the *Hanshu* opines that the key source for many historical events mentioned in the *Hanshu* omen treatise is the *Shiji* 史記 by Sima Qian 司馬遷 (d. ca 86 BCE),[204] but is Yan correct in his surmise, or does the *Hanshu* omen treatise cite many unidentified "historical records" (lowercase *shiji* 史記)? One *Hanshu* entry, for example, is loosely tied in date to a passage in *Shiji* 34.1554, but the analysis offered by the *Hanshu* omen treatise diverges from that in today's *Shiji* text. The as-yet unanswerable questions are, How loose could "citations" be in manuscript culture, and how much do the key texts that we hold in our hand differ from those known to Yan in early Tang times?

(3) While the connection with the "Great Plan" chapter in the *Documents* is obvious, the omen treatise seldom mentions other chapters in the same classic when its analyzes. As there

are but two citations of the Han-era Postface to the *Documents*,²⁰⁵ precursor to today's so-called "Minor Preface,"²⁰⁶ were the Xiahou traditions in the predictive arts premised on this single chapter in the *Documents*?

(4) Can we be sure what value and meaning to assign the *wuxing* 五行 in each passage where it is found? In other words, was the final compiler of the Han omen treatise as aware as modern native-speakers of English must be of the different connotations and traditions surrounding the distinct usages of the term?²⁰⁷

(5) We may note, but we do not always understand the reasons why a certain ominous incident was judged "nearly" or "approximately" (*jin* 近) to fit within a certain category but not to belong to it.²⁰⁸ Further work or newly excavated manuscripts may help us here.

(6) To some degree, many of the omen theories (e.g., those by Dong Zhongshu) posit an interventionist Heaven acting as judge and juror, while others imply an equation between cosmic balance and perfect justice (as in Jing Fang). It is fair to ask, then, what picture of retributive justice does the *Hanshu* treatise presuppose when read against or with Ban Gu's treatise on penal laws in the same history?²⁰⁹ (That chapter opens with Heaven's conferral of the *wu cai* 五才 [Five Material Resources],²¹⁰ just as the omen treatise begins with Heaven's conferral of signs.) Thanks to recent pathbreaking work, we know that the Han administration divided civil from penal law, with each to be adjudicated by separate experts. By the prevailing notions of Han justice (*gongxin* 公心, *yi* 宜 or 義, *ping* 平), assessed from the twin goods of reciprocity (*bao* 報) and hierarchy (*fen* 分),²¹¹ civil laws primarily dealt with voluntary transactions of goods, where compensation typically entailed simple monetary restitution for an unjust loss; by contrast, the penal laws judged involuntary loss through destructive interactions, for which there exists no simple way to determine just requital, because the situation can never be returned to the status quo ante, with murder, rape, maiming, or adultery.²¹² Is

the sheer impossibility of predetermining the proper punishment, absent precise information about the involuntary interaction(s), one reason why Ban Gu supplies a range of authoritative readings as precedents to be weighed by courts when disaster threatens the court? (Ban's treatise on the laws suggests that just decisions in legal cases are not enough; people must made to see how fitting those sentences are, and often enough fit sentences do not rule entirely in favor of the defendant or accept the entirety of the initial report of the crime.)[213]

The eminent *Shiji* expert Shi Ding 施定 has opined that later generations can glimpse Ban Gu's principles and ideas mainly through his "Wuxing zhi."[214] To my mind, the degree to which the omen treatise reflects Ban Gu's personal beliefs remains to be seen. That said, this technical treatise (as one of the *Hanshu*'s major departures from the *Shiji*) unquestionably constitutes a major contribution to the Han court's efforts to study the distant past as history, so as to provide a more accurate mirror for policy debates at the time, a contribution all the more intriguing since its multivalent readings provide a stark contrast to Ban Gu's "Table of Figures, Past and Present," which mandates a single, explicit evaluation of each "personage" in persuasion pieces.

As I see it, the chief contribution of this chapter of mine may be to ask us to recast the omen treatise as revelation text offering testimony on extra-human but quasi-legal judgments. Meanwhile, my chapter queries some earlier assumptions, chiefly the presumption of Fu Sheng's strong influence on omen theories.[215] It further questions the facile transfer of insights from Biblical studies to research on the unseen powers in China, observing that the authoritative writings in classical Chinese never commanded the extraordinary degree of reverence that the scriptures did, as the "[direct] word of [a single] God" who engendered the Good Book shared by Jews, Christians, and Muslims.[216] Even the most sacred revelation texts in Chinese were transmitted through fallible people and interpreted by experts prone to error: priests, wizards, classicists, and so on, as Ban Gu showed.[217] (One notable feature of the *Annals*, from which so many omen records are drawn, is that its very text and commentaries drew attention to the fact that reasonable people might dispute its pronouncements.)[218] One may hope that continuing research on Liu Xin's and Ban Gu's place in Han history will lead to a more fine-grained approach regarding their roles in the world-making glimpsed in this *Hanshu* chapter.

Appendix: A Brief Think Piece on Resonance Theory

For nearly a decade now, I have sought to better understand resonance theory because practically all Han-era writings are predicated on it, and particularly those we tend to label as "technical writings." Spurred by Nathan Sivin, who once remarked that new scientific knowledge was often parked in the apocrypha, as well as by Grègoire Espesset's painstaking work on that set of texts, I began to make a bit of headway, thanks in large part to seemingly unrelated work on Han exegetical traditions. (Let this be a lesson to all of us undertaking research: frontal attacks on a problem often prove far less useful than wide roaming.) I write this, knowing that future scholars will doubtless improve upon my musings, possibly by overturning them. Still, the fundamental insight informing my mode of thinking—that the early Chinese thinkers were often far more sophisticated than we give them credit for—has always been proven right.

As I see it, the central problem with resonance theory is one of scale: how does a single person (if generally the ruler) so affect the cosmos that the order exhibits visible changes? Seen and unseen must be factored into the equation, with few changes entirely legible transfers of *qi* in the *quid pro quo* manner. Why does imbalance in one aspect of the universe manifest itself in another in different form and frame? I have begun to analogize this to wave theory (light wave or electromagnetic), where one wave (albeit present) may appear to cancel or be canceled by another wave, so that no obvious change occurs. (From high school physics classes one can imagine the complex impact of multiple waves.)

Nathan Sivin's seminars taught me about Jupiter (aka the Year Star, *suixing* 歲星) and the invisible counter-Jupiter in Chinese astronomy that passes through *Sheti ge* 攝提格), deemed responsible for some irregular motions of the five visible planets, which do not always travel neatly around their orbits, but instead can be stationery or retrograde.[219] Note in this connection the continual interplay of visible and invisible here, whose interactions intermittently become visible at due times. Next I think of the *Annals* classic, believed in Han to be operating simultaneously on two levels: the visible world of events (which ultimately become invisible as history) and the ideal, visionary world as yet unrealized. As Gary Arbuckle, Joachim Gentz, and Newell van Auken have said,[220] by the *Gongyang Tradition* (*Gongyang zhuan*), the Master of legend had two antithetical aims in compiling his history: he wanted the history at once to accurately reflect past events, and to show the ideal principles that *should* operate in history.

To accomplish both antithetical aims simultaneously, the language of the *Annals* compiled by Confucius had to be perfectly gauged to convey both overt truths and subtle messages (*wei yan* 微言) unnoticed by all but the most adept. And because historical events have the force of precedents, they operate in present and future, in visible and in unseen ways. The expectation, sometimes made explicit, was that only highly trained specialists can decode the hidden message and thereby extract a correct interpretation for present and future, often by "taking deviations from an ideal pattern as expressions of [significant] meaning," following well-established divinatory and juridical traditions.[221] Again, real and ideal worlds are intertwined in the same arena or plane of existence, with the cosmic and sociopolitical confounded. We meanwhile confront what appears to us, the untrained, as inexplicable, strained, and incoherent,[222] for example, the sheer impossibility of keeping the actual events under Zhou correlated with what He Xiu, the *Annals* exegete, calls "New Zhou" 新周 (the ideal dispensation under which future dynasties are to rule).[223]

One final example comes readily to mind: in the mid-1930s, Gu Jiegang 顧頡剛 and Tong Shuye 童書業 tried to account for discrepancies among the different accounts of the Nine Provinces, the territorial borders of the realm under the sway of elites whose writings used Chinese script.[224] The "Tribute of Yu," for example, imagines a vast empire, but since it focuses on the waterways of China, and secondarily on mountains, its frontiers remained credible, even as over time new geographical knowledge came to light. By contrast, the picture presented in the "Terrestrial Terrain" chapter of the *Huainanzi* locates phenomena within two nested sets of eight regions (*ba hong* 八紘, *ba ji* 八極) placed "beyond" those "distant" eight regions (*ba yin* 八殥, probably still inhabited by mortals) that lie "beyond" the Nine Provinces. What happens in the furthermost "beyond-beyond" lands matters a great deal to ordinary people, for it is those lands that supposedly generate the climate and weather upon which sedentary farmers and the early empires depend. And yet some significant features of the "beyond-beyond" plainly lie outside the mundane realm: Heaven's gates (e.g., Changhe 閶闔 Gate), for example, and the outer edge of the cosmos (Buzhou 不周 Mountain),[225] not to mention the realm of the dead (Gate of Darkness 幽都). So while a few names listed in both the "Tribute" and "Terrain" chapters coincide (e.g., Youdu), their referents must differ, for all place names in the "Tribute" chapter refer to earthly sites, whereas this is not the case with all place names listed in "Terrain." For Gu and Tong, the primary interest of their work lay in correlating the names for sites with fixed geographic locations;

John Major says nothing in this connection.[226] My interest lies elsewhere, for here again the verbal maps invite us to see the ways in which the interplay between the seen and unseen worlds affects the palpable in the real time and space. Put another way, we discover Han thinkers mapping the real (to us) onto the imaginary (to us), weighing the readily calculable factors against the unpredictable and the unknown.

I write this as my university has closed all libraries for the foreseeable future, due to the pandemic. There is much more work to undertake, but for now I perforce embrace the epistemological modesty that undergirds the writings by the smartest and most erudite thinkers of the two Han dynasties, with respect to omens and other vital matters.

Notes

1. *Hanshu* 82.3371, said by Wang Feng 鳳. I dedicate my essay not only to Michael Loewe, but also to three other teachers of mine: Nathan Sivin, Paul Serruys, and William Hung, all of whom gave me a taste for the technical.

2. For Bielenstein, see "Interpretation of the Portents" and "Han Portents and Prognostications." For Eberhard, see "Political Function of Astronomy and Astronomers." For Kern, see "Religious Anxiety and Political Interest." For Yang Shao-yun, see "Politics of Omenology." Hans Bielenstein ("Interpretation of the Portents," 128, 143), noted, there were "many eclipses which are not recorded by Ban Gu, in spite of the fact that they were often clearly visible." (Bielenstein discounts local conditions entirely.) Upon careful investigation, Bielenstein concluded that "all the portents recorded in the *Ts'ien-Han-shu* form a homogeneous material, influenced by one and the same motive power," namely "the desire to level indirect criticism against the ruler." This line of reasoning has been followed to this day by nearly all scholars, including Kern.

3. See Fan Ye, on Ban Gu's biography, in Nylan, "Historians Writing about Historians."

4. Schwermann ("Anecdote Collections") has shown that even where large parts of a work derive from a second work (as some 60 percent of the anecdotes in *Shuoyuan* have counterparts from *Hanshi waizhuan*), the second work is not necessarily derivative; its compiler has his own distinct style and points to register.

5. As a historian who usually insists on distinguishing Western Han from Eastern Han, not to mention specific administrations within each dynasty, I would emphasize that attempts to deepen our understanding of our sources will inevitably fail unless we situate those compilations as precisely as possible within specific times and places.

6. These crop up in connection with the *Odes* most often when discussing failings in speech or song; see *Hanshu* 27A.1339–40 (twice), 27B(a).1358, 1360, 1362, 1373; 1376–77 (three times), 1394, 27B(c).1405 (twice); 27C(a).1451, 1465, 1468, 1472, 1476; 27C(c).1494 (three times, in connections with eclipses?), 1511. NB: the whole segment of the treatises devoted to *shi yao* 詩妖 (sprung from yang *qi*, manifested as locusts 陽氣所生也, 於春秋為螽, 今謂之蝗, 皆其類也).

7. *Hanshu* 75 is a collective biography of six omen experts: Sui Hong/Meng 眭弘/孟, Xiahou Shichang 夏侯始昌 and Xiahou Sheng 夏侯勝 (the "Two Xiahous"), Jing Fang 京房, Yi Feng 翼奉, and Li Xun 李尋. For example, Sui Meng's expertise and theories are cited in Hanshu 27A.1317, 27B(a).1400, 27B(c).1412; Xiahou Sheng appears in *Hanshu* 27A.1317, 27B(a).1367, 27C(a).1459. Many of these experts also figure in the "Rulin zhuan" chapter; for example, Sui Meng is called a "great master" of the *Gongyang* (*Hanshu* 88.3617).

Note the chapter begins with two figures tied to the later period of Wudi's reign, but the significance of Xiahou Shichang, Xiahou Sheng, and Sui Meng is tied up with the deposal of Changyi wang/Haihun hou. Xiahou Shichang was Senior Tutor to Liu He "when young" (i.e., before 74 BCE). Xiahou Sheng early on condemned the failings of the new emperor. *Hanshu* 75.3155.

8. For example, Yang Xiong's reputation as poet outstripped his reputation as omen expert, and Liu Xiang's quadruply honored standing as noble scion, minister, major literary figure, and director of the imperial clan outweighed his reputation as omen expert, no matter how important that role was to his reputation at any given time.

9. In *Hanshu* 72, we have biographies for Wang Ji 王吉, Gong Yu 貢禹, the two Gongs 兩龔 (Gong Sheng 勝 and Gong She 舍, two friends from Chu), and Bao Xuan 鮑宣. Wang Ji is said to have been well versed in all Five Classics and to have seen that his son was taught by Liangqiu He, the *Changes* expert. Gong Yu was on the lookout for disasters and anomalies (*Hanshu* 72.3071). Gong Sheng talked of cosmic imbalances after injustices had occurred (*Hanshu* 72.3081). Bao Xuan "read" an eclipse as Heaven's warning against Dong Xian. Even the very best of the last Western Han officials in Ban Gu's view (the subjects of *Hanshu* 71) shared an interest in omens, as when Yu Dingguo (*Hanshu* 71.3045) considers disasters and anomalies, and Ping Dang considers the potential cosmic fallout shown through disasters and anomalies (*Hanshu* 71.3049).

Usually, it is wise not to posit a "public/private" divide in early China, although a "court/domestic life" divide exists.

10. Ban Gu here puts together two phrases from the "Shuo gua" of the *Changes* (昔者聖人之作《易》也, 幽贊於神明而生蓍, 參天兩地而倚數) and the *Annals* (子產曰, 天道遠, 人道邇, 非所及也, 何以知之), which describe the work the sages did in matching Heaven and earth. He also evokes Yang Xiong's *Taixuan jing*, which made similar rhetorical moves.

11. *Analects* 5/12.

12. We know little about Tian Zhongshu, except that he, like Liu Xin, studied with the master Zhai Fangjin, who was also chancellor for the Han court. See *Hanshu* 84.3421, which says Tian was a calendrical expert.

13. *Hanshu* 75.3194.

14. Contrast *Hanshu* 56.2515, which has Dong Zhongshu operating from this single starting point in the Dao.

15. *Hanshu* 75.3195.

16. Literally, "shallow and deep."

17. Again, we have the repetition of the "Shuo gua" phrasing, to which this time is added a reference to "Xici, xia" 繫辭下: 又明於憂患與故 … 苟非其人, 道不虛行.

18. Tentative translation, given *Analects* 2/18.

19. *Hanshu* 100.4261.

20. For those who do not know Chen Lai's writings, he would style all "Confucian" thinkers secular humanists, although this construction is deeply anachronistic for the antique period.

21. *Hanshu* 75.3154.

22. *Hanshu* 28B.1645 is the other instance, in Ban Gu's geographic treatise.

23. In English, the two standard works remain Queen, *From Chronicle to Canon*, and Loewe, *Dong Zhongshu*.

24. *Hanshu* 56.2515, where the Chinese reads: 故春秋之所譏, 災害之所加也; 春秋之所惡, 怪異之所施也。書邦家之過, 兼災異之變, 以此見人之所為, 其美惡之極, 乃與天地流通而往來相應, 此亦言天之一端也.

25. *Hanshu* 72.3097, with the last line citing *Analects* 15/36. Even with the citation, this must be a tentative translation, for the last line could also mean "firm in his principles but not stubborn to a fault" (as Legge, *Documents*, 305, reads it), even "firm in his principles but not closed-mouthed" (as per the *Documents* classic). The passage goes on to discuss men of standing who were either willing or unwilling to serve Wang Mang: "Guo Qin and Jiang Yu preferred to flee the filth. In this they were utterly different from Ji Yuan and Tang Lin and Tang Zun, who served Wang Mang." For this last group of characters, see *Hanshu* 99C.4149. Tang Lin 唐林 and Tang Zun 唐尊 ("the two Tangs" 兩唐) appear in *Hanshu* 72.3095, 86.3485.

26. In using this phrase, I do not deny that what constitutes the "public interest" (aka the common good) was hotly debated during the two Han dynasties (as today): some argued that it was in the public interest to serve the ruler's interests, while others insisted that the public interest was best served by courts acting in the best interests of the lowliest subjects of the realm, regardless of the ruler's avowed aims. Mark Csikszentmihalyi notes that besides Xue Fang who is credulous, several other people are credulous, including Wang Mang.

27. Ban in *Hanshu* 72.3056–57 alludes to Yang Xiong's *Fayan* 6/19 regarding Yan Junping, his teacher: "Yan of Shu would not act improperly just to gain

visibility, nor would he become a bureaucrat for unauthorized gain. Long years he spent in obscurity, yet he never changed his principles. How could even Marquis Sui's pearl and He Bian's jade be of greater value than this? We should uphold such examples as standards. Were they not rare treasures also?"

28. This refers to a complex omenological theory injected into *Odes* theory by the end of Western Han; it does not appear to have been there earlier.

29. But, as Gentz ("Confucius Confronting Contingency," 65) reminds us, in only two of the "approximately 140 instances" does the *Annals* classic itself comment or reflect upon the cause of the calamities and ominous events.

30. *Hanshu* 75.3172.

31. See Lee Chi-hsiang's chapter in this volume for the fictive and real relation between Luo and ChengZhou.

32. Gu Yong's memorials are the subject of Liu Tseng-kuei's extraordinary essay "A Study of Gu Yong's Three Troubles Theory."

33. *Hanshu* 75.3153–54.

34. Their condemnation may have been so vociferous because their grandfather, Ban Zhi 穉, in particular, had been part of the circle of Wang Mang's admirers in late Western Han. See Clark, *Ban Gu's History*, esp. 79–82.

35. Presumably, because his clansman (his uncle?) Xiahou Shichang had been Senior Tutor to Changyi wang, he was alerted to his alleged depravities.

36. Xiahou Sheng had been trained in many classical traditions, but most especially the *Documents* classic (both the Ouyang and Xiahou traditions) and the *Hongfan wuxing zhuan* 洪範五行傳. See *Hanshu* 75.3155.

37. *Hanshu* 75.3159.

38. *Hanshu* 27A.1317, 27B(a).1353, 27B(a).1367, 27C(a).1459. Xiahou Sheng was also an expert in the rites, and he paid particular attention to clothing. Both Xiahou Shichang (who compiled something titled or related to the *Hongfan wuxing zhuan*) and Xiahou Sheng are named as *the* experts on certain interpretations regarding bearing (*mao* 貌) (1353).

39. Ban was hardly alone in this way of thinking. Germane here is the *Bohu tong* section devoted to "remonstrance" (*jian* 諫), which seems to register the court consensus of the time.

40. Su, *Hanshu Wuxing zhi yanjiu*, 155; Loewe's *Biographical Dictionary* says "over sixty" (200). Jing Fang is the subject of a recent book by Guo Yu, *Jing shi Yi yuanliu*.

41. I say this because the order in which Ban Gu lists the Five Classics varies. In Eastern Han it is by no means standard, though it was to become standard in the post-Han period. The Western Han order put the *Changes* third, after the *Odes* and *Documents*, based on their presumed age of transcription.

42. Under Yuandi, Ban Gu's maternal grandfather, Jin Chang 金敞, was working hand-in-glove with Liu Xiang, Xiao Wangzhi, and Zhou Kan, against the powerful imperial favorite Shi Xian. Again, Ban Gu's disdain doesn't seem to have

been motivated by a disagreement over policies; instead, it was a philosophical disagreement over how to conduct court discussions.

43. An atmospheric effect related to constant yin, which does not dissipate; a threat to yang *qi* and so to the ruler, also a swipe at Shi Xian as the imperial favorite.

44. *Hanshu* 75.3164–66, dated Jianzhao 2, second month, and later (3 memorials altogether).

45. I note that Liu Xiang sharply distinguished the *Changes* teachings that derived from Tian He 田和 from teachings traceable to Jing Fang. Loewe (*Biographical Dictionary*, 199–200) rightly notes that there are two Jing Fangs, which fact undoubtedly complicates our interpretations of the distant past.

46. Possibly relevant in this connection are Dong Zhongshu's omen theories, because Ban Gu approved so much of Dong Zhongshu. Dong, by his theories, attempted to persuade Wudi that he could become another sage-king Yao, and Dong had seen history as moving in a two-stage, six-part cycle. Yao had inaugurated a first stage of order (where the throne passes according to merit), while Yu's decision to pass the throne on to his son inaugurated the Three Dynasties of Xia-Shang-Zhou (all of which relied upon hereditary succession). As both Stages 1 and 2 divide into three, by Dong's theory, Dong's Western Han court was at the beginning of a whole new cycle.

47. I use these terms because that is what the Eastern Han sources use, although Hans Bielenstein's magisterial three-volume work, "The Restoration of the Han Dynasty," amply demonstrates how inappropriate those terms are, given that most officials who served Guangwu and Mingdi had once served Wang Mang, and the dynasty was never really "restored" to its former glories. Instead, a strong oligarchy began to rule from the time of the Eastern Han founder, a distant descendant of the Western Han ruling house who functioned as *primus inter pares*, except for certain ritual matters. Doubtless that is why the Eastern Han focused so much on imperial rituals, which many historians have (mis)taken as proof of their profound interest in Confucian moral precepts.

48. *Hanshu* 100.4261. Jack Dull asserts that a "feeling" had been "developing for quite some time that the [Western] Han house was about to end," but this is to simplify Ban Gu's complex analysis (see Dull, "Historical Introduction," 305). Certainly, Sui Hong and Gan Zhongke had predicted the downfall of the Western Han.

49. Trenton Wilson and I have coauthored a forthcoming essay titled "A Brief History of Daring" that discusses *shan* and *zhuan*. We are currently writing a longer monograph titled "An Early History of Daring," Ban Gu's thoughts on when and how to dare to remonstrate figure largely in both works.

50. This eventually became Nylan, "Mapping Time," for a special volume on early tables.

51. The work on the Tables and on the *Suishu* omen treatises is now published, as Nylan, "Mapping Time," and "The *Suishu* 'Wuxing zhi.' "

52. E.g., Zhou Zhenhe, *Xi Han zhengqu dili*.

53. Nylan, "Manuscript Culture." Similar points have been raised by Charles Sanft and many others.

54. I give but one example of the importance of placement within the two histories, even when it comes to accepted "facts": the *Shiji* tells the story of Dowager Empress Lü's illness in her "Basic Annals" chapter, but the *Hanshu* treatise treats her fatal illness as an omen, rather than as a necessary part of a chronicle that details her rule of the empire.

55. The five-*juan* work occupies 181 pages, from *Hanshu* 27A.1315 to 27C(c).1506. The next-longest work (Wang Mang's three-*juan* biography) occupies 54 pages, from *Hanshu* 99A.4039 to 99C.4194. Obviously not all pages are equal in length, but this gives us a rough idea of the length of the omen treatise. Although *Hanshu* 26 ("Treatise on the Celestial Patterns") is likewise an omen treatise, as Chapman's chapter in this volume shows, when I speak of the "omen treatise," I refer solely to *Hanshu* 27, if only because "Treatise on the *wuxing*" is a mouthful.

56. One would expect Lisa Raphals's book on divination, for example, to discuss this omen treatise, but it does not (pp. 50, 91, 347).

57. See Su Dechang in the note above. Meanwhile, Su's work underscores much of Liu Tseng-kuei's work on omens and charts, another marvel of erudition.

58. See Mansvelt Beck, "Treatises of Later Han." The thesis was later published as a book by Brill.

59. *Hanshu* 27B(a).1353.

60. See Nylan, "Yin-yang, Five Phases, and *qi*." For the incoherence of talk about "correlative cosmology," please consult the appendix to that piece.

61. See Miu Fenglin, "*Hanshu* 'Wuxing zhi' fan li." Miu's work is reviewed in Xu Xingwu, *Liu Xiang ping zhuan*, a good introduction to the two Lius, father and son.

62. For the formula *yi yue* 一曰, see *Hanshu* 27A.1320, 1324 (twice), 1329, 1330, 1334, 1343, 1351, 1353, 1366 (twice), 1370, 1374, 1377; *Hanshu* 27B(a).1384, 1391 (twice), 1400; *Hanshu* 27B(b).1406, 1407, 1410, 1411 (three times), 1412 (twice), 1416, 1419, 1421, 1425, 1426, 1430, 1434 (three times); *Hanshu* 27C(a).1441 (three times), 1442 (three times), 1447, 1450, 1452, 1455, 1458 (twice), 1459, 1460, 1465, 1471 (twice), 1472, 1473 (three times), 1477; *Hanshu* 27C(b).1494, 1498, 1511, 1514, 1516 (twice), 1517. For the formula *huo yue* 或曰 (some say), see *Hanshu* 27A.1320, 27B(b).1416, 27C(b).1508, 1516. (A few additional objections are registered in the commentaries to the omen treatise. By contrast, in the *Suishu* omen treatise, *all* authorities for the omen readings are cited by name.)

63. See *Hanshu* 27B(c).1354; 27B9c.1406 非是.

64. Section 9 details the Five Blessings that ensue from virtuous conduct. However, the *Hanshu* does not always cite a Blessing that followed an ominous event. For an example where it does, see *Hanshu* 27B(a).1377; and where not, 27B(a).1352.

65. At this remove, it is not always possible to tell, but Su Dechang, *Hanshu Wuxing zhi yanjiu*, chapter 7, also registers his idea that the merging of *wuxing*

as Five Resources and as Five Phases is incomplete and inelegant, suggesting the difficulties of combining the two fundamentally disparate theories. As Su writes, to adapt an old system stipulating Five Material Resources to a new system about Five Phases *qi* was far from easy. The main Five Phases theorizing comes up in the analysis in the fifth *juan* discussions of *wuxing xiangli* (where one or more type of *qi* harms another).

66. The former category is extensively treated; the second appears twice in the *Hanshu*, once in the omen treatise (*Hanshu* 27B(c).1411, 1411n6), and once in the omen expert Gu Yong's biography (*Hanshu* 85.2450).

67. Mansvelt Beck, "Treatises of Later Han," 155.

68. As the identical term represents the Five Material Resources on earth (as it does in both the "Great Plan" chapter and in Jing Fang's writings), long before signifying the Five Phases, should we not regard the standard translation of the title as "Treatise on the Five Phases" as deeply misleading? For a recent study on Jing Fang (d. 37 BCE), showing his usage of the term, see Guo Yu, see note 40 above.

69. Hirasawa, "Kansho gogyō to Ryū Kō Kōhan gogyō ten," 25.

70. *Hanshu* 27A.1316 defines *huangji* in its own way, as a mode of governance that encompasses the Three Treatments (section 6 of the "Plan") and Five Modes of [Virtuous] Engagement (section 2): 黃極統三德五事. One could argue that this definition is implied by the "Hongfan" chapter, but I would argue otherwise.

71. These remarks accept Liu Xin's new theory equating the "He tu" with the "Luo shu," and thus the *Changes* with the *Documents* traditions. There are some oddities with Ban Gu's summary of Liu Xin's theory (as discussed below).

72. See, e.g., Saso, "What is the Ho-t'u?" and Nylan, *Shifting Center*.

73. Niu Jingfei, "Lun Handai Xibu bianjiang shang de 'Yu gong' diming."

74. Revelation texts appear to some scholars to signal ruptures. Espesset ("La religiosité en Chine") argues strongly that all revelations occasion ruptures. Certainly, this is true in the case of the Christian revelations, which in theory supersede those of the Old Testament figures, but I wonder whether this holds true in early China.

75. *Hanshu* 27A.1316 reads: 箕子對禹得雒書之意也. . . . [Liu Xin] 以為河圖、雒書相為經緯，八卦、九章相為表裏.

76. See *Shiji* 2.49. Evidently the controversial association stemmed from the story cycle that had Yu's successful flood-quelling activities following the Yellow River water systems.

77. For the *Suishu* objections to Liu Xin's theories, see *Suishu* 32.948. Worth review is You Ziyong, "Shi lun zheng shi Wuxing zhi de yanbian." *Tongzhi* 通志, *juan* 14.1905, ascribes this theory to Liu Xiang, instead of his son Liu Xin; it is the outlier here.

78. The earliest record purports to interpret a "dragon omen" that it traces back to the entirely legendary Xia dynasty (tradit. 2205–1766 BCE).

79. *Hanshu* 27B(c).1458 begins the *huangji* section. In yet another sign of adaptation, Su Dechang's chart in *Hanshu Wuxing zhi yanjiu*, 111, shows that the Six Extremities are not listed in the same order in the *Hanshu* "Wuxing zhi";

drawing apparently upon Liu Xin's "Wuxing zhan," they are consequently assigned to different characteristic failings.

80. The Five Resources and Five Modes of Engagement correspond to sections 1 and 2 of the nine-section "Great Plan." *Huangji* is section 5 in the same work. Section 8 consists of "Signs" in the form of climate, weather, and minor irregular events in the sky. See below.

81. Revelation texts appear to some scholars to signal ruptures. See note 74 above.

82. On the Five Phases as a late Western Han innovation, see Nylan, "Yin-yang, Five Phases, and *qi*."

83. As when only the Five Phases in operation *li* 沴 ("destroy") each other singly and in groups.

84. This, of course, has serious implications. The Generative Order supposedly may be traced back to Dong Zhongshu, who preferred it to the older Conquest Order. See also note 46 above.

85. For the changing reputation of Dong Zhongshu over the course of the two Han dynasties and beyond, see Michael Loewe, *Dong Zhongshu*.

86. In the "Wuxing zhuan," three terms signal blame (*jiu* 咎, *fa* 罰, and *ji* 極) in stages that are increasingly baleful, moving from situations that can be read and corrected through misbehavior that requires concerted reforms to inevitable (or nearly inevitable) calamities.

87. Elsewhere these vocabulary items are used with considerably more inexactitude, as pointed out by Su Dechang, *Hanshu Wuxing zhi yanjiu*, 594. Su presumes the sequence of stages cannot predate Liu Xiang and possibly dates to the time of the final compiler.

88. Here Chengdi's inability to bear healthy children is one relevant example.

89. *Hanshu* 27C(a).1448. Judging from the *Shuowen*, this technical language may reflect the *Hongfan wuxing zhuan*. See Yang Tianyu 2011, 91.

90. Mark Csikszentmihalyi comments (personal communication, August 2020), "This is not Evans-Pritchard's view of the Azande having a view of a magical 'second spear' that explains when something happens that can't be explained by physics, but rather a notion that there is a certain amount of noise built into the system." Yes, precisely.

91. One may compare, for example, *Hanshu* 27 for omens relating to Chengdi's reign (e.g., that for 建始元年, reported in *Hanshu* 27A.1336), with the counterpart passage in *Hanji*.

92. *Hanshu* 27B(a).1353.

93. The treatise posits the superiority of the Central States over the "barbarian" groups (mostly semi-nomadic), which are once described as being "of different stock," species, or category (*yi lei* 異類), in *Hanshu* 27B(a).1398.

94. Su Dechang, *Hanshu Wuxing zhi yanjiu*, 103, for the first observation; Nylan, for the second. This, of course, is a hope expressed down through the ages (e.g., Lü Nangong, *Guanyuan ji*, 5/9b, 8/13b–15b).

95. Espesset, "Portents in Early Imperial China," 29, 29n117.

96. Here the work of Gerard Colas on divination as a discipline has influenced me, for Colas shows how porous are the semantic boundaries between sign and cause in many languages. In Indian terms, for example, single words can mean "sign," "mark," or "causes" (linga, nimitta, laksana). Colas notes a "certain disrespect toward icons," which he thinks is intended to cast doubt on divination as producing reliable signs. I argue below that Ban Gu's seeming indecisiveness when it comes to choosing among rival omen theories may be rooted in his habitual regard for court deliberations. Note that I say "may." At this remove, only a fool would claim certainty.

97. See, e.g., *Hanshu* 27B(a).1352, where the binomial expression occurs five times on a single page. The phrase 時則 is mentioned twenty-one times in the *Hanshu* omen treatise, in connection with every stage of omen production, usually in clusters of five or six occurrences. *Hanshu* 27B(a).1353 clearly defines the expression, as something like "betimes" (not always occurring, occurring in unpredictable ways).

98. *Hanshu* 100.4243, cited by Wang Xianqian. Cf. *Hanshu* 27A.1316, for one example.

99. *Hanshu* 27B(b).1409 presents one omen reading for several events under Wudi's reigns: disastrous wars of aggression, but also gifts to the underprivileged and disadvantaged, the last of which ended in "all the realm feeling delight" (*tianxia xian xi* 天下咸喜). Auspicious events are rarely mentioned in the omen treatise, but this instance is not unique. *Hanshu* 27B(b).1412 lists another "mixed" omen, citing the initial investiture of the King of Changyi 昌邑 as Han emperor, followed by his deposal, and the installation of a more suitable replacement, Liu Bingyi 病已 (the future Xuandi).

100. As readers may recall, Ban Gu's father, Ban Biao, asserted in his "On the Destiny of Kings" (Wang ming lun 王命論) that the dynastic fortunes did not go to the merely lucky, but only to those with the right heredity and moral standing. This piece has been translated in de Bary and Bloom, *Sources of Chinese Tradition*, vol. 1 (all editions).

101. Wang Mingsheng 王鳴盛 (1722–1798), in *Shiqi shi shang que* 十七史商榷, *juan* 13, theorized that when the omen treatise cites the *jing* 經 ("the classic"), it refers to the *Documents* chapter itself; when it next cites *zhuan* 傳 ("tradition"), it refers to Fu Sheng's *Hongfan wuxing zhuan*; and when it cites what "the explication says" (*shuo yue* 說曰), it refers to the court-sponsored Ouyang and Xiahou interpretations (i.e., official learning at the court). Compare the 1984 study by Wang, in the second volume of *Ershiwu shi bubian*. Certainly, the word "explication" is often linked with the Academicians' court-sponsored teachings. For details, see the Shanghai shudian 2005 edition of Wang's work, punctuated by Huang Shuhui. Wang Xianqian followed Wang Mingsheng's assumptions. More recently the scholar Chen Kanli, in his book *Ruxue, shushu yu zhengzhi*, argues that the "explications" were based, in the main, if not in toto, on Liu Xiang's *Hongfan wuxing zhuan*.

Hirasawa, among others, disputes this, saying the treatise is too multilayered and complex to convey one message from Liu Xiang.

102. *Hanshu* 27A.1318, following Yan Shigu's glosses, which say, 田獵不宿, 飲食不享, 出入不節, 奪民農時, 及有姦謀.

103. Alternate explanations for the single case of "wood freezing" are given in *Hanshu* 27A.1320. The point is, however, that none of the alleged political misdeeds had anything to do with either the *zhuan* or the *shuo* preceding the entry particulars.

104. The phrase *zhuan yue* 傳曰 ("the tradition says") probably refers to the teachings of Xiahou Shichang 始昌, transmitted to Xiahou Sheng 夏侯勝, and then on to other *Documents* experts.

105. See Su Dechang, *Hanshu Wuxing zhi yanjiu*, 207.

106. Su Dechang (*Hanshu Wuxing zhi yanjiu*, 592–94) makes this point.

107. *Hanshu* 27A.1346 rails against "a female ruler who is governing on her own" (*nü zhu du zhi* 女主獨治), and reports of her misconduct follow, offering a far less balanced view of her reign than that given in the *Shiji* "Basic Annals."

108. See, e.g., *Hanshu* 27A.1331, 27B(a).1457. The compiler casts Jia Yi not as a brilliant advisor but as a troublemaker, possibly a rebel. The same is said of Chao Cuo 晁/鼂錯.

109. Mention of the usurpation of power by Huo Guang and outright treason by his wife and family members begins with *Hanshu* 27A.1335. Cf. *Hanshu* 27B(b).1409.

110. Mention of misgovernment under Yuandi's administration begins with *Hanshu* 27A.1335, 1347.

111. See, e.g., *Hanshu* 27B(a).1368.

112. See, e.g., *Hanshu* 27B(b).1416.

113. Mention of the Wang family usurpation begins with *Hanshu* 27A.1334, 1336–37, and continues throughout the omen treatise. See, e.g., *Hanshu* 27B(a).1370, 1395 (with the latter mentioning Wang Mang's birth during Yuandi's reign). For Xuandi's and Yuandi's part in the Wang family rise, see *Hanshu* 27A.1336, which concurs with *Hanshu* 98.4014–15.

114. See, e.g., *Hanshu* 27A.1327, *Hanshu* 27B(a).1391, 1393. *Hanshu* 27B(a).1396 characterizes the Xiongnu incursions on the Central States as contact with *yi lei* 異類 (of a different species or type, not comparable). The Central States are said to be *yang* 陽 to the semi-nomadic groups, who are yin 陰, in *Hanshu* 27B(b).1414.

115. *Hanshu* 27A.1334.

116. Mansvelt Beck, "Treatises of Later Han," 155. Needless to say, I part company with him when he says that "Ban Gu constructed a system."

117. That said, it is true that the lightning omens are mainly blamed on women or eunuchs, because Liu Xiang associates "bad behavior connected with women" with fires/Fire.

118. In Ban Gu's own time, we learn from HHS 79A.2554, one Wei Man 魏滿, Governor of Hongnong during the reign of Mingdi (r. 57–75), was a cele-

brated expert in Jing Fang learning. Wei is mentioned nowhere else in *Hanshu* or *Hou Hanshu*, though this may signify nothing.

119. Liu Zhiji, *Shitong*, *juan* 19.10/1ff. contains many of these criticisms (see note 131 below). However, the "Tianwen zhi," compiled by Ban Zhao, Ma Xu, and possibly Ma Rong, roughly twenty years *after* Ban Gu's death, treats solar eclipses quite differently, presumably to avoid duplication (see *Hanshu* 26.2391 [3x]; 26.1300 [2x]). Ban Gu mentions ten comets and Ma Xu, an additional twenty-two, for a total of thirty-two.

120. We have plenty of evidence of fierce and open debates in Han times, though some presume the court debaters were irenic. For abundant evidence, one may the consult the (forthcoming) paper by Nicholas Constantino and Michael Nylan ("On the Rites") on rites controversies during Zhangdi's reign (i.e., precisely when Ban Gu is compiling his history). This does not mean that ambiguity is *not* a feature of the epistemic realm here. It could be that *any* past occurrence in the category is an admissible precedent and therefore a possible outcome of a future occurrence.

121. For Ban Zhao and Deng Sui, see Nylan and Wilson, "Dowager Empress Deng Sui" (2021).

122. Michael Loewe is currently writing a discussion on Kongzi's role in Western and Eastern Han, but already, in *Men Who Governed*, chap. 10, he showed that it was only in the first years of the first century that regular sacrifices to Confucius were made, under imperial patronage. That the *Hanshu* omen treatise includes so many pre-Qin omens only underscores the morality of listing such warnings in writing, for it was a prime contention of the Eastern Han court, in particular, that Kongzi had ingeniously devised the "praise and blame" style for omens in the *Annals* classic to provide a blueprint for Han imperial policy-making and thus a preemptive set of warnings for the Eastern Han.

123. Like the *Siku* editors, I question the traditional ascription of the current *Bohu tong* to Ban Gu. However, the *Bohu tong* compilation must date to the milieu in which Ban Gu operated. On omen blame, see Kageyama Terukuni, "Kandai ni okeru saigi to seiji."

124. For this reason, Su Dechang offers another idea (see below).

125. This incident and the three dominant explanations are given in Mansvelt Beck, "Treatises of Later Han," 152. Many other examples of conflicting expert readings could be adduced.

126. However, we can see that Dong does not invariably represent the dominant *Gongyang* readings in the *Hanshu*'s second example of lightning striking a building, where Dong's view is not identical with the dominant tradition. On the other hand, some of the opinions registered by Liu Xiang must represent *Guliang* traditions, because we find them so labeled in Xu Shen's *Wujing yi yi* 五經異義.

127. But Liu Xiang adopted the Xiahou readings for the *Documents*, including the "Great Plan," if tradition is correct. See below.

128. See *Hanshu* 27A.1322–23.

129. This is why this chapter consistently speaks of omens, as many come after the event, rather than portending it.

130. For this reason, Su Dechang (*Hanshu Wuxing zhi yanjiu*, 144) reviews disparate views regarding an ominous event blamed on an emperor's favoring of Shi Xian at the expense of loyal officials at his court. *Hanshu* 27B(a).1393 lists an ominous occurrence during the reign of Chengdi, which it does not analyze; there are only a few comparable cases.

131. For Liu Zhiji's complaints, see chap. 19 of the *Shitong* (with the complaints summarized on p. 506). That the omen treatise contains entries on events in the sky means that it sometimes duplicates material presented in the *Hanshu* astronomical treatise, purportedly compiled by Ban Zhao, a fact that elicited sharp criticism from Liu Zhiji, an early Tang proponent of economy in historical writings. Liu's original complaint also notes that the treatise does not assess all the ominous events in the *Annals* classic or in the *Shiji* and *Hanshu*. Liu's original note (speaking only of a few categories) says: "Of the thirty-six solar eclipses recorded for the Chunqiu, in two cases, there are no responses. Of the fifty-three solar eclipses recorded for Han, for forty there are no responses given. Furthermore, earthquakes took place in the second year of Emperor Hui [193 BCE], the second year of Zhenghe of Emperor Wu [91 BCE], the fourth year of Benshi of Emperor Xuan [70 BCE], the third year of Yongguang [41 BCE] of Emperor Yuan, and the second year of Suihe [7 BCE] of Emperor Cheng. Meteorites fell a total of eleven times, and no responses are given for any of them."

132. Contra Su Dechang, who definitely regards Ban Gu as "author" of the treatise, discounting the strong possibility that Ban added only some finishing touches to previously compiled materials, following Mansvelt-Beck, "Treatises of Later Han," 154. For an example of the rare remarks addressed directly to the reader, see, e.g., *Hanshu* 27B(a).1354.

133. For Bielenstein, Eberhard, and Kern, see note 1 in this chapter. Similarly, the PRC expert Shi Ding (see the conclusion below) speaks of the *Hanshu* treatise as mainly, if not entirely, Ban Gu's own creation, and thus a reflection of his own idealism (as opposed to materialism). I demur, finding the antonyms "idealism" and "materialism" anachronistic.

Of great use were three secondary sources: Sakamoto, "Kanjo gogyōshi sainan setsu"; Hirasawa, "Kansho gogyō to Ryū Kō Kōhan gogyō ten"; and Su, *Hanshu Wuxing zhi yanjiu*. Also of use was Mansvelt Beck, "Treatises of Later Han." I came to the Japanese sources early, but learned of the Taiwanese volume only shortly before finishing the first draft of this chapter. Michael Loewe in August of 2016 was kind enough to lend me some handwritten notes he had taken on the "Wuxing zhi" in the course of researching Dong Zhongshu. I have also consulted Tanaka Masami's 田中麻紗巳 authoritative study of Liu Xiang's portent theory titled "Ryū Kyō no sai-i setsu ni tsuite: Gen Kan sai-i shiso no ichimen," as the aforementioned scholars were all cognizant of Tanaka's theories, as was Zhang Shuhao's 2015 essay "Shitan Liu Xiang zaiyi lunzhu de zhuanbian."

134. Grégoire Espesset was kind enough to share with me two of his writings, one in draft form and one forthcoming: "Portents in Early Imperial China" and "Sketching out Portents: Classification and Logic." I learned from each of them. For Yang Shao-yun, see "Politics of Omenology."

135. *Hanshu* 27C(A).1459 associates Xiahou Sheng 夏侯勝 with compiling such a work; *Hanshu* 36.1950 credits Liu Xiang with compiling such a work, in 11 *juan*; and *Hanshu* 75.3155 seems to make Xiahou Shichang 始昌 the compiler of a set of writings by the same title. Xiahou Shichang was an expert in the rites and the *Odes* who held a position as Academician. It was he who taught Kuang Heng, Yi Feng, and Xiao Wangzhi (in other words, all the main reformers in the 30s BCE). Some early sources claim that Liu Xin was author or compiler of a *Hongfan wuxing zhuanlun* 洪範五行傳論 (now no longer extant) or a *Huangji zhuan* 皇極傳, as in *Hanshu* 27C(a).1459. However, this title may be the same as a third title ascribed to Liu Xin: *Wuxing zhuan shuo* 五行傳說.

Note meanwhile that the *Hanshu* bibliographic treatise makes no mention of these works, though it does catalogue works on the Five Phases (*Hanshu* 30.1769). For an attribution to an unnamed author whose work Liu Xiang glossed, see *Suishu* 32.913 (光祿大夫劉向注).

136. E.g., Ma Shiyuan, *Liang Han Shangshu xue yanjiu*. I narrowly avoided this mistake myself in *The Shifting Center*, and like Ma, I misattributed the *Shangshu dazhuan* to Fu Sheng in earlier publications.

137. *Hanshu* 27B(a).1353. Creating some confusion is *Hanshu* 30.1705, which conflates two titles.

138. *Hanshu* 27B(b).1429. Li Xun wrote an omen text of unknown title that was organized around the "Great Plan" chapter of the *Documents*.

139. *Hanshu* 100A.4203. Ban You 斿/游, Ban Gu's great-uncle, was a disciple of Yang Xiong, one of whose omen readings appears in the treatise.

140. See, e.g., *Hou Hanshu* 24.860 for Ban Bo's patronage by Wang Feng 鳳. Full details of the Ban family appear in ibid., 40A–B, and in Clark, *Ban Gu's History*, passim.

141. *Hanshu* 27B(a).1353.

142. Rpt. in *Ershiwu shi bubian* 二十五史補編.

143. Yao cites Zhao Chusheng, saying, 自大小夏侯明五行之後，劉向遂著為《洪範五行傳論》，其書不可而見於班固《漢書·五行志》者，皆其遺法也。

144. Wang's *Han Wei yishu chao*, in its preface to the fragments of Liu Xiang's 洪範五行傳, concurs, saying that Ban Gu, as final compiler of the *Hanshu* omen treatise, most substantially used Liu Xiang's work: 謨按班固《前漢書·五行志》原本伏生《尚書大傳》，兼采董仲舒、劉向、向子歆及睢孟、夏侯勝、京房、谷永、李尋諸

家之說。而劉知幾《史通》乃云班史《五行志》出劉向《洪范》, 趙樞生亦云劉向《洪範五行傳》其書不可見, 班書《五行志》乃其遺法.

145. See Chen's *Ruxue, shushu yu zhengzhi*.

146. *Hanshu* 21B.1101. Recall that *Hanshu* 27B(a).1353 says that "only Liu Xin's tradition of reading differed [from those of Xiahou Shichang and Liu Xiang]" 唯劉歆傳獨異.

At first, it seemed easy to find a good reason why Liu Xin would not be named as principal contributor to the *Hanshu* omen treatise, given his support of the usurper Wang Mang, yet Liu Xin's contribution is foregrounded from the beginning and acknowledged in other very important parts of the *Hanshu*.

147. *Hanshu* 27A.1334, *Hanshu* 27B(a).1353. Xiahou, who flourished during the reign of Han Wudi, was characterized as particularly well versed in all the Five Classics, an erudition he brought when extrapolating (*tui* 推) his omen readings: 善推五行傳 (1353).

148. Jing Fang plainly equates the *wuxing* with the earth's material resources, as the most recent edition of his writings shows; see Guo Yu, *Jing shi Yi yuanliu*.

149. Readers will profit from consulting the chart prepared by Su Dechang, *Hanshu Wuxing zhi yanjiu*, 161.

150. See Zhou Zhenhe, *Xi Han zhengqu dili*.

151. Personal communication, July 2016. We know that the omen readings offered in the *Hanshu* omen treatise did not forestall further re-readings. Liu Tao 劉陶, fl. ca. 157–184, expert in the Ouyang and Xiahou schools of *Documents* classics, correlated an unidentified *guwen Shangshu* with the Ouyang and Xiahou readings, and then re-interpreted thirty-one events.

152. See Mansvelt Beck, "Treatises of Later Han," 150.

153. To my way of thinking, it's significant that the treatise appears to be intent upon proving that the contents of the entire Five Classics corpus—not just the *Annals, Changes,* or *Documents* classic—represents a single vast repository of omen readings, because each of the five contributes to the overall plan of the treatise, through citations and careful correlations, even if the omen treatise's most obvious source of inspiration is the "Great Plan" chapter of the *Documents*. Thus, by the time of this compilation of the omen treatise, many once-separate passages bearing upon events had been made to articulate with one another as seamlessly as possible. Cf. Su Dechang, *Hanshu Wuxing zhi yanjiu*, 112: 王鳳薦班伯於成帝, 宜勸學, 召見宴昵殿是也.

154. Su Dechang, *Hanshu Wuxing zhi yanjiu*, 32. Su's theory is designed to answer the question he poses there, and my purposes differ from Su's. If we assume that there was a purpose behind the compilation of the "Wuxing zhi," how was that purpose served by *not* confining remarks to a single dynasty? In one conspicuous departure from pre-Qin omen readings, the Han readings do not include a single instance of contemporary experts predicting that trouble would come to those in

power on account of their "failures to be dutiful or reverent" (*bu jing* 不敬) conduct. Han experts were ready enough to offer post-facto analyses, however.

155. *Hanshu* 27A.1315.

156. I think here, for example, of Hans van Ess's painstaking comparison of the *Shiji* and *Hanshu*, which demonstrates beyond a doubt that small changes in wording in the chapters that overlap sometimes signal major interpretive shifts. See *Politik und Geschictsschreibung*. My comparison of the *Hanshu* and *Hanji* prompts the same conclusion. Nominally a mere "digest" of the *Hanshu*, the *Hanji* project has entirely different rhetorical aims, judging by the remarks Xun Yue directs to his readers and also Xun's deletions. The *Hanji* alters the focus of the account, so that the emperor always remains front and center, regardless of the identities of the actors who foment a change.

157. If Ban Gu was the compiler, his family members would still get a great deal of credit. "Originality" of view was far less valued in Ban's day and in manuscript culture in general. See Nylan, "Manuscript Culture."

158. *Hanshu* 27C(a).1445. Liu Xiang and Liu Xin disagreed both in their assignments of individual portents to specific omen categories and in their selections of historical events to tie to the portents. For example, in one case, Liu Xiang assigns an omen to the category of "speech that is discordant" 言之不從, whereas Liu Xin assigns the same portent to the category "hearing that is not acute" 聽之不聰. This represents a reversal of directionality, with the first focusing on the agent's inappropriate speech to others and the second on the agent's reception of others' speech. Liu Xin also seems to have been more preoccupied with the place of the Duke of Zhou (Zhougong) than his predecessors, perhaps as a model for Liu Xin's patron, Wang Mang. See also *Hanshu* 27A.1343, *Hanshu* 27B(a).1372.

159. *Hanshu* 21A.985.

160. Chapter 14 in *Chang'an 27 BCE*, Michael Loewe's "Liu Xiang and Liu Xin," begins this exploration. Relevant to my remarks is the chart in Hirasawa, "Kansho gogyō to Ryū Kō Kōhan gogyō ten," 12, which shows that, in terms of omenological categories, Ban Gu was inclined to adopt those categories that Liu Xin attached to the Five Modes (or failures thereof), not those proposed by Liu Xiang. Thus, Ban Gu, who had more or less the same motivation and orientation as Liu Xiang (to warn the ruler about the dangers attached to certain acts), retains references to both Liu Xin and Liu Xiang, without always explicitly ascribing the source of his views to Liu Xin. One would hardly be able to guess the pervasiveness of Liu Xin's theories by reference to the *Hou Hanshu* counterpart to the *Hanshu* omen treatise, because Liu Xin is mentioned by name in the later compilation but once. Even in the *Hanshu* omen treatise, at one point, Ban Gu makes a point of saying that he tends to favor the omen readings of Dong; he does not specify whether this preference applies to all cases or just the one. See Mansvelt Beck, "Treatises of Later Han," 152, citing *Hanshu* 27A.1317. Cf. ibid., 160ff., argues that the *Han Houshu* omen treatise by Sima Biao represents a simplification in organization from

the *Hanshu* model. But contrast *Jinshu* 27.800, which argues that the later standard history "did not go beyond the model of the *Hanshu*."

161. Twice the compiler (Ban Gu?) says of Liu Xin's explanation that it is "not right" (*shuo fei shi* 說非是). See *Hanshu* 27B(a).1354 and the discussion above.

162. Comparable success rates for court physicians are seen in Chunyu Yi's twenty-five descriptions of encounters with the ill and his replies to the court inquiries, and also in the "Yao" 要 manuscript from Mawangdui. For more on Chunyu Yi, see Brown, *Art of Medicine*.

163. Cf. *Shiji* 130.3319: 舊聞有遺失放逸者, 網羅而考論之也.

164. This preoccupation of Ban Gu's becomes readily apparent when the *Hanshu* entries are read against the *Hanji* and other Eastern Han histories. See note 160.

165. For the formulae *yi yue* 一曰 and *huo yue* 或曰 (some say), see part II above. The latter formula appears more often in the related "Tianwen zhi."

166. The omen treatise mentions Dong Zhongshu, Liu Xiang, Liu Xin, Jing Fang, Xiahou Shichang 夏侯始昌, Xiahou Sheng 勝, Sui Meng 睢孟, Xu Shang 許商, Li Xun 李尋 (Xu's student), Yang Xiong 揚雄, Gu Yong 谷永, and his contemporary Zhou Kan 周勘, among others. Jing Fang's *Yi zhuan* 易傳 is cited as a textual authority repeatedly; for Jing Fang's theories, see Guo Yu's *Jing shi Yi yuanliu*. At the same time, its pronouncements usually occur at the end of the omen readings and do not always seem relevant to the incident under discussion. Yang Xiong's opinion on an omen was solicited, we are told in *Hanshu* 27B(b).1429.

Moreover, through irregular resort to the formulae *yi yue* 一曰 ("one says") and *huo yue* 或曰 ("some say"), the *Hanshu* omen treatise in effect cites the opinions of additional unnamed experts, presumably too influential to omit, which may include Ban Gu, the members of his family, or circle. There are other formulae, *shuo yue* 說曰 ("the explication says"), *ruo yue* 若曰 ("this seems to say"), and *ji yue* 記曰 ("the record says"), but these seem different, though Su Dechang tends to lump them all together in his treatment of the *Hanshu* omen treatise. *Shuo yue* apparently refers to one or more authoritative (i.e., court-sponsored) explications of the text (as noted above), and *ji yue*, by analogy, to one or more authoritative records of events, presumably.

167. One should note, without approval or disapproval, what Su Dechang believes to be the current scholarly consensus: that when the *Hanshu* omen treatise supplies conflicting omen analyses, the particular analysis favored by the final compiler appears first in the group. Su Dechang (*Hanshu Wuxing zhi yanjiu*, 167) tries to prove this. Obviously, I feel less certain that this is right.

Unlike other *Hanshu* chapters, the omen treatise never mentions any early Eastern Han advisors and exegetes such as Huan Tan 桓譚, Huan Rong 榮, or Fan Sheng 范升, whose views were still contested in Ban Gu's time.

168. *Hanshui* 30.1745, citing *Analects* 19/4, where the speaker is Zixia (italics mine).

169. Genette, *Paratexts*.

170. *Hanshu* 27B(b).1418: 天尚不能感動陛下, 臣子何望? Wang served as Prefect of the Secretariat from 22–15 BCE; he also held the position of *da sima*, roughly Commander-in-Chief. Cf. *Hanshu* 85.3473–74.

171. From the modern vantage point, one would hesitate to call Wang Mang a usurper, but it is clear that the Eastern Han texts did not hesitate to apply that label (and worse!) to him.

172. As readers may recall, such cases celebrating good omens become more common by early Tang, when the *Suishu*'s omen treatise was put together, and they are most common in the omen treatises of late imperial China. For comparison, see Nylan, "*Suishu* 'Wuxing zhi.'"

173. It might well be that Ban Gu reserves the "Tian wen" treatise for the inner court circle, whereas the *wuxing* can relate to much larger circles.

174. In the case of the *Hanshu* omen treatise, certainly, the principal compiler (unknown) is unlikely to be the same person as the final compiler (presumably Ban Gu?). The omen treatise makes explicit mention of a range of experts, but *not* the Academician Fu Sheng 伏生 in early Western Han, the man credited with assembling the twenty-eight or twenty-nine chapters of the Han-era *Documents*.

175. See *Shiji* 121.3128. See also Loewe's *Dong Zhongshu*. Both Ban and Xu were said to be experts in the *Documents* (*Shangshu*).

176. That said, remarks ascribed to Dong in *Hanshu* 27A.1332 tally well enough with the main message conveyed by Dong's three attested memorials in the *Hanshu*, insofar as the memorials register complaints that the Han has cleaved too closely to the methods of governance that it adopted from Qin, which critics claimed built on the worst of the Zhanguo impulses toward Realpolitik. For Qin and Han methods of governance, see Nylan, "Han Views of the Qin Legacy."

177. As readers will doubtless recall, in Dong's own writings, he only mentions yin and yang *qi*, and not the Five Phases.

178. These are my numbers, but they match those of Su Dechang, *Hanshu Wuxing zhi yanjiu*, 143.

179. Solar eclipses (one of five events that are *always* dated in the *Annals* and *Zuozhuan*) are cited in both the "Celestial Patterns" treatise and "Wuxing zhi" chapters (e.g.), presumably because they are too significant to ignore. E.g., *Hanshu* 27C(a).1452; 27C(c).1459, 1479–80, 1482–1503, 1504–7, 1510. To a far lesser but still significant extent, fires at significant sites (mainly temples and palaces) and other portents and anomalies tend to be recorded, between 33 and 67 percent. Meanwhile, one must realize that four-fifths of the entries in the *Annals* and *Zuozhuan* are never assigned day-dates in the sexagenary cycle. I owe the information in the last two sentences to Newell Ann van Auken (private communication).

180. Su Dechang, Hanshu Wuxing zhi yanjiu, 139–41.

181. See note 32 for Liu Tseng-kuei's contribution on this.

182. *Hanshu* 36.1972.

183. See below.

184. The *pian* count is explicitly mentioned in *Hanshu* 36.1950, as is Liu Xiang's motivation in compiling his work.

185. At this remove, it is often difficult to ascertain whether *tian* should be read as the "cosmic powers" or as an anthropomorphic god, Heaven. I have argued that in the *Documents* chapters themselves, *tian* usually refers to the ancestors residing in heaven. However, my reading of those chapters need not coincide with Han readings known to the omen experts cited in the *Hanshu* treatise.

186. Mansvelt Beck, "Treatises of Later Han," 153, unaccountably speaks of five times, rather than thirty-one. There are, in addition, other formulae that say more or less the same thing, as with *Tian zai ruo yu bixia* 天災若語陛下 ("Heaven-sent calamities seemed to say to the emperor"), as in *Hanshu* 27A.1332.

187. Multiple translations of this formula have been offered over the years, but the Han readings seem to take it to mean "The king spoke to this effect." For further information, see the forthcoming translation of the Han *Documents* classic prepared by Nylan and He Ruyue.

188. Thanks to Wang Chong 王充 (d. 97), a student of Ban Gu's father, we know that such problems preoccupied those in Ban's circle, and, notably, in the omen treatise the strong forces in the cosmos typically but do not invariably conduce to good order. Wang Chong's *Lunheng* ("Wu shi" 物勢 chap.) suggests that Heaven does not intentionally give birth to things, for if it did so, there would not be so much chaos and disorder in the universe. But the Five Phases do destroy each other, by this *Hanshu* treatise's account.

189. See, e.g., *Hanshu* 27A.1321, for Dong's talk of such warnings.

190. See *Mencius* 3B/2 for talk of floods as Heaven's warnings. *Mencius* 3B/9 cites a four-character phrase presumed to be from a lost *Documents* chapter: 洚水警余.

191. Contra Su Dechang (*Hanshu Wuxing zhi yanjiu*, 137), who ascribes the one phrase to both Dong Zhongshu and Liu Xiang.

192. This to some degree confirms the relative unimportance that Professor Loewe assigns to Dong in Western Han.

193. See the discussion of Sakamoto Tomotsugu, "Kanjo gogyōshi sainan setsu," 50, regarding this and noting similar language used in the treatise and in Liu Xiang's *Hanshu* biography.

194. Thanks to the Fan Ning commentary to the *Guliang Tradition* of the *Annals*, we are certain that with respect to incident 4, under Lord Li, the main part we are reading is the theory of Liu Xiang. See Sakamoto Tomotsugu, "Kanjo gogyōshi sainan setsu," 51, for details.

195. Liu Xiang spoke out against Wudi's wars of aggression in Central Asia and maltreatment of his officials, for example.

196. For the *haogu* movement, see Nylan, *Yang Xiong and the Pleasures of Reading*.

197. For examples of one *haogu* reformer's disparagement of Han Wudi and Gongsun Hong, his classicist-sycophant and henchman, see *Fayan* 11/18.

198. See Gu Yong's memorial for an example of astonishing forthrightness; this is translated for Liu Tseng-kuei's essay, as in note 32. Liu Zhiji, *Shitong* 3.8/22 says baldly that Liu Xiang is the principal compiler of the *Hanshu* "Wuxing zhi": 〈五行〉出劉向〈洪範〉.

199. The treatise mentions ten comets and Ma Xu, an *additional* twenty-two comets, which means that the list of ominous events is very incomplete or selective. I surmise, but cannot prove, that the treatise chooses omens easily tied to key events in Han history, whenever possible, while disregarding the others. For an example, *Hanshu* 27B(b).1410, essentially tells a tale drawn from the "Gaozong rongri" chapter of the *Documents*, wherein a crowing pheasant becomes an omen. NB: I suspect that the so-called "Minor Preface" to the *Documents* was originally a separate treatise or essay.

200. Cf. Mansvelt Beck, "Treatises of Later Han," 153.

201. *Hanshu* 27B(a).1330, cited in Su Dechang, *Hanshu Wuxing zhi yanjiu*, 493–94.

202. A very high number of bad omens in Western Han occurs at the beginning of the reign of Emperor Cheng (r. 33–7 BCE), with its seventeen bad omens. Compare this with Andi in Eastern Han for whom 130 bad omens are recorded for his single reign (r. 107–125) or the 120 recorded during Huandi's reign (147–167). Major epidemics broken out in 37, 38, and 50. Plague hit the capital in 125 and revisited the populace in 151, 161, 171, 173, 179, 182, 185, and 217. See Waley's *Three Ways of Thought*, 93.

203. The phrase is not often used, but it does appear in *Hanshu* 36.1972 (Liu Xiang's biography). More often we see instead 天人之際.

204. See *Shiji* 1355n1. Neither does the text on *Hanshu* 27B(a).1377 appear to come straight from *Shiji*, so again Yan seems to be moving outside Sima Qian's *Shiji* for his information.

205. *Hanshu* 27B(b), 1410, 1411.

206. He and Nylan, "On the Han-era Postface (*xu* 序)."

207. Elsewhere, following the advice of Nathan Sivin, I have devised a test for deciding when Five Phases *qi* maps onto yin and yang *qi*. See my "Yin-yang, Five Phases, and *qi*."

208. For three examples, see *Hanshu* 27B(a).1365, *Hanshu* 27B(b).1415, 1416.

209. So far as we know, Ban Gu himself never served as judge. In the main, his treatise on penal laws follows Xunzi.

210. *Hanshu* 23.1901.

211. For Yang Lien-sheng, *bao* was the basis of just ethical and legal reasoning. See his *"Concept of Pao."*

212. For civil law in Han, see Zhang Zhaoyang, "Civil Laws and Civil Justice." My thinking here has been sharpened by the arguments lodged in Brickhouse, "Aristotle on Corrective Justice."

213. Yu Dingguo (fl. 69–52 BCE), for this reason, in Ban's view, was a better judge than Zhang Shizhi (d. ca. 135 BCE); see *Hanshu* 71.3043. Cf. for Ban's interest in *jiaohua* (civilizing influence), *Hanshu* 76.3211; 74.3137, 3139; 72.3090 (plainly shown to be a joint enterprise involving the ruler and his officials). Similarly, the *Lunheng*, "Jiang Rui" 講瑞 chapter, by Ban's contemporary, compares assessment of omens with legal assessment by men of the time. Mark Csikszentmihalyi has dealt with these issues in "Severity and Lenience."

214. Shi Ding, "Ma Ban yitong san lun," 234.

215. Hirasawa ("Kansho gogyō to Ryū Kō Kōhan gogyō ten," 5) has simply noted some discrepancies between what the fourth-century CE *Shangshu* "Minor Preface" ascribes to the *guwen Shangshu* or the *Shangshu dazhuan*, and what Hirasawa identifies as Liu Xiang's project in the *Hanshu* omen treatise. Accordingly, my own chapter contradicts Wang Mingsheng's conclusions, as summarized in note 101 above.

216. As Barrett ("Reading the *Liezi*," 25) says, "the Bible has been read and reread, and incessantly commented on, in Christendom. But it has never been treated as just a book, a book among other books."

217. Here I am very much thinking of the Daoist revelation texts, including the Maoshan revelations issued between 364 and 370.

218. As is true with the famous judgment rendered about Zhao Dun 趙盾 by Jin scribes following the "rules for documenting" (*shu fa* 書法) events. In other words, even the words of the "noble man" (possibly Kongzi?) are not infallible. See also Gentz, "Confucius Confronting Contingency."

219. Chen Jiujin, "Suixing jinian," 339; Wang Jianmin, "Shi'er ci," 318. The twelve-year cycle is actually 11.86 years; the counter-Jupiter moved from east to west. In English, see Ho Peng-yoke, *Chinese Mathematical Astronomy*, 32–33. As Ho notes (33), about the counter-Jupiter, "Being an imaginary heavenly body, no observation was needed or could be made on the movement of Counter-Jupiter and it was not even necessary to know the actual position of Jupiter." There are some indications that the Counter-Jupiter was important to retain to determine the reigning deity at any given time (115).

220. All work on the *Gongyang zhuan* in Western languages reflects the work of Gary Arbuckle, in particular his thesis: "Restoring Dong Zhongshu." For Gentz, see "Language of Heaven"; "Past as a Messianic Vision." For van Auken, see her *Commentarial Transformation*.

221. Gentz, "Language of Heaven," 818. The *jie* 詰 in the title of the *Gongyang* scholar He Xiu refers to "interrogations" in the courts and "inquiries" of the spirits. The close connections between religious and legal vocabulary in received and excavated texts is notable (as with Yunmeng Shuihudi).

222. For example, the impossibility of squaring the Four Rules 四法 (Shang-Xia-Basic-Refined, most associated with *Chunqiu fanlu*) with the the "Four Rules" outlined in Liu Xiang's *Shuoyuan* (chap. 19), Shang/basic versus Xia/refined, with

the Three Standards of Liu Xiang or the Three Eras of Dong Zhongshu and He Xiu's Black-Red-White sequence. Equally conflicting, but understandable is the phrase "The *Annals* classic in fact has its constant significant, and in addition its responsive changes" 春秋固有常義, 又有應變.

223. Most clearly explained in Huang Pumin, *He Xiu ping zhuan*, 186–92. As Huang remarks, as soon as one imagines interventionist deities, one accepts the interplay between the visible and unseen realms.

224. Gu Jiegang and Tong Shuye, "Handai yiqian Zhongguo ren de shijie guannian yu yuwai jiaotong de gushi." These views follow three main lines of transmission.

225. Changhe is the Gate of Heaven; Buzhou Mountain is "Not-Orbiting Mountain," as the pivot around which the entire cosmos circles. Youdu is the land where half-man, half-bird creatures dwell, but no human beings, according to Jia Yi's *Xinshu*, *juan* 9 ("Xiu zheng yu, shang" 脩政語上 chapter); as beyond "sun and moon," in *Han Feizi*, *juan* 10 ("Shi guo" 十過 chapter); and as the land of the dead in *Chuci* ("Zhao hun" 招魂 chapter).

226. Major et al., *The Huainanzi*.

Asian-language Bibliography

Ban Gu et al. 漢書. Compiled by Ban Gu 班固, 32–92, Ban Zhao 班昭, 45–ca. 116, et al. 12 vols. All refs. to the punctuated edition. Edited by Yan Shigu 顏師古. Beijing: Zhonghua shuju, 1962.

Chen Jiujin 陳久金. "Suixing jinian" 歲星紀年. In *Zhongguo da baike quanshu* 中國大百科全書, *Tianwen xue* 天文學. Beijing and Shanghai: Zhongguo da baike quanshu chubanshe, 1980.

Chen Kanli 陳侃理. *Ruxue, shushu yu zhengzhi: Zhongguo gudai zaiyi zhengzhi wenhua yanjiu* 儒學, 數術與政治: 中國古代災異政治文化研究. Beijing: Beijing Daxue chubanshe, 2015.

Ershiwu shi bubian 二十五史補編.

Fayan 法言. Compiled by Yang Xiong 揚雄. All refs. to *Exemplary Figures (Fayan)*. Seattle: University of Washington Press, 2014 [using standard paragraphing].

Gu Jiegang, and Tong Shuye. "Handai yiqian Zhongguo ren de shijie guannian yu yuwai jiaotong de gushi" 漢代中國人的世界觀念與域外交通的故事. *Yugong banyue kan* 禹貢半月刊 5, nos. 3–4 (April 1936): 97–120.

Guo Yu 郭彧. *Jing shi Yi yuanliu* 京氏易源流. Beijing: HuaXia chubanshe, 2007.

Hanshu: see Ban Gu et al.

Hirasawa Ayumu 平澤步. "Kansho gogyô to Ryû Kô Kôhan gogyô ten" 漢書五行と劉向洪範五行傳 [English title: "On *Hanshu* 'Wuxing zhi' and Liu Xiang's 'Hongfan Wuxing zhuan'"]. *Chugokû tetsugaku kenkyû* 中國哲學研究 25 (2011): 1–65.

Huang Pumin 黃朴民. *He Xiu ping zhuan* 何休評傳. Nanjing: Nanjing daxue chubanshe, 1998.

Kageyama Terukuni 景山輝國. "Kandai ni okeru saigi to seiji" 漢代における災異と政治. *Shigaku zasshi* 史學雜誌 90, no. 8 (1981), 46-68.

Liu Zhiji 劉知幾. All refs. to *Shitong tongshi* 史通通釋. Edited by Pu Qilong 浦起龍. Taipei: Yiwen, 1974.

Lü Nangong 呂南公 (1047–1086). *Guanyuan ji* 灌園集. Shanghai: Shangwu yinshu guan, 1935.

Ma Shiyuan 馬士遠. *Liang Han Shangshu xue yanjiu* 兩漢尚書研究. Beijing: Zhongguo shehui kexue chubanshe, 2014.

Miu Fenglin 繆風林. "*Hanshu* 'Wuxing zhi' fan li" 漢書五行志凡例. *Shixue zazhi* 史學雜誌 1, no. 2 (May 1929): 1–3.

Niu Jingfei 牛敬飛. "Lun Handai Xibu bianjiang shang de 'Yu gong' diming" 論漢代西部邊疆上的禹貢地名. *Xueshu yuekan* 學術月刊 2108, no. 3 (March 2018): 125–38.

Sakamoto Tomotsugu 坂本具償. "Kanjo Gogyôshi no sainan setsu" 漢書五行志の災異 [English title: "On the theory of *Zaiyi* (Portents) in the *Hanshu* 'Wuxing zhi'"]. *Nippon Chugoku gakkai ho* 日本中國學會報 40 (1988): 47–59.

Shi Ding 施定. "Ma Ban yitong san lun" 馬班異同三論. In *Sima Qian yanjiu xinlun* 司馬遷研究新論. Edited by Shi Ding and Chen Keqing 陳可青. N.d.: Henan Renmin chubanshe, 1982.

Shiji: see Sima Qian.

Sima Qian. 史記. Compiled by Sima Qian 司馬遷 et al. 12 vols. All refs. to the punctuated edition. Beijing: Zhonghua shuju, 1972.

Su Dechang 蘇德昌. *Hanshu Wuxing zhi yanjiu* 漢書五行志研究 [author's English title: A Study on "The Treatise of the Five Elements" in the Book of Han]. Taipei: Taida chubanshe, 2014.

Tanaka Masami 田中麻紗巳. "Ryū Kyō no sai-i setsu ni tsuite: Gen Kan sai-i shiso no ichimen" 劉向の災異說について：前漢災異思想の一面 [English title: "About Liu Xiang's Theory of Anomalies"]. *Shūkan tōyōgaku* 集刊東洋学 24 (October 1970): 173–85.

Tongzhi 通志. Compiled by Zheng Qiao 鄭樵. All refs. to Wang Shumin 王樹民, annot. *Tong zhi, ershi lue* 通志二十略, "Zai xiang lue" 災祥略. Beijing: Zhonghua shuju, 2004.

Wang Chong 王充. *Lunheng* 論衡. All refs. to Huang Hui 黃暉. *Lunheng jiaoshi* 論衡校釋. With collected commentaries by Liu Pansui 劉盼遂. 3 vols. Beijing: Zhonghua shuju, 1990.

Wang Jianmin 王健民. "Shi'er ci" 十二次. In *Zhongguo da baike quanshu* 中國大百科全書, *Tianwenxue* 天文學. Beijing and Shanghai: Zhongguo da baike quanshu chubanshe, 1980.

Wang Mingsheng 王鳴盛. "Ershiwu shi bubian jiaozheng, *juan er*" 二十五史補編校證, 卷二, *Han Ru yan zaiyi* 漢儒言災異 (Beijing: Zhonghua shuju, 1984), 38–40 [annotation of Zhao Yi's 趙翼 essay].

———. *Shiqi shi shang que* 十七史商榷. Punctuated by Huang Shuhui. Shanghai: Shanghai shudian, 2005.

Wang Mo 王謨. *Han Wei yishu chao* 漢魏遺書鈔. In *Han Wei congshu* 漢魏叢書. Shanghai: n.p., 1895 [includes "Han-Tang dili shu chao" 漢唐地理書鈔].

Wang Xianqian 王先謙. *Hanshu buzhu* 漢書補注. Changsha (preface 1900). Reprint, Taipei: Yiwen chubanshe, 1955; Beijing: Zhonghua shuju, 1983. Also available in the *Basic Sinological Series*.

Xu Xingwu 徐兴无. *Liu Xiang ping zhuan, fu Liu Xin pingzhuan* 刘向评传：附刘歆评传. Nanjing: Nanjing University Press, 2005.

Yao Zhenzong 姚振宗. *Hanshu Yiwen zhi tiaoli* 漢書藝文志條理. In *Ershiwu shi bubian* 二十五史補編, ed. Ershiwu shi kanxing weiyuan huibian 二十五史刊行委員會. Shanghai: Kaiming shudian, 1935.

Yang Tianyu 杨天宇, "Luelun Xu Shen zai Handai jinguwen jingxue ronghe zhong de zuoyong" 略论许慎在汉代今古文经学融合中的作用, *Zhengzhou daxue xuebao (Zhexue, shehui, kexueban)* 郑州大学学报 (哲学社会科学版) 40, no. 6 (November 2011): 90–93.

You Ziyong 游自勇. "Shi lun zheng shi Wuxing zhi de yanbian" 式論正史五行志的演變. *Shoudu Shifan daxue xuebao (Shehui kexue ban)* 首都師範大學學報 (社會科學版) 169 (February 2006): 1–6.

Zhang Shuhao 張書豪. "Shitan Liu Xiang zaiyi lunzhu de zhuanbian" 試探劉向災異論著的轉變. *Guowen xuebao* 國文學報 57 (2015): 1–28.

Zhou Zhenhe 周振鶴. *Xi Han zhengqu dili* 西漢政區地理. Beijing: Renmin chubanshe, 1987.

Western-language Bibliography

Arbuckle, Gary. "Restoring Dong Zhongshu (195–115 BCE): An Experiment in Historical and Philosophical Reconstruction." PhD diss. University of British Columbia, 1991.

Barrett, T. H. "Reading the *Liezi*: The First Thousand Years." In *Riding the Wind with Liezi: New Perspectives on the Daoist Classic*, edited by Ronnie Littlejohn and Jeffrey Dippmann, 15–30. Albany: State University of New York Press, 2011.

Bary, Wm. Theodore de, and Irene Bloom. *Sources of Chinese Tradition, Volume I: From Earliest Times to 1600*. New York: Columbia University Press, 1999.

Bielenstein, Hans. "An Interpretation of the Portents in the *Ts'ien-Han-Shu*." *Bulletin of the Museum of Far Eastern Antiquities* 22 (1950): 127–43.

———. "Han Portents and Prognostications." *Bulletin of the Museum of Far Eastern Antiquities* 56 (1984): 97–110.

———. "The Restoration of the Han Dynasty, Volume I." *Bulletin of the Museum of Far Eastern Antiquities* 26 (1954): 82–165.

———. "The Restoration of the Han Dynasty, Volume II: The Civil War." *Bulletin of the Museum of Far Eastern Antiquities* 31 (1959): 11–256.

———. "The Restoration of the Han Dynasty, Volume III: The People." *Bulletin of the Museum of Far Eastern Antiquities* 39 (1967): 1–198.
Brickhouse, Thomas C. "Aristotle on Corrective Justice." *The Journal of Ethics* 18, no. 3 (September 2014): 187–205.
Brown, Miranda. *The Art of Medicine in Early China: The Ancient and Medieval Origins of a Modern Archive*. New York: Cambridge University Press, 2015.
Clark, Anthony. *Ban Gu's History of Early China*. Amherst, NY: Cambria Press, 2008.
Colas, Gérard. *Penser l'icône en Inde ancienne*. Bibliothèque de l'École des Hautes Études. Sciences Religieuses. Turnhout: Brepols, 2012.
Constantino, Nicholas, and Michael Nylan. "On the Rites and Rites Controversies in Mid-Eastern Han." In a publication edited by Anne Cheng and Stephane Feuillas for the Collège de France. Forthcoming.
Csikszentmihalyi, Mark. "Severity and Lenience: Divination and Law in Early Imperial China." *Extrême-Orient, Extrême-Occident* 21 (1999): 111–30.
Dull, Jack L. "A Historical Introduction to the Apocryphal (Ch'an-wei) Texts of the Han Dynasty." PhD diss. University of Washington, 1966.
Eberhard, Wolfram. "The Political Function of Astronomy and Astronomers in Han China." In *Chinese Thought and Institutions*, edited by John K. Fairbank, 33–70. Chicago: University of Chicago Press, 1957.
Espesset, Grégoire. "La religiosité en Chine au haut Moyen-Âge: données chinoises et interprétation sinologique." Habilitation for UMR 8582 "Groupe sociétés, religions, laïcités" (GSRL). Paris, 2019.
———. "Portents in Early Imperial China: Observational Patterns from the 'Spring and Autumn' Weft *Profoundly Immersed Herptile*." *Divination and the Strange: Constructing Fate in Pre-and early Modern East Asia and Europe*, edited by Sophia Katz. Leiden: Brill, forthcoming.
———. "Sketching Out Portents: Classification and Logic in the Monographs of Han Official Historiography." *Bochum Yearbook of East Asian Studies*/BJOAF 39 (2016): 5–38.
Genette, Gérard. *Paratexts: Thresholds of Interpretation*. Cambridge: Cambridge University Press, 1997.
Gentz, Joachim. "Confucius Confronting Contingency in the *Lunyu* and the *Gongyang zhuan*." *Journal of Chinese Philosophy* 39, no. 1 (March 2012): 60–70.
———. "Language of Heaven, Exegetical Skepticisms, and the Re-insertion of Religious Concepts in the *Gongyang Tradition*." In *Early Chinese Religion, Part One: Shang through Han (1250 BC–220 AD)*, edited by John Lagerwey and Marc Kalinowski, 813–38. Leiden: Brill, 2009.
———. "The Past as a Messianic Vision: Historical Thought and Strategies of Sacralization in the Early *Gongyang Tradition*." In *Historical Truth, Historical Criticism, and Ideology: Chinese Historiography and Historical Culture from a New Comparative Perspective*, edited by Helwig Schmidt-Glintzer, Achim Mittag, and Jörn Rüsen, 227–54. Leiden: Brill, 2005.

He Ruyue, and Michael Nylan, "On the Han-era Postface (*xu* 序) to the *Documents Classic*." *Harvard Journal of Asiatic Studies* 75, no. 2 (Summer 2015) [actual date: Summer 2016]: 377–427.

Ho Peng-yoke. *Chinese Mathematical Astronomy*. London: Routledge Curzon, 2003.

Legge, James. *The Chinese Classics, with a Translation, Critical and Exegetical Notes, Prolegomena, and Copious Indexes*, Vol. 3 (*Shoo jing*, i.e., *Documents* classic). London: Trübner, 1861–1872.

Liu Tseng-kuei 劉增貴. "A Study of Gu Yong's Three Troubles Theory." In *Chang'an 26 BCE: An Augustan Age in China*, edited by Michael Nylan and Griet Vankeerberghen, 292–322. Seattle: University of Washington Press, 2015.

Loewe, Michael. *A Biographical Dictionary of the Qin, Former Han, and Xin Periods: 221 BC–24 AD*. Leiden: Brill, 2000.

———. *Dong Zhongshu: A "Confucian" Heritage and the Chunqiu fanlu*. Leiden: Brill, 2011.

———. "Liu Xiang and Liu Xin." In *Chang'an 26 BCE: An Augustan Age in China*, edited by Michael Nylan and Griet Vankeerberghen, 369–89. Seattle: University of Washington Press, 2015.

———. *The Men Who Governed Han China: Companion to a Biographical Dictionary of the Qin, Former Han and Xin Periods*. Handbuch der Orientalistik IV.17. Leiden: Brill, 2004.

Kern, Martin. "Religious Anxiety and Political Interest in Western Han Omen Interpretation: The Case of the Han Wudi Period (141–87 BC)." *Studies in Chinese History* 10 (2000): 1–31.

Major, John S., Sarah A. Queen, Andrew Seth Meyer, and Harold D. Roth, eds. and trans. *The Huainanzi: Liu An, King of Huainan. A Guide to the Theory and Practice of Government in Early Han China*. New York: Columbia University Press, 2010.

Mansvelt Beck, B. J. "The Treatises of Later Han." PhD diss. Leiden University, 1986.

Nylan, Michael. "Han Views of the Qin Legacy and the Late Western Han 'Classical Turn.'" *Bulletin of the Museum of Far Eastern Antiquities* 79–80 [December 2013/actually 2020]: 51–98.

———. "Historians Writing about Historians: Fan Ye 范曄 (398–446), Liu Zhiji 劉知幾 (661–721), and Ban Gu 班固 (32–92)." *Monumenta Serica* 67, no. 1 (June 2019): 183–213.

———. "Manuscript Culture in Late Western Han and Authors' Authority." *Journal of Chinese Literature* 1, no. 3 (2014–2015): 155–85.

———. "Mapping Time in the *Shiji* and *Hanshu* Tables." *East Asian Science, Technology, and Medicine*, Special Issue on Numerical Tables and Tabular Layouts in Chinese Scholarly Documents: Part I: On the Work to Produce Tables and the Meaning of their Format, edited by Karine Chemla (2016): 1–65.

———. *The Shifting Center: The Original "Great Plan" and Later Readings*. Monumenta Serica Monograph Series 24. Nettetal: Steyler, 1992.

———. "The *Suishu* 'Wuxing zhi' and Resonance Theories." In *Monographs in Tang Official History*, edited by Karine Chemla, Damien Chaussende, and Daniel Patrick Morgan. Dordrecht: Springer, 2018.

———. *Yang Xiong and the Pleasures of Reading and Classical Learning in Han China*. New Haven: The American Oriental Society, 2011.

———. "Yin-yang, Five Phases, and *qi*." In *China's Early Empires, supplement to The Cambridge History of China, vol. 1, Ch'in and Han*, edited by Michael Nylan and Michael Loewe, 398–413. Cambridge: Cambridge University Press, 2010.

Nylan, Michael, and Ruyue He. Translation of *The Documents*. Seattle: University of Washington Press, forthcoming.

Nylan, Michael, and Trenton Wilson. "A Brief History of Daring," accepted for a volume comparing the antique Greek and Chinese emotions, edited by David Konstan and Huang Yang (to be published in Chinese and English).

———. "Dowager Empress Deng Sui 鄧綏 (d. 125), a Study of Legal, Cosmological, and Gender Theories in Eastern Han," *Frontiers of Chinese Philosophy* (March 2021), 1–37.

Queen, Sarah A. *From Chronicle to Canon: The Hermeneutics of the Spring and Autumn, According to Tung Chung-shu*. Cambridge: Cambridge University Press, 1996.

Raphals, Lisa. *Divination and Prediction in Early China and Ancient Greece*. New York: Cambridge University Press, 2013.

Saso, Michael. "What is the Ho-t'u?" *History of Religions* 17, nos. 3/4 (February–May 1978): 399–416.

Schwermann, Christian. "Anecdote Collections as Argumentative Texts: The Composition of the *Shuoyuan*." In *Between History and Philosophy: Anecdotes in Early China*, edited by Paul van Els and Sarah A. Queen, 147–92. Albany: State University of New York Press, 2017.

van Auken, Newell Ann. *The Commentarial Transformation of the* Spring and Autumn. Albany: State University of New York Press, 2016.

van Ess, Hans. *Politik und Geschictsschreibung in alten China: Pan-Ma i-t'ung*. 2 vols. Wiesbaden: Harrassowitz, 2014.

Waley, Arthur. *Three Ways of Thought in Ancient China*. London: G. Allen & Unwin, 1953.

Yang Lien-sheng. "The Concept of Pao as a Basis for Social Relations in China." In *Chinese Thought and Institutions*, edited by John K. Fairbank, 291–309. Chicago: University of Chicago Press, 1957.

Yang Shao-yun. "The Politics of Omenology in Chengdi's Reign." In *Chang'an 26 BCE: An Augustan Age in China*, edited by Michael Nylan and Griet Vankeerberghen, 323–46. Seattle: University of Washington Press, 2015.

Zhang Zhaoyang. "Civil Laws and Civil Justice in Han." PhD diss. University of California at Berkeley, 2010.

6

Western Han Sacrifices to Taiyi 泰一

TIAN TIAN 田天

TRANSLATED BY MICHAEL NYLAN

Worship of Taiyi first became important during Western Han (206 BCE–9 CE). Although the god Taiyi was already recognized in the pre-unification period, no evidence prior to the reign of Han Wudi (r. 141–87 BCE) attests Taiyi's worship among the *chief* gods during Qin and early Western Han. In the early days of unified empire, in Qin and in early Western Han, the sacrifices to the southern suburbs were not the highest form of the imperial sacrifices; those took place at the four altars at Yong, near present-day Baoji (Shaanxi province), and at other shrines distributed throughout the country. Some of these shrines belonged to the original Six Kingdoms conquered by Qin, and even more of them came from the pre-dynastic Qin's own traditions. But Han Wudi deliberately altered many long-standing precedents with the aim of making a new set of practices that would set his own Han dynasty apart. Known to be inordinately interested in sacrifices and the immortality cult, he introduced a number of innovations in objects and methods of sacrifice, the most important of which was his decision, in 112 BCE, to make Taiyi his chief object of worship, at an altar devoted solely devoted to Taiyi in Yunyang, on the grounds of his Sweet Springs palace complex. This *first* reorganization of cult practices, midway through Wudi's reign, became a new binding precedent, with the result that that splendid altar was maintained until the reign of Han Aidi (7–1 BCE). In 5 CE, a *second* major reorganization of the gods took place, by which a single imperial pantheon was forged, to be worshipped in the suburban sacrifices at the capital with the emperor as its chief officiant. Although one can trace the theories underlying the suburban sacrifices to the ritual Classics, there is no sign anywhere that a major imperial sacrifice took place in the suburbs before 5 CE. For the second time, however, the court and ruling house adopted the new organization as the norm.

As the worship of Taiyi during that last century in Western Han departed so significantly from Qin practices, this chapter explains the evolution of the cult, based on the extant sources from the period, most substantially the "Treatise on the *Feng* and *Shan* Sacrifices" (Fengshan shu 封禪書) from Sima Qian's *Shiji* (Historical Records) and its counterpart chapter from Ban Gu's *Hanshu* (History of Han), the "Treatise on the Suburban Sacrifices" (*Jiaosi zhi* 郊祀志), whose judgments of events differed sharply. By Ban Gu's era, in the early reigns of Eastern Han, the southern suburban sacrifices had already become the only correct form of court sacrifice on behalf of the ruling house. Because of this, Ban Gu's evaluation of the Taiyi cult fundamentally differed from that of Sima Qian and his peers two centuries earlier. As the differences between Sima Qian and Ban Gu are noteworthy, instead of conflating the two accounts (as most have done), we may use their different stances to reveal a process of innovation and change.

I. Early Sitings of Taiyi

a. Taiyi before Han Wudi's Reign

The early history of the Taiyi cult is complicated and somewhat murky. As many other secondary studies have been devoted to it, only a summary of their findings is presented here.[1] The pre-unification Chu bamboo strips often mention worship of a god named Tai, and the scholarly consensus today presumes that this god is Taiyi.[2] But because of the limitations of these sources, we cannot say much about the character and iconography of this god.[3] We can only say that Tai was a god in the common religion shared alike by members of the governing elite and lower-ranking men and women, and Tai probably was regarded as head of the gods, as the character Tai 泰 means "great." From two sources, the *Huainanzi* (compiled ca. 138 BCE) and the *Shiji* (finished ca. 90–86 BCE), we learn that Taiyi was an astral deity,[4] who supposedly resided in the Central Palace in the brightest constellation corresponding to the North Pole Star, the heavenly counterpart to the supreme ruler on earth.[5]

Aside from this, we have the Guodian manuscript (ca. 300 BCE) titled "Taiyi Gives Birth to Water," which seems to equate Taiyi with the Dao, the cosmic Way and origin of phenomenal existence. Professor Li Ling (Peking University) has written that Taiyi had three aspects simultaneously: first, as an astral deity; second, as a star; and third, as ultimate being in phenomenal existence.[6] But how these three aspects were related and subsumed in the same god Taiyi is quite unknown. As we will see, these three aspects all come into play in Western Han, but are superseded by other aspects.

b. The Establishment of the Western Han Cult to Taiyi

Taiyi made its next appearance during the reign of Han Wudi. A miracle worker (*fangshi* 方士) named Miu Ji 謬忌, from Bo 亳 county, came to Wudi's court and proposed that, as Taiyi was head of the gods, he should be worshipped alongside the Five Lords (Wudi 五帝).[7] Finding Miu's claim credible, Wudi ordered an altar to Taiyi to be erected in the southeastern suburbs of the capital, outside the city walls, and there cult was offered under Miu Ji's explicit directions. Soon afterward, again at the urging of Miu Ji, worship of the Triune Gods (*sanyi* 三一) began at the same altar complex with the Triune Gods the Heavenly Unity, the deified Earthly Unity, and Taiyi. The first of the three gods corresponded to three small *yinde* 陰德 stars at the "mouth" of the Northern Dipper, whose light, argued Sima Qian, Director of Astronomy for Wudi, was seen as one. The eighth-century Tang commentator Zhang Shoujie 張守節, however, thought the Heavenly Unity was a single small star, and Taiyi a star to the south, and most star maps have followed Zhang Shoujie's account. This problem arose because the Pole Star that signified the Han emperor was Ursa Minor, 小熊座 β, while the Tang Pole star had moved to another position in Ursa Minor (小熊座 γ), which necessitated the reassignment of all the stars and constellations. If we follow the lead of Sima Qian, as he is our main source here, the star for Heavenly Unity ought to be three small stars in the Dipper.

As for the cult to "Earthly Unity," it is only known from brief mentions in the two treatises in the official histories for Western Han.[8] Judging from the current evidence, it is entirely possible that it was brought in simply to make a third with the other powers. Certainly, as Professor Li Ling has suggested, the term is hard to square with what we know of the astral deities and astronomy of the time.

By Miu Ji's plan, there were other gods besides the Triune Gods who should be worshipped at the altar of Taiyi, but about these we know almost nothing. In connection with these other cults, the *Shiji* treatise mentions the "Sons of Heaven [i.e., supreme rulers] in antiquity" (*guzhe Tianzi* 古者天子). Probably Miu Ji wanted to give his new cult greater legitimacy by bringing in such old associations, although we cannot entirely rule out the possibility of some deeper plan behind Sima Qian's account of Miu, and we can probably trace this kind of linkage to the late Zhanguo eastern cults.

Wudi had a Taiyi altar erected in the suburbs to the southeast of his capital at Chang'an.[9] His attitude seems rather passive, in that he seems mainly to have hoped to avoid offending a powerful god by a refusal to erect such an altar. And Miu Ji, for his part, seems not to have had a systematic theology driving his proposals to Wudi. In other words, Wudi may not have had a larger design in mind when erecting the Taiyi altar, and it seems likely that Miu Ji didn't need to suggest that Taiyi would take precedence over the other gods. For these reasons, we can identify the Taiyi cult during Wudi's reign as "phase 1" in the establishment of the cult.

Not long after the Taiyi altar was set up, a *fangshi* from Qi named Shao Weng 少翁 proposed to Wudi that in the palace there be hung a portrait of all the gods, for by this means the emperor might summon the gods to his presence. This suggestion Wudi followed, and he had a portrait of Taiyi especially hung in his Sweet Springs Palace. Then, when Wudi contracted an illness, he sent around an order to the commanderies to find a healer who might be able to call "the gods down" (*xia shen* 下神) to cure his illness, and a Master was found who was credited with such powers.[10] The list of the healing gods included the name of Taiyi, presumably the "highest" of these gods who were to be summoned. When Wudi's illness was cured, the fortunes of Taiyi were suddenly ascendant.

At this time, the visage of the god Taiyi becomes clearer, and not only because there are images that depict the god.[11] Taiyi could be summoned by a *wu* 巫 (a wizard),[12] and he could be conversed with; moreover, other, lesser gods were said to be in attendance on him. So we know that when Wudi worshipped Taiyi, it was not merely that he worshipped the North Pole Star, but rather that this god had evolved from an astral deity and in the process acquired some human properties. In the end, promotion of the cult was always to promote the status of Wudi himself, so that he could join the ranks of the gods, with his personal immortality assured, so long as he made Taiyi the most powerful god among the gods.

c. Reasons for the Great Altar at Ganquan (Sweet Springs) Palace Complex

In 113 BCE, when Wudi performed the suburban sacrifices to Heaven at Yong, some at court suggested that the object of the cult be changed to Taiyi, with the Five Lords reconfigured as his attendants. Wudi hesitated and could come to no decision. But before very long, the *fangshi* Gongsun Qing 公孫卿 advanced the argument that a precious tripod had come to light on a certain date on the winter solstice (冬辛巳朔旦冬至), the very date that tradition had the Yellow Emperor or Huangdi finding one or more precious tripods. Gongsun Qing then supplied the throne with further details to corroborate his claims. For example, he reported that he had learned these things from his teacher, Master Shen 申, who had found a tripod whose inscription read "The Han will rise when it restores the era of Huangdi" 漢興復當黃帝之時. Meanwhile, Gongsun Qing suggested that perhaps Wudi could assure his immortality if he performed the *feng* sacrifice at Mount Tai, because the Yellow Emperor had done this at the same site.[13]

Gongsun Qing had managed, quite masterfully, to weave together several of Wudi's preoccupations: the change to a new reign era, performance of the *feng* and *shan* sacrifices at Mount Tai, and Wudi's longing for immortality. Once he had tied these things successfully to the Yellow Emperor, he really captured Wudi's full attention. By the *Shiji* narrative, Wudi decided to set up a Grand Altar to Taiyi at Sweet Springs Palace very soon after Gongsun Qing made his proposal regarding this. So while we know that there was some discussion at court about setting up a new altar of this sort as soon as the precious tripod was unearthed, we know, too, that action had been temporarily tabled until the consideration of Gongsun Qing's proposal. Perhaps we must understand the association between Fenyin 汾陰, the location which yielded the precious tripod, and the Taiyi cult, if we hope to understand this particular timing.

Actually *two* precious tripods came to light around the same time: one in 113 BCE at Fenyin and a second, specifically labeled as an "antique," in the same year, which was found in the vicinity of the Houtu 后土 altar. According to the records, this second tripod was quite unlike the other. The experts whom Wudi consulted regarding the provenance of the

Fenyin bronze argued that it was related somehow to the Great Lord's 泰帝 bronze tripod.[14] Report of this consultation is the first time a Great Lord is mentioned in the *Shiji* chapter devoted to the *feng* and *shan* sacrifices. In 111 BCE, just as Wudi had men preparing the songs and music to be offered to the Great Lord—who may or may not be identical with Taiyi—an unknown person mentioned a legend that featured the god. Supposedly, the Great Lord instructed the Plain Girl 素女 to strum a fifty-string lute (*se* 瑟), and when she did so, he found the tune so terribly sad that he ordered her to stop. When she ignored his plea, the Great Lord in a rage broke her lute into two, with the result that lutes even down to Wudi's era only had twenty-five strings.[15] According to this tale, the Great Lord was a ruler of great antiquity, like Taiyi himself.[16] But the relation between the Great Lord and Taiyi still requires more explanation, as it is not entirely self-evident.

Apparently, the Great Lord is just another name for Taiyi, or so the apocryphal sources say,[17] for those apocryphal sources identify the Great Lord with the Pole Star, while we know that Taiyi corresponds to a star within Ursa Minor. The *Wuxing dayi* 五行大義 (compiled before 614) cites Zheng Xuan 鄭玄 (127–200) to the same effect,[18] a further corroboration of this identity. This identity apparently came about as the astral deity during Western Han slowly acquired the attributes of a sage-king in antiquity. For example, one chart issued by Wang Mang and recorded in the *Hanshu*, the Map of the Purple Pavilion 紫閣圖, talked of the Great Lord and the Yellow Emperor as two supreme rulers of antiquity who eventually achieved immortality.[19] (Supposedly, the Great Lord was an emperor preceding the Yellow Emperor—the very same ruler who was credited with casting the precious tripod and inventing the *se*.) That Taiyi is indeed depicted in Han-era materials.[20] Once we understand the dual roles played by the Great Lord Taiyi, we understand Wudi's special interest in the god. One of the precious tripods associated with the Yellow Emperor gave a specific date, and so it figured as a kind of prediction or "divine instruction" (*shen ce* 神策) revealing the end date of the old cycle and the start date of the new. Legends held that the only emperor before the Yellow Emperor who could have cast a precious tripod was the Great Lord.[21] In Gongsun Qing's construction of history, the precious tripod that the Yellow Emperor had gotten could only have been commissioned and cast by the Great Lord. By the cosmic cycle, the precious tripod had now made its third appearance, this time to Han Wudi. So long as Wudi performed the same sacrifices as the Yellow Emperor, he could expect to become an immortal, too. This "logic" was the last piece of encouragement that Wudi needed to erect the altar.

Han Wudi explained his motivation very clearly the first time he worshipped at that new altar: "at the start of a new Heavenly cycle, the precious tripod, with its divine inscription, was conferred upon the emperor; the cycle had gone round, and the current emperor respectfully bows low to acknowledge its auspicious reappearance" (天始以寶鼎神策授皇帝，朔而又朔，終而復始，皇帝敬拜見焉).²² This praise song clearly states the reasons for Wudi's ritual observances: that he felt sure that he had received sufficient powers to start a new cosmic cycle. Later, in 104 BCE, Wudi carried out the sacrifices on Mount Tai, on the winter solstice ("in the eleventh month, the *jiazi* day [i.e., the first day in the sixty-day cycle], at dawn" 十一月甲子朔旦冬至). There he offered cult in a Worship Hall²³ (Mingtang 明堂), and there his message noted that he had paid his dues to Taiyi,²⁴ in language similar to that used seven years earlier. From all this, we can see that Wudi's primary motivation for starting the ceremonies at Mt. Tai was his interest in starting the cosmic cycle anew, in the hopes that the renewal would assure his capacity to join the ranks of the immortals. Wudi was absolutely convinced that he had received messages from the highest gods to proceed in this way, and he, in consultation with his court, had been planning for a long time related policy moves (e.g., the change in the color of court robes), not just because of Gongsun Qing's advice. Nonetheless, the flurry of activities was certainly promoted by Gongsun Qing.

Let us recapitulate the Western Han history of the Taiyi imperial cult, as we know it from the extant records: Miu Ji, the *fangshi*, was the first to promote the Taiyi cult. In 124 BCE, in Chang'an's southeastern suburbs, an altar was erected. When Wudi fell ill in 118 BCE, greater attention went to one avatar of Taiyi, in connection with Wudi's pursuit of longevity, if not immortality (either in 118 or 117 BCE). In 112 BCE, the Sweet Springs altar was erected to Taiyi. So altogether the cult's elevation to the premier rank of state sacrifices had taken a mere twelve years. Once it had achieved that status, the emperor himself participated in the sacrifices. Below, we examine the ceremony itself to gain a greater understanding of the cult from our sources.

II. Cult Ceremonies at the Great Altar at Sweet Springs 甘泉泰畤

a. The Material Form of the Ceremonial Site

In 112 BCE, as noted above, Wudi erected an altar at Sweet Springs dedicated to offerings to Taiyi (and, so far as we know, solely dedicated to that purpose). According to the *Shiji* and *Hanshu* treatises,[25] the Great Altar there had three distinctive features: (1) it was a three-tiered structure erected on a square base, with an octagonal altar sitting atop the tiers; (2) the iconography of that octagonal altar, also named the Purple Altar 紫 壇, tied it to the Big Dipper constellation; and (3) the altar was heavily decorated (more on this below).

In Chinese tradition, we have round altars dedicated to the worship of heaven/Heaven (tian/Tian)[26] and square altars dedicated to the worship of earth, but only very rarely does the extant literature mention an eight-sided worship site. In the Sui and Tang periods, we know of two eight-sided cult sites, and these we may use for comparative purposes. According to the *Great Tang Record of the Suburban Sacrifices* (*Da Tang jiaosi lu* 大唐郊 祀錄, compiled 793), to the north of the walled palace compound was a square mound with just such an octagonal site,[27] which was dedicated to the worship of all the various earth gods. In addition, the Tang-era Mingtang was octagonal in design.[28] For Northern Zhou, the same sources mention a similarly configured altar, and Watanabe Shin'ichirō 渡辺信一郎 believed that the Sui dynasty "hill of earth" (*diqiu* 地丘) was also octagonally shaped, with a two-tiered altar.[29] Thus, based on some evidence, we may postulate the existence in these three dynasties of such eight-sided structures meant to symbolize the eight directions of the earth below the skies. Evidently, the Northern Zhou, Sui, and Tang all erected octagonal altars because the same cosmological thinking informed their architectural designs; because of the close connections between the ruling houses of these three dynasties, they shared many ideas in common.[30]

In all probability, the Tang Mingtang shared certain design features with the Western Han altar to Taiyi at Sweet Springs, because religious structures tend to be conservative in their design. That said, the Tang-era octagonal altars, including the Mingtang, were dedicated to Earth's eight directions, while the Han-era Great Altar was erected for the express purpose of worshipping Heaven. This important difference aids us when we reimagine the distinctive features of the Han Great Altar and the cult to Taiyi, and we argue that the Han *imaginaire* (but not that of Tang) is based on the *shipan* 式盤 or "diviner's board."[31] A diviner's board found at another site, in Fuyang, Anhui province, illustrates this tie-in beautifully (fig. 6.1),[32] for even if the Anhui diviner's board reproduced in figure 6.1 has no direct connection with the Western Han Great Altar, the board and the altar share a vision of the cosmos. The four directions and eight subdivisions of the heavens (each 45 degrees) correspond to the domain of

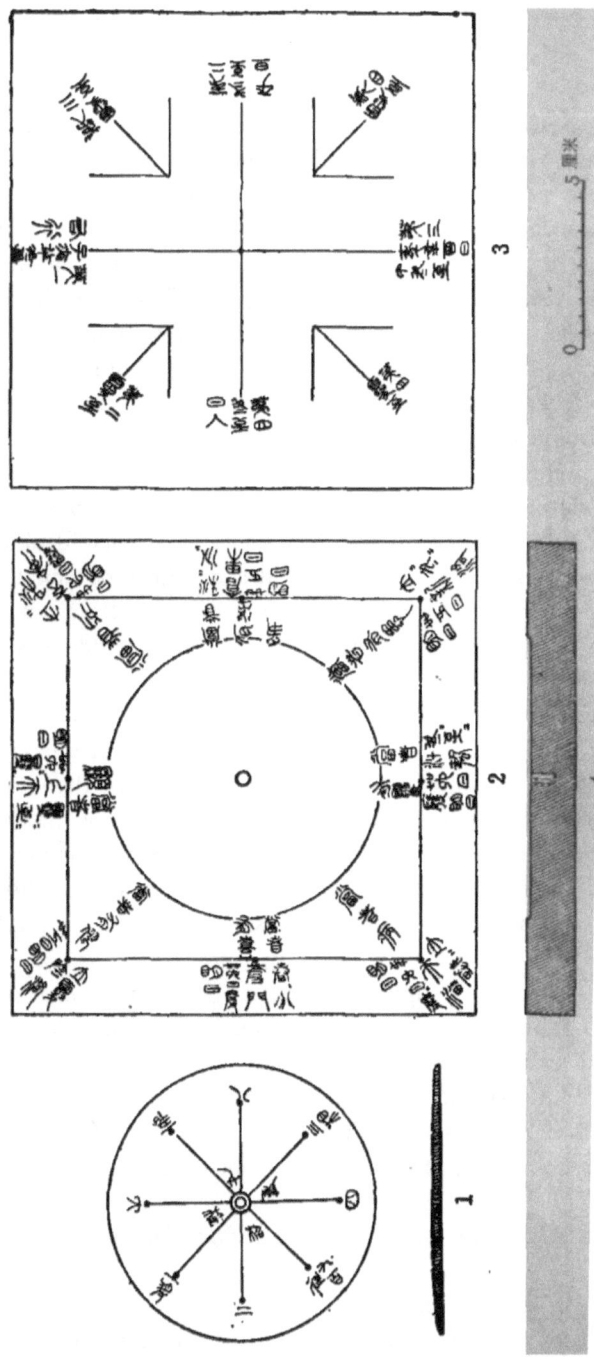

6.1. Fuyang Shuanggudui Xi Han qimu shi 阜陽雙古堆西漢漆木式 [Lacquer diviner's board excavated from Shuanggudui in Fuyang, Anhui, from *Wenwu* 文物 1978: 8, 25.]

Taiyi, by implication on the diviner's board and explicitly at the octagonal altar erected in honor of Taiyi.[33]

Second, the most sacred spot in the Great Altar complex was the altar called the "Purple Altar." An excavation carried out in 1980 at Sweet Springs led to the discovery of an altar composed of four rows of eight brown stones some 250 meters to the south of a western foundation of pounded earth. Each stone measured 1 square meter, with a thickness of 0.45 meter. In the exact middle, there was a round "nest" marked with engraved grooves. The original excavation report called this a "gate hinge stone" (*men shu shi* 門樞石). Three years previously, in 1977, to the northeast of the western foundation of pounded earth, another slab was unearthed. It, too, was of brown stone, and, though partially broken, it measured 170 centimeters long, 73 centimeters wide, and some 50 centimeters thick.[34] One face of the stone had carving on it, but these markings were illegible.[35] Lin Meicun 林梅村 believed that this stone came from the Purple Altar at Sweet Springs.[36] This remains a possibility. As mentioned above, the site of the Great Altar had a "stone altar, dedicated to the cult of the immortals" 石壇、仙人祠, and it is entirely possible that the altar used stone paving or great stone slabs as decoration. Third, we believe the Purple Altar was heavily decorated; also that there were offerings buried there. If these surmises are correct, then there must have been a burial cavity of some size atop the Great Altar.[37] According to the *Shiji* "Treatise on the *Feng* and *Shan* Sacrifices," the Qin buried huge quantities of offerings and sacrificial animals at the four Qin altars at Yong. At or near the Qin altars at Yong there must have been pits for such materials, insofar as the Qin was merely following earlier traditions. Thus the most important features of the Great Altar are shared with altars from high antiquity.[38] Judging from the Liangzhu 良渚 altar at Hangzhou, Huanguanshan 匯觀山; from the altar at Lingjiatan 凌家灘, in Hanshan 含山 county, Anhui, and the slightly later earthen altar at Chengdu's Yangzishan 羊子山土壇,[39] south China early on had developed an altar form composed of three levels. In addition, already at Lingjiatan, there were decorative stone slabs and an offering pit, just as at the Great Altar at Ganquan. The form of the Qin and Han altars under review seem not to have differed all that much from those of the Neolithic periods, even if the naming conventions were not the same.

b. The Liturgies and Ceremonies at the Great Altar Erected to Taiyi

For the first time at the winter solstice sacrifice on the *xinsi* day of the eleventh month of 112 BCE, Wudi personally conducted a sacrifice to

Taiyi.⁴⁰ When Wudi did so, he used the rituals associated with the Suburban Sacrifices, whose liturgies and standards were the same as those of the Five Altars at Yong. Below we examine every aspect of the sacrifices, from the place, the time, and the ceremonial itself, plus the procedures, in order to lay out the Great Altar at Sweet Springs (Ganquan). Of course, Guanquan is well-known. But the *Shiji* and *Hanshu* accounts do not tell us whether or not Wudi personally offered cult when he ascended the altar to Taiyi. We can only consult the records of later eras to see what they say, which may be speculation. The *Beitang shuchao* 北堂書鈔 (compiled before 638) cites the *Old Ceremonies of the Han* (*Han jiu yi* 漢舊儀), which says, perhaps on some good authority, that Wudi personally worshipped Taiyi at the Purple Altar there.⁴¹ Besides worshipping Taiyi, Wudi carried out a ceremony called Chaori xiyue 朝日夕月 at Sweet Springs. He rose at dawn to worship the sun, and very late, at midnight, he carried out a ceremony in honor of the moon.⁴² Wudi apparently believed the sun and moon to be aides to Taiyi, and thus in need of proper sacrifices. His "Basic Annals" records his edict, which says, "I look from afar at Taiyi. I repair the Altar to Heaven's Starry Signs" (Tianwen shan 天文禮).⁴³ The phrase "Tianwen" here refers to sun and moon, as adjuncts to Taiyi.

Each time, the sacrifices offered at the Great Altar were elaborate and time-consuming. The *Shiji* treatise on music tells us that the sacrifice to Taiyi began at dusk, with an Evening Cult Offering (*ye ci* 夜祠), and it ended only with dawn's first light.⁴⁴ The "Treatise on the *Feng* and *Shan* Sacrifices" in the same history also speaks of Wudi participating in a dusk-to-dawn offering on the *xinsi* day, which means that at dawn, in the suburbs, he was to commune with Taiyi. From this we see that the emperor's personal participation in the sacrifices was only one of several forms of sacrifice scheduled in the sacrificial calendar, as many other sacrifices were conducted by the Grand Invocator and other imperially appointed prayer-masters.⁴⁵

The Taiyi cult at Wudi's court was not limited to Sweet Springs. In addition to the Great Altar there, a Taiyi altar was built in the southeastern suburbs at Chang'an under the supervision of Miu Ji, and a Mingtang/Worship Hall was erected at Mount Tai to Taiyi. Moreover, the *Hanshu* geographical treatise lists yet another altar to Taiyi under the entry for Langya commandery.⁴⁶ From these, we say that there were multiple Taiyi altars, but perhaps the other sites were less elevated than that at Ganquan. So, although we may characterize Taiyi as the highest god in a certain period of the Western Han's pantheon, strictly speaking, such a claim only has relevance to the cult offering made to Taiyi at the single location at Sweet Springs.

III. The *Shiji* and *Hanshu* Treatises on Sacrifice

a. The Nature of the Sacrifice at the Great Altar

Before the erection of the Great Altar at Sweet Springs, the highest sacrifice personally conducted by the Western Han emperors was paying cult at the Five Altars at Yong inherited from Qin. The sources sometimes conflate the Sweet Springs Great Altar and the Suburban Sacrifice, for reasons that can be easily imagined, given that both were imperial cults offered to an astral deity of the highest level. The conflation was even more likely once Wang Mang proclaimed the Great Altar ceremonies to be the Suburban Sacrifices,[47] at which point the regular sacrifices dedicated to the four seasons at the Great Altar included various songs whose names (e.g., Bright Yang or Qingyang 青陽, Red Light or Zhuming 朱明) clearly show their derivation from the nineteen-part song cycle known as the Suburban Sacrifice Songs (*Jiaosi ge* 郊祀歌).[48]

Looking backward to Wudi's era, the sacrifices at Yong had not yet been abandoned, however. It was simply that the two pairs of sacrifices at Yong and at Sweet Springs were roughly similar, we think, as were the ceremonies used to worship Heaven (Tian). For example, before Wudi carried out the *feng* and *shan* sacrifices at Mount Tai, he offered cult to the Taiyi god in rites called the *leici* 類祠 offering.[49] Later, for the *feng* sacrifice at Mount Tai, Wudi used a ceremony very much like the one that he had used earlier for the worship of Taiyi.[50] And when, on an autumn day, Wudi carried out the *feng* and *shan*, a comet made its appearance in the sky, which the relevant officials considered an auspicious omen sent from Heaven. After Wudi in the next year carried out the Suburban Sacrifice at Yong, he then carried out a special sacrifice to Taiyi a second time.[51] That Wudi carried out this second sacrifice to Taiyi in requital for Heaven-sent omens strongly suggests that the attributes of the god Taiyi approximated those of Heaven. And when Wudi erected the Mingtang, dedicated in part or *in toto* to Taiyi, he also made Gaozu, the Han founder, the co-adjutor and fellow recipient of sacrifices there.[52] His way of conducting sacrifices, not coincidentally, was very much like the pairing of Gaozu with Heaven in the Suburban Sacrifices. So in many features, these imperial sacrifices all conveyed a single message: that by the second century of Western Han, specifically at the time of Wudi, the worship of Taiyi was, to all intents and purposes, conflated with the worship of Heaven itself, as both were thought of as a high god.

That said, we must stress the following: although Taiyi approximated Heaven, strictly speaking, in the Western Han imperial sacrifices, Taiyi cannot be considered identical with Heaven in every respect. Among the songs sung for Wudi's ceremonies were two titled "Wei Taiyuan" 惟泰元 and "Tiandi" 天地. "Wei Taiyuan" is dedicated to the worship of the Heavenly gods and the Earthly spirits. One line in it reads *Wei tai yuan zun / Ao shen fan li* 惟泰元尊 / 媼神蕃釐. Li Qi 李奇, writing before Yan Shigu in early Tang, parsed the phrasing so that the binomial expression *yuanzun* referred to Heaven: "It is precisely the Great One, who is the Origin, the Honored. . . ." Yan Shigu 顏師古 (581–645) objected to Li's rendering, for he took Taiyuan 泰元 instead as the epithet for Heaven, as in the translation "Only the Great Origin is to be honored. . . ."[53] Actually, both constructions proposed by these scholars are off, by my construction, although Li Qi's punctuation is correct, insofar as *wei* is a particle and Tai means Taiyi, while *yuanzun* is an adjectival phrase describing Taiyi. I would then read the lines, "It is Taiyi who is the Origin Honored. . . ." The fine rhetoric here is smooth and elegant.

From the one song "Wei Taiyuan" we can see that Taiyi and Heaven were not tied together in a simple relation. That Taiyi has his counterpart in an earth deity named Lady Ao 媼神 clearly means that Taiyi is a god in heaven. At the same time, the paired deities Taiyi and Lady Ao pervade Heaven and Earth and make the four seasons. Taiyi enjoys power over the sky and thus is higher. And turning to the second of the two songs, "Tiandi," we also find information on Taiyi at the end. From the last, we learn that the song has been offered in the Taiyi sacrifice, and earlier lines say, "With due attention we reverently regard the generous gifts from Heaven and Earth, / and so have erected this Purple Altar, / to seek to communicate with the remote powers."[54] (Recall that the Purple Altar is the Great Altar.) This song pairs Taiyi as dedicatee with Houtu 后土, Lord of Earth, quite exceptionally, and the spirit of Taiyi apparently pervades all of the cosmos ("heaven-and-earth"). Taiyi is the highest god in heaven, it seems. What we need to add is simply this: despite all this language, the god Taiyi cannot be regarded as identical to Heaven itself. But in Han theories, Taiyi is the highest god in heaven, and sometimes, in a shorthand way of speaking, becomes Heaven. Just as Du Ye 杜鄴 (d. ca. 2 BCE), in late Western Han, complained, "Today, there are Suburban Sacrifices to Heaven and Earth offered at Sweet Springs and at Hedong; these have lost their correct directional allocation."[55] The sacrifices mentioned here by Du Ye are the Sweet Springs Great Altar and the altar erected to Houtu at Fenyin. Ban Gu offers a similar criticism, that "over the course of thirty-plus years, the shrines to Heaven and Earth were moved no fewer than five times."[56] Wang Shun 王舜, Liu Xin 劉歆, and others, when speaking of Wudi's erecting shrines to Heaven and Earth, always meant the Great Altar at Sweet Springs and the altar to Houtu (Lord Earth, sometimes cast as Lady Earth) at Fenyin. We can see that sacrifices at these two altars stood for sacrifices to Heaven and Earth.[57]

Under the leadership of Kuang Heng 匡衡, who was Chancellor between 36–30 BCE, reforms to the sacrificial schedule were urged upon the court by many. The main motivation behind such reformist programs was to do away with the Great Altar at Sweet Springs and the sacrifices to Houtu in Fenyin in order to move the entire pantheon of gods to the southern suburbs of Chang'an.[58] As Kuang Heng himself noted, the reform would relocate and rebuild the altar erected to Taiyi. Actually, when the Suburban Sacrifices were conducted at Chang'an during Chengdi's reign, the Great Altar at Sweet Springs was disbanded and ousted from the sacrificial schedule.[59]

Ostensibly, the classicists led by Kuang Heng were simply urging the relocation of the Great Altar to the Suburban Sacrifices, so as to bring the imperial rituals more in conformity with those they deemed "early" and in line with the antique sage-kings. In actuality, by this set of reforms, the classicists were urging a substantial revolution in ritual practice.[60] Taiyi exited from the Western Han sacrificial space, and Heaven represented the highest deity or deities. By such reforms, the worship of Taiyi was removed from the official sacrifices. This also led to changes in the beliefs of those who came later.

b. Discrepancies in the Shiji and Hanshu Accounts

When the *Shiji* "Treatise on the *Feng* and *Shan* Sacrifices" was compiled by Sima Qian, the Taiyi cult was at its most influential, with the cult paid to Taiyi the very highest. When the *Hanshu* treatise on sacrifices was written, some two centuries later, the Suburban Sacrifices had become the only imperial ritual. So while the *Hanshu* treatise, in its first half, copies the *Shiji* account fairly closely, in both narrative and vocabulary, there are still places where the two treatises clearly reflect two different experiential worlds. The biggest discrepancy between the two concerns the way the two treatises deal with the term "suburban."

Both the *Shiji* and *Hanshu* employ the term "suburban," but only the *Hanshu* describes sacrifices following the Yuanshi 元始 reforms of 5 CE as the imperial sacrifices. Naturally enough, then, their theories, structures, and contents constitute major differences in usage. Historians up to now have tended to take any reference to *jiao* 郊 ("suburban") to indicate the Suburban Sacrifices. Many of them base their understanding on the middle and later periods of Western Han history, and specifically on the late Western Han reformers' discourse. But if we wish to understand the meaning of *jiao* in early to mid-Western Han, the time up to and including Han Wudi's reign, we must look to the discourse dating to the time *before* the reforms, obviously enough.[61]

The *Shiji* treatise uses the term *jiao* in a way closer to Qin and early Western Han usage, where *jiao* and *jiaosi* refer to the highest sacrifices to be offered.[62] Note, however, that *jiao* in the pre-Qin Classics and masterworks is deployed differently, but when the early Han classicists used the term *jiaosi*, they thought they were referring to Western Zhou ritual practices. As scholars today know so little about the Western Zhou, and cannot even say what rituals were ever practiced in early Western Zhou, we will not try to analyze their imagined use of the term *jiaosi* here. But happily, there are in the received literature some few clues in the form of texts mentioning a suburban ritual in the pre-unification state of Lu in the *Zuozhuan* or *Zuo Traditions*.[63] Lu, by this tradition, every year, in spring, in the first month, in the suburbs of its capital conducted a sacrifice. Yet another form of record relating to suburban rituals is preserved in the pre-Qin ritual classics, which are more preoccupied with formal hierarchies in ritual settings. So, the passages in the pre-Qin ritual Classics do not coincide with the entries in the *Zuozhuan* as it pertains to the Lu kingdom rituals.

That said, in bureaucratic documents and in the contemporary discourse dating from Qin and Western Han, the highest sacrifices all are called *jiao* or *jiaosi*, regardless of the changes the high gods underwent. For example, when the Qin and Han emperors worshipped at Yong, the *Shiji* says that "in the *jiao*, they saw Shangdi, the High Lord." Wendi, when he worshipped at Weiyang, at the ancestral temple for Wudi, supposedly "visited for the *jiao*."[64] Wudi, when he offered cult at the Great Altar to Taiyi, reportedly "conducted the *jiao* at the Great Altar."[65] So *jiao* is deployed in the *Shiji* in the same way as in the pre-Qin ritual classics dating to the Zhanguo period. On the one hand, this usage makes the later sacrifices seem in conformity with hallowed traditions, stretching back to the Three Dynasties with their legendary sage-kings. On the other hand, people of the time really did not think that the highest sacrifices they were offering were precisely the same; instead, they thought to give a nod to their presumed origins, and so they would call them *jiao*. The Western Han people did not feel the need for the same precision and "conformity with [antique] rituals" as later scholars would have liked.

Having gone through the multiple changes witnessed by the mid- to late Western Han ritualists, Wang Mang, in connection with the Yuanshi 5 CE reforms, under Pingdi, raised the issue of standardizing the entire imperial pantheon, beginning with implementing the contemporary classicists' desire to institute a suburban sacrifice in the southern suburbs of the capital. It was that innovation by Wang Mang that would set the pattern for the succeeding two millennia of Chinese history. It was the Yuanshi ceremonies that constructed the Suburban Sacrifices, conceived of as an attempt to "restore Antiquity," that is, an attempt to get back to the Western Zhou practices, as configured in the *Rites* classics and in the imagination of the learned people of the time. From that point on, scholars decided that the term *jiao* in every case meant *jiaosi*, or Suburban Sacrifice. Beginning with the Yuanshi ceremonies, then, the Suburban Sacrifices began to be conflated with the sacrifices offered in the suburbs that were mentioned in the ritual classics favored by the classicists. By the late Western Han reforms, the most important features to be noted are these: (1) the emperor, as the Son of Heaven, was to personally conduct these sacrifices; (2) the location of the sacrifices was to be south of the city walls of the capital; (3) the principal recipient of the sacrifices was to be Heaven itself; (4) worship of Heaven was to take place at the Circular Moat 圜丘; (5) the Circular Moat was to serve as the reception site or audience hall for the entire pantheon of gods.[66] Later, the Suburban Sacrifices of Eastern Han followed the design of the Yuanshi ceremonies.

We need to pay attention to this: when Sima Qian was compiling his treatise on the *feng* and *shan*, the place and recipients of the *jiao* had not yet undergone massive changes. But by the time of the compilation of the *Hanshu* treatise devoted to the sacrifice, the notion of *jiao* had completely changed. So although the *Hanshu* treatise borrows a great deal of vocabulary from the *Shiji*, its theoretical stance has already undergone a fundamental alteration. At the end of the *Shiji* treatise, Sima Qian speaks in his role as Senior Archivist, concluding that he has analyzed the "inner and outer [visible and invisible] aspects of the service to the spirit world since antiquity."[67] The purpose of compiling the *Shiji* treatise is clear: Sima Qian wants to register the ideas underlying the sacrifices that have evolved from the kings and emperors of the antique past down to his present day. But when we come to the *Hanshu* Appraisal, the main story line concerns the charismatic virtue of the Western Han kings and court. As what Ban deems important is merely or mainly the legitimacy of the Han ruling house, the *Hanshu*

Appraisal for the treatise on sacrifice, naturally enough, is preoccupied with the changes made to the *fangshi* and "shrine officials." Ban Gu approves of a position that was expressed by Gu Yong 谷永 in 14 BCE.

For Gu Yong had berated all the *fangshi* 方士 for using illicit means, even black magic, to cheat the emperor ("They hold skewed ways, they harbor treachery and deceit" 挾左道, 懷詐偽). To Gu's way of thinking, the worst mistake of the *fangshi* was that they did not honor the Five Classics' teachings, and instead "turned their back of the right Way of humaneness and duty" (背仁義之正道). Not only this. Gu Yong also criticized them by name, speaking of the *fangshi* in the reign of King Huai of Chu (d. 290 BCE), in Wendi's era and in Wudi's era. Although Gu Yong never explicitly brought up the subject of Taiyi, he hardly needed to, because everyone at court would have known that the altar was set up at the urging of the *fangshi* Miu Ji (as above). As some may recall, during Western Han the *fangshi* had gained influence at the court as holders of the title "candidates awaiting appointment" in a range of court bureaus. If they were able to secure the emperor's trust they then could become chief officiant at the cult, and sometimes they were even ennobled for such services. Several major sacrificial reforms in the Western Han Dynasty, including the *feng* and *shan* ceremonies and the establishment of the temple for Taiyi, represent the joint work of the officers in the regular officialdom and the *fangshi*. Emperor Wu even instructed his regular officials to learn from the *fangshi*.[68] Many high-ranking officials in late Western Han were somewhat dubious of Han Wudi's activities, but no one before Gu Yong had denied the value and worth of the *fangshi* as strongly as Gu Yong did, and none had demanded that all persons at the court act in accordance with the Five Classics. According to *Hanshu*, Gu Yong did not want his ruler to be deluded about the gods or prodigies; he did not want others to be deceived regarding their superior status in a supramundane world; and he did not approve of any search for requital to secure immortality or extraordinary blessings.[69] Such concerns as Gu Yong's derived from a particular small group of classicists, who were willing, implicitly and forthrightly, to criticize Han Wudi's beliefs and deeds. Thus the *Hanshu* critique of Wudi, registered in the concluding Appraisal, sharply diverges from the *Shiji* treatise in this one respect. And this discrepancy between the *Shiji* and *Hanshu* accounts, in both content and ultimate purport, signals the major change that had occurred in connection with the great ritual reforms of the Qin and Han times.

One other thing: the *Hanshu* treatise reflects an Eastern Han understanding of events, but not the Eastern Han understanding of events. Scholars often seem to believe that Ban Gu and Gu Yong represent the *only* classical understanding of *jiao* and *jiaosi*, rather than *one* such understanding at a certain point in time. They then, on that basis, conclude that the sacrifices to such gods as Lady Chen Bao must be "illicit," when that conclusion does not fit the *Shiji* treatise at all. Instead, we should embrace the notion that *jiao* originally meant no more than "conducting the highest sacrifice."

IV. Conclusion

The Qin imperium originally conducted its highest sacrifices at the altars at Yong, and these sacrifices the early Western Han emperors continued. But in the first year of the Yuanding era, Han Wudi instituted a new set of sacrifices to the high god Taiyi, and he performed these sacrifices to Taiyi at the Great Altar at Sweet Springs. From that time on, the worship of Taiyi became one of the most important sacrifices of the Han ruling house. Taiyi, as a high god, belonged to the heavens, and sometimes even was taken to represent Heaven itself. After the institution of the Suburban Sacrifices in 5 CE, however, Heaven (not Taiyi) became the high god in the imperial pantheon, Heaven being conceived as a somewhat abstract moral force. As Taiyi had been an astral deity and also, supposedly, one who enjoyed the status of an antique sage, his position at points seemed higher than that of heaven or *tian*, in the sense of "sky." Still, the position of Taiyi would be altered forever with the ritual reforms undertaken during Pingdi's reign and Wang Mang's regency, and those reforms ushered in a set of cult practices that was neither like that of Qin and early Western Han nor like that of Western Zhou.

The *Shiji* "Treatise on *Feng* and *Shan*" expended a great deal of time recording Wudi's worship of Taiyi, the establishment of the altar at Sweet Springs, and the relation between the Yellow Emperor and the Taichu calendar, while giving readers to understand that the Taiyi sacrifice was an innovation. These two last points were Sima Qian's preoccupations, and for that reason, he was quite clear in his narrative about the order and substance of successive events. But this was hardly just Sima Qian's preoccupation. Roughly a century later, Liu Xin was determined to prove that one of Han Wudi's most important achievements was this erection of a cult site to heaven-and-earth. For at the end of Western Han, with the establishment of the Suburban Sacrifices at Chang'an, the chief object of worship became Tian or Heaven, a high god. The position of Taiyi retreated from the stage of the imperial sacrifices, as it was negated, denied, or trivialized in the *Hanshu* account. When Sima Qian and Liu Xin disagreed, Ban Gu had no way to understand and adjudicate their different views of Taiyi; still less did he have a way to identify the suitability of Taiyi as the supreme object of cult for Han. Thus consideration of the changes in the status of Taiyi allows us also to see something unexpected in the rise of Heaven to preeminence within the pantheon of gods. We must forget what the classicists have tried to persuade us of: that Heaven was also the main object of worship in Antiquity, for we have no evidence to support such a belief.

Notes

1. The earliest study of Taiyi was that by Qian Baocong 錢寶琮 (1892–1974), "Taiyi kao" 太一考 ("Research on Taiyi"), published in 1932 and reprinted in 1983. My work here builds upon an earlier piece, "The Suburban Sacrifice Reforms and the Evolution of the Imperial Sacrifices." Editors' note: we believe that the *Yili* dates to the pre-Qin period; we know that the *Liji* compilation dates to Western Han, although it contains earlier materials. The *Zhouli* was regarded by many in Eastern Han as a possible forgery. The *Zuozhuan* is based on the Han-era *Zuoshi chunqiu*, but we do not know how much the two compilations differed.

2. Scholars, when researching the Chu slips from Baoshan, pointed out that the "Tai" mentioned there is "Taiyi." See Liu, "Baoshan Chujian shenming yu Jiuge shenqi"; Li, "Baoshan Chujian yanjiu, zhanbu lei," 425–48. Yan Changgui spoke to the debate over this identification in his 2010 *Wugui yu yinsi: Chujian suo jian fangshu zongjiao kao*, 80–81n2.

3. In addition to those cited above, in note 2, one may mention in relation to "Taiyi" the following works: Harper, "The Nature of Taiyi," and, more recently, Dong Shan's rather complete ordering of the Baoshan slips, in his 2014 "Chujian zhong cong 'da' zhi zi de dufa." Dong (143–73) believes that the character *da* 大 refers to the "plague demons" (*li gui* 厲鬼), something worth further consideration.

4. *Shiji* 27.1289n4. All references to *Shiji* are to this edition.

5. On Beichen 北辰, see He, *Huainanzi jishi* [hereafter HNZ], 3.200–2; Qian, "Taiyi kao," 216.

6. Li, "Taiyi chongbai de kaoguxue yanjiu."

7. See He, HNZ, 3.200–2.

8. Rao Zongyi wrote that in the Mawangdui work titled "Taiyi Chart for Avoiding War" (Taiyi bi bing tu 太一避兵圖), the green and yellow dragons represent Heaven's Unity and Earth's Unity, and the dragon with the yellow head and green body represents Taiyi. See Rao, "Tushi yu cifu." Lai renames this work "Diagram of the Taiyi Incantation" (Taiyi zhu tu 太一祝圖). See Lai, "Mawangdui 'Taiyi zhu tu' kao."

9. Li, "Sanyi kao" 三一考, 244–46.

10. *Shiji* 28A.1388.

11. For a synopsis of the pre-Qin and Han statements on Taiyi, see Hsing, "'Taiyi sheng shui,' 'Taiyi chuxing' yu 'Taiyi zuo.'"

12. This translation of *wu* is borrowed from Sivin, *Health Care in Eleventh-Century China*, 94–95.

13. *Shiji* 28.1393.

14. *Shiji* 28.1392.

15. *Shiji* 28.1396.

16. See Gu and Yang, "Sanhuang kao," which registers the belief that "The Tai Lord 泰帝, for a time, came down from heaven" (這個人是臨時由天上拉下來的). Qian, "Taiyi kao," 225 (reprinted ed.), registers some problems with this account. As a group, the *Gushi bian* 古史辨 scholars devoted a great deal of attention to the problem of the relation between "Taiyi" 太一 and the Great Lord or Taidi 泰帝. Besides Gu, Yang, and Qian, Lü Simian wrote *Sanhuang wudi kao*.

17. The *Shiji suoyin*'s 史記索隱 sources (now in fragments) include *Chunqiu wenyao gou* 春秋文耀鉤 and *Chunqiu hecheng tu* 春秋合誠圖. See *Shiji* 27.1289n1, 1290n3. However, it is extremely hard to date these apocryphal sources, and they may date to any time up to the Tang, when the *Suoyin* cites them.

18. See *juan* 5 of the work by Xiao Ji 蕭吉.

19. *Hanshu* 99C.4154.

20. Huang (*Heguanzi huijiao jizhu*, 222–25) believed that Taiyi represented the emperors and kings from remote antiquity, such as those who appear in the *Heguanzi* 鶡冠子 ("Taihong" 泰鸿 chapter).

21. *Shiji* 28.1392.

22. *Shiji* 28.1395.

23. This is Loewe's translation of the Hall that is often called the Bright Hall or Hall of Light.

24. *Shiji* 28.1401.

25. *Shiji* 28.1395; *Hanshu* 25B.1256.

26. While the Chinese sources do not distinguish the two possible meanings, in English the uppercase signifies an anthropomorphic god, while the lowercase typically signifies the cosmic processes, which, insofar as they are self-generated, do not require the presence of an anthropomorphic god.

27. See "Da Tang jiaosi lu," *juan* 8 ("Xiazhi ji huangdi zhi" 夏至祭皇地祇); *Da Tang Kaiyuan li*, *fu* (photographic reprint), 782 下欄; *Da Tang Kaiyuanli*, 782 上欄.

28. It may be important to say here that many academics have reconstructed an eight-sided structure for the Han Mingtang, retrojecting this back into the Han dynasties. This information about the Tang Mingtang is found in the "Treatise on Ceremonials" in the *Jiu Tangshu* 22.857.

29. See Watanabe, *Chūgoku kodai no ōken to tenka chitsujo: Nitchū hikakushi no shiten kara*, translated by Xu Chong as *Zhongguo gudai de wangquan yu tianxia zhixu*, 147–148n9.

30. Above, the *Jiu Tangshu* 舊唐書 also cites a Han history (又按漢書, 武帝立八觚壇以祀地. 登地之壇, 形象地, 故令為八方之基, 以象地形), ascribing the eight-sided altar platform to the era of Han Wudi. We know from the "Fengshan shu" chapter that the Han leaders worshipped Earth via Houtu 后土; also that the altar to Houtu was not octagonal. See *Shiji* 28.1389. The somewhat misleading citation in the *Jiu Tangshu* is actually to the Great Altar dedicated to the worship of Taiyi at Sweet Springs.

31. On the diviner's board, see Li Ling, *Zhongguo fangshu kao*, chap. 2 ("Shi yu zhongguo gudai de yuzhou moshi"); also Kalinowski, "Xianqin suili wenhua jiqi zai zaoqi yuzhou shengchenglun zhong de gongyong."

32. Anhui sheng Wenwu gongzuodui, "Fuyang Shanggudui Xi Han Ruyin hou mu fajue jianbao."

33. In addition, one might also consult the "Treatise on Sacrifice" in the *Hou Hanshu*, zhi 7.3158n2. From its pages, we find that the Yuanshi altar dedicated to the worship of Heaven is round, rather than octagonal. Possibly this is mistaken.

34. Those last two dimensions are rough estimates, as the stone was broken in parts.

35. See Yao, "Guanyu Han Ganquan zhuti jianzhu weizhi wenti."

36. Lin, "Xi Han diguo daxing shidiao yishu de fazhan," 116.

37. Yao, "Guanyu Han Ganquan," 95.

38. For research on the earliest forms of altars, see Li, "Shuo jitan he jisikeng," in his 2004 *Rushan yu Chusai*, and Jing, "Woguo shiqian jisi yiji chutan."

39. The Liangzhu altar was built atop a mountain. It was rectangular, with three levels, like an overturned dipper. See Zhejiang province, Wenwu kaogu yanjiusuo, "Liangzhu wenhua Huanguanshan yizhi dierci fajue jianbao." Similarly, a three-level altar is erected at the highest point in the Lingjiadui tomb complex; it is an irregularly shaped rectangle with rounded corners 圓角長方形. Its height is approximately one meter; its top level has three rectangular offering pits. See Anhui province, "Anhui Hanshan Lingjiatan yizhi disanci fajue jianbao." For the Yangzishan altar, see Sichuan province, "Chengdu Yangzishan tutai yizhi qingli baogao."

40. *Shiji* 28.1395.

41. Yu Shinan (d. 638), "Jisi shijiu," *juan* 90, p. 371 (bottom).

42. *Hanshu* 6.185–86.

43. *Hanshu* 6.185.

44. *Shiji* 24.1178.

45. *Shiji* 28. 1395.

46. *Hanshu* 28A.1585.

47. *Hanshu* 25B.1265.

48. *Hanshu* 22.1054–55.

49. *Shiji* 28.1396.

50. *Shiji* 28.1398.

51. *Shiji* 28.1399. Altogether Wudi visited Yong ten times, and during nine of those times, he worshipped the Five Lords at the Five Altars 五畤, judging from the extant records.

52. *Shiji* 28.1401.
53. *Hanshu* 22.1057n1.
54. *Hanshu* 22.1057–58.
55. *Hanshu* 25B.1262.
56. *Hanshu* 25B.1266.
57. *Hanshu* 73.3126.
58. *Hanshu* 25B.1256.
59. *Hanshu* 10.304.
60. Michael Loewe and Lü Min 呂敏 both note this point, with Loewe stressing that greater emphasis be put on Heaven and Earth. See Loewe, *Crisis and Conflict in Han China*, 171. For Lü Min, see Bujard, "State and Local Cults," 793. These two scholars point out fairly clearly where the problems in such analyses lie, when scholars ascribe, anachronistically, more systematicity than existed during the two Han dynasties, since changes at the highest levels of the imperial cult were made very gradually.
61. A more detailed survey of the early meaning of the term is found in chapter 1 of Tian, *Qin Han guojia jisi shigao*.
62. Editor's note: The *Annals* mentions *jiao* sixteen times, and typically these events are assigned day-dates, signifying that great import was attached to them in Lu. That *jiao* seems only gradually to have gained significance in Western Han suggests that the *Annals* ascribed to Confucius was not consulted for ritual precedents early on. Also, many records in the *Annals* concerning the *jiao* (typically conducted in summer, but once in the autumn) concern a problem with the sacrifice (an inauspicious divination result, a damaged sacrificial victim). There is some dispute as to what the *jiao* sacrifice was. See Yang Bojun's extensive remarks to Lord Huan, Year 5, 106–7; Mao Qiling's *Chunqiu Maoshi zhuan*, *juan* 136.11a.
63. Editors' note: There is, of course, the problem of how to date the *Zuozhuan*, with most scholars dating it to ca. 300 BCE, but others putting its composition as late as mid-Western Han. This is not the place to resolve this controversy. All we can say at this point is that the *Zuoshi chunqiu* (which may or may not have been in circulation before Liu Xin's time) is not today's *Zuozhuan*.
64. The Chinese is 郊見渭陽五帝廟.
65. See *Shiji* 28.1382, where the Chinese is 郊泰畤.
66. Gan, *Zhongguo zhonggu jiaosi li de yuanliu yu tezhi*, 6.
67. *Shiji* 28.1404. This is a loose translation of the lines 退而論次自古以來用事於鬼神者, 具見其表裏.
68. *Shiji* 28.1386: 使黃錘史寬舒受其方.
69. *Hanshu* 25B.1270, 1260.

Asian-language Bibliography

Anhui province, Wenwu kaogu yanjiusuo; Hanshan county, Wenwu guanlisuo. 安徽省文物考古研究所、含山縣文物管理所. "Anhui Hanshan Lingjiatan yizhi disanci fajue jianbao" 安徽含山凌家灘遺址第三次發掘簡報. *Kaogu* 考古 (1999.11): 1–12.

Anhui sheng Wenwu gongzuodui 安徽省文物工作隊. Fuyang diqu bowuguan 阜陽地區博物館. Fuyang xian wenhuaju 阜陽縣文化局. "Fuyang Shuanggudui Xi Han Ruyin hou mu fajue jianbao" 阜陽雙古堆西漢汝陰侯墓發掘簡報. *Wenwu* 文物 (1978.8): 12–32.

"Da Tang jiaosi lu" 大唐郊祀錄. *Da Tang Kaiyuan li* 大唐開元禮, *fu* 附 (photographic reprint 适園叢書). Beijing: Minzu chuabanshe, 2000.

Dong Shan 董珊. "Chujian zhong cong 'da' zhi zi de dufa" 楚簡中從"大"之字的讀法. *Jianbo wenxian kaoshi luncong* 簡帛文獻考釋論叢. Shanghai: Shanghai guji chubanshe, 2014.

Fan Ye 范曄 (398–446), *Hou Hanshu* 後漢書, and Sima Biao 司馬彪 (d. 306). *Xu Han zhi* 續漢志. References are to the punctuated edition. Beijing: Zhonghua shuju, 1965.

Gan Huaizhen 甘懷真. "Zhongguo zhonggu jiaosi li de yuanliu yu tezhi" 中國中古郊祀禮的源流與特質." In *Zhonggu shidai de liyi, zongjiao, yu zhidu* 中古時代的禮儀、宗教与制度, ed. Yu Xin 余欣. Shanghai: Shanghai guji chubanshe, 2012.

Gu Jiegang 顧頡剛, and Yang Xiangkui 楊向奎. "Sanhuang kao" 三皇考. *Gushi bian* 7, 20–275, 337–90. Beijing: Pushe, 1926–1941. Reprint, Shanghai: Shanghai guji chubanshe, 1982.

Hanshu 漢書. Compiled by Ban Gu 班固 (32–92). Annotated by Yan Shigu 顏師古 (581–645). 12 vols. Beijing: Zhonghua shuju, 1962. Reprint, 1996.

He Ning 何寧. *Huainanzi jishi* 淮南子集释. Beijing: Zhonghua shuju, 1998.

Hou Hanshu: see Fan Ye.

Hsing I-t'ien 邢義田. "'Taiyi sheng shui,' 'Taiyi chuxing' yu 'Taiyi zuo': du Guodian jian, Mawangdui bohua he Dingbian, Jingbian Hanmu bihua de lianxiang" 太一生水、太一出行与太一坐：讀郭店簡、馬王堆帛畫和定邊、靖邊漢墓壁畫的聯想. *Meishu shi yanjiu jikan* 美术史研究集刊 30 (2011): 1–34, 351.

Huang Huaixin 黃懷信. *Heguanzi huijiao jizhu* 鶡冠子匯校集注. Beijing: Zhonghua shuju, 2004.

Jing Zhongwei 井中偉. "Wo guo shiqian jisi yiji chutan" 我國史前祭祀遺跡初探. *Beifang wenwu* 北方文物 (2002.2): 6–15.

Jiu Tang shu 舊唐書. Compiled by Liu Xu 劉昫 et al. Beijing: Zhonghua shuju, 1975.

Kalinowski, Marc. "Xianqin suili wenhua jiqi zai zaoqi yuzhou shengchenglun zhong de gongyong" 先秦歲曆文化及其在早期宇宙生成論中的功用. *Wenshi* 文史 (2016.2): 5–22.

Lai Guolong. "Mawangdui 'Taiyi zhu tu kao'" 馬王堆《太一祝圖》考. *Zhejiang University Journal of Art and Archaeology* 浙江大學藝術與考古研究 1 (2014): 1–27.

Li Ling 李零. "Baoshan Chujian yanjiu, (zhanbu lei)" 包山楚簡研究（占卜類）. In *Zhongguo dianji yu wenhua luncong* 中國典籍與文化論叢, vol. 1, 425–48. Beijing: Zhonghua shuju, 1993.

———. *Zhongguo fangshu kao* 中國方術考, 215–35. Beijing: Zhonghua shuju, 2006.

———. *Zhongguo fangshu xukao* 中國方術續考, 182–92. Beijing: Zhonghua shuju, 2006, which includes "Sanyi kao" 三一考; "Taiyi chongbai de kaoguxue yanjiu."

———. "Shuo jitan he jisikeng" 说祭壇和祭祀坑. *Rushan yu chusai* 入山与出塞, 17–40. Beijing: Wenwu chbuanshe, 2004.

Lin Meicun 林梅村. "Xi Han diguo daxing shidiao yishu de fazhan" 西漢帝國大型石雕藝術的發展. *Gudao xifeng: kaogu xin faxian suo jian Zhong-Xi wenhua jiaoliu* 古道西風——考古新發現所見中西文化交流, 112–37. Beijing: SDX Joint Publishing Company, 2000.

Liu Xinfang 劉信芳. "Baoshan Chujian shenming yu Jiuge shenqi" 包山楚簡神名與九歌神祇. *Wenxue yichan* 文學遺產 (1993.4): 11–16.

Lü Simian 呂思勉. *Sanhuang wudi kao* 三皇五帝考. *Gushi bian* 7, 350–60. Beijing: Pushe, 1926–1941. Reprint, Shanghai: Shanghai guji chubanshe, 1982.

Mao Qiling 毛奇齡. *Chunqiu Maoshi zhuan* 春秋毛氏傳. Taipei: Taiwan shangwu yinshuguan, 1983.

Qian Baocong 錢寶琮. "Taiyi kao" 太一考 ("Research on Taiyi"). *Yanjing xuebao* 燕京學報 8 (January 1932): 2449–78. Reprinted in *Qian Baocong kexueshi lunwen xuanji* 錢寶琮科學史論文選集, 207–34. Beijing: Kexue chubanshe, 1983.

Rao Zongyi. "Tushi yu cifu: Mawangdui xinchu Taiyi chuxing tu sijian" 圖詩與辭賦——馬王堆新出《太一出行圖》私見. In *Hunan sheng bowuguan sishi zhounian jinian lunwenji* 湖南省博物館四十周年紀念論文集, 79–82. Changsha: Hunan jiaoyu chubanshe, 1996.

Shiji 史記 [*Writings by the Senior Archivist*, aka *Historical Records*]. Compiled by Sima Qian 司馬遷 (?145–?86 BCE). 10 vols. Beijing: Zhonghua shuju, 1959. Reprint, 1982.

Sichuan province. Wenwu guanli weiyuanhui 四川省文物管理委員會. "Chengdu Yangzishan tutai yizhi qingli baogao" 成都羊子山土臺遺址清理報告. *Kaogu xuebao* 考古學報. (1957.4): 17–31.

Sima Biao: see Fan Ye.

Tian Tian 田天. *Qin Han guojia jisi shigao* 秦漢國家祭祀史稿. Beijing: SDX Joint Publishing Company, 2015.

Watanabe Shin'ichirō. *Zhongguo gudai de wangquan yu tianxia zhixu: Cong Ri-Zhong bijiaoshi de shijiao chufa* 中國古代的王權與天下秩序：從日中比較史的視角出發. Translated by Xu Chong 徐沖. Beijing: Zhonghua shuju, 2008.

Xiao Ji 蕭吉. *Wuxing dayi* 五行大義. Shanghai: Shanghai shudian chubanshe, 2001.

Xu Han zhi: see Fan Ye.

Yan Changgui 晏昌貴. *Wugui yu yinsi: Chujian suojian fangshu zongjiao kao* 巫鬼與淫祀——楚簡所見方術宗教考. Wuhan: Wuhan daxue chubanshe, 2010.

Yao Shengmin 姚生民. "Guanyu Han Ganquan gong zhuti jianzhu weizhi wenti" 關於漢甘泉宮主體建筑位置問題. *Kaogu yu wenwu* 考古與文物 (1992.2): 25.

Yu Shinan 虞世南. *Beitang shuchao* 北堂書鈔. Tianjin: Tianjin guji chubanshe, 1988.

Zhejiang province Wenwu kaogu yanjiusuo 浙江省文物考古研究所. "Liangzhu wenhua Huiguanshan yizhi dierci fajue jianbao" 良渚文化匯觀山遺址第二次發掘簡報." *Wenwu* 文物 (2001.12): 36–40.

Zuozhuan, Yang Bojun 楊伯峻. *Chunqiu Zuozhuan zhu* 春秋左傳注. Beijing: Zhonghua shuju, 1990.

Western-language Bibliography

Bujard, Marianne. "State and Local Cults in Han Religion." In *Early Chinese Religion, Part One: Shang through Han (1250 BCE–220 AD),* edited by John Lagerwey and Marc Kalinowski, 833–68. Leiden: Brill, 2009.
Loewe, Michael. *Crisis and Conflict in Han China, 104 BC to AD 9.* London: George Allen & Unwin Ltd, 1974.
Harper, Donald. "The Nature of Taiyi in the Guodian Manuscript *Taiyi shengshui*: Abstract Cosmic Principle or Supreme Cosmic Deity?" *Zhongguo chutu ziliao yanjiu* 中國出土資料研究 5 (2001): 1–23.
Sivin, Nathan. *Health Care in Eleventh-Century China.* Archimedes 43. Cham, Switzerland: Springer, 2015.
Tian Tian. "The Suburban Sacrifice Reforms and the Evolution of the Imperial Sacrifices." In *Chang'an 26 BCE: An Augustan Age in China,* edited by Michael Nylan and Griet Vankeerberghen, 263–91. Seattle: University of Washington Press, 2015.

7

Writing Abstractly in Mathematical Texts from Early Imperial China

KARINE CHEMLA 林力娜

Introduction

In the nineteenth century, information about the history of mathematics in China became more systematically available to European readers, thanks to the works of such China specialists as the Paris-based Edouard Biot (1803–1850) and the Protestant missionary Alexander Wylie (1815–1887).[1] In European discussions drawing contrasts between "the Chinese" and "the Europeans," knowledge about the history of mathematics was regularly adduced to support views about the "Chinese people" (as they were called, taken as a single entity, past and present): that they purportedly lacked intellectual abilities, for instance. One of the recurring themes in these discussions was the fact that "the Chinese" lacked a capacity for abstraction or, similarly, that their practical bent made them incapable of theorizing.[2] Views of this kind had been widespread since the end of the eighteenth century, and they more generally opposed "the Europeans" and all the other "peoples."[3] The publication throughout the nineteenth century of information about past mathematics in China did not significantly alter these views, which, unfortunately, are still quite widespread among academics today. In this chapter, my aim is to establish that such views fly in the face of the evidence at hand, based on the mathematic writings from early China. To achieve that goal, I concentrate on epistemological values to which writings from early China attest.

In previous publications, I have argued that the practitioners of mathematics in early China whose writings were handed down prized generality as a key epistemological value. I have also put forward the thesis that, in their eyes, generality mattered more than abstraction.[4] These conclusions, however, do not by any means imply that the same practitioners had no interest in, or use for, abstraction. So, in much the same way that I have described the meaning and practices of generality, I employ the earliest extant mathematical writings in Chinese to demonstrate and examine the formulation and the practice of abstraction in technical discussions found in the mathematical documents from Western and Eastern Han (206 BCE–220 CE).

Such a project confronts us with a major methodological problem. It is easy to approach abstraction in these contexts as an observers' category.[5] However, it requires more subtlety to deal with abstraction as an actors' category. The reason for this is simple. The mathematical writings that have survived from Han display mathematical knowledge in the form of problems, algorithms (or, to use the actors' term, "procedures" *shu* 術), and numerical tables. These writings apparently add no second-order comment on the related practice of mathematics. In particular, they formulate no assertions that allow us to capture how they understood and practiced abstraction. This holds true for the Han classic that was handed down through the written tradition, *The Nine Chapters on Mathematical Procedures* (九章算書, a title hereafter abbreviated to *The Nine Chapters*),[6] as well as the mathematical manuscripts excavated from Western Han tombs, such as *Writings on Mathematical Procedures* (筭數書).[7]

We thus have to devise a strategy to approach abstraction as an actors' category, in a situation where actors left no clear and unambiguous evidence for it. In the first part of this chapter, I present the strategy whose adoption, I suggest, will address this problem of approaching mathematical abstraction in the Han. I then outline how and where in *The Nine Chapters* we can spot abstraction, so as to interpret which benefits actors might have expected from their use of abstraction. In a third part, I turn to a manuscript, *Writings on Mathematical Procedures*, to show that the same form of abstraction that I have been describing occurs therein. Notably, one cannot find any trace of abstraction in the same sense in the (presumably) older manuscript originally titled *Mathematics* (數), whose text has been published.[8] Thus, this discrepancy, and what it suggests about the qualitative difference between the two manuscripts, needs further elucidation. In the fourth and final part of the chapter, I suggest that we can do more than simply perceive in Han dynasty works a notion of abstraction understood

Writing Abstractly in Mathematical Texts from Early Imperial China | 309

in the terms I have proposed. In fact, we have evidence of actors' work to achieve a greater abstraction of this kind. I conclude by describing what this approach can tell us about technical writing on mathematics in Han.

I. Abstraction as a Commentator's Category

As its title makes explicit, *The Nine Chapters* is composed of nine chapters. This point, which we might think unworthy of notice, will become interesting from our perspective. Each chapter contains a list of mathematical problems, most of which appear to be practical and particular. For example, problems about the area of a field give the magnitudes of length and width expressed as numbers of *bu*. However, the text of the procedures that follow these problems and their respective answers refers to length and width without assigning numerical values to them. (We return to this appearance below.) The mathematical problems are of three types. In some cases, a problem is given in relation to a "procedure" (this is how I translate *shu* 術 in this context, which in other contexts might be interpreted as "artful application"). In other cases, a set of problems is attached to a single procedure. In yet other cases, as we will see, procedures are given alone, without any mathematical problems. Let me provide an illustration by way of a problem that opens Chapter 8, which bears the title "Measures in Square" (*fangcheng* 方程) and which is entirely devoted to the procedure of the same name. (The capital letters indicate here that this is part of the canon):

(**8.1**) SUPPOSE THAT THREE *BING* OF HIGH-QUALITY MILLET, TWO *BING* OF MEDIUM-QUALITY MILLET AND ONE *BING* OF LOW-QUALITY MILLET PRODUCE (*SHI*) THIRTY-NINE *DOU*; TWO *BING* OF HIGH-QUALITY MILLET, THREE *BING* OF MEDIUM-QUALITY MILLET AND ONE *BING* OF LOW-QUALITY MILLET PRODUCE THIRTY-TWO *DOU*; ONE *BING* OF HIGH-QUALITY MILLET, TWO *BING* OF MEDIUM-QUALITY MILLET AND THREE *BING* OF LOW-QUALITY MILLET PRODUCE TWENTY-SIX *DOU*. ONE ASKS HOW MUCH IS PRODUCED RESPECTIVELY BY ONE *BING* OF HIGH-, MEDIUM- AND LOW-QUALITY MILLET.

今有上禾三秉，中禾二秉，下禾一秉，實三十九斗；上禾二秉，中禾三秉，下禾一秉，實三十四斗；上禾一秉，中禾二秉，下禾三秉，實二十六斗。問上、中、下禾實一秉各幾何。⁹

a. Evidence for Ancient Actors' Expectation of Abstraction

My reason for choosing this particular mathematical problem is that it allows us to immediately establish an essential point: the issue of abstraction is not one that only modern readers raise with respect to *The Nine Chapters*. It was meaningful for the most ancient readers we can observe.

The Nine Chapters partly owes its survival in the written tradition to its inclusion with nine other mathematical classics in the anthology *Ten Canons of Mathematics* (算經十書) in the earlier decades of the Tang dynasty (640s–650s). A group of scholars, commissioned to prepare an edition of these classics under the supervision of Li Chunfeng 李淳風 (ca. 602–670), presented it to the throne in 656. The same group also decided which ancient commentaries on the mathematical classics to include in the anthology, and, in the case of *The Nine Chapters*, the ancient commentary selected was the one completed by Liu Hui 劉徽 in 263. In addition, Li Chunfeng and his associates composed subcommentaries on all these texts. This anthology played a key role in the education in mathematics delivered within the context of state institutions.[10]

All of the earliest editions of *The Nine Chapters* that have survived contain both Liu's commentary and what, for the sake of simplicity, I refer to as "Li's subcommentary." This suggests the enormous impact that the Tang edition of the anthology had on the textual history of *The Nine Chapters*, as well as on the textual history of the other mathematical classics. What is of the utmost importance here is that *The Nine Chapters* was thus handed down with a carefully selected commentary and a subcommentary, through which the written tradition was received, and commentary and subcommentary yield evidence about how the oldest readers whom we can observe interpreted *The Nine Chapters*. Thus, they give us an essential insight for the issue addressed in this chapter.

The key point for us is that in relation to the procedure "Measures in square" given after the problem 8.1, the commentary ascribed to Liu Hui comments in the following way:

> This is a *universal procedure*. It would be difficult to *understand* [the procedure] with *abstract expressions* [*kongyan*]. This is why it was *deliberately linked* to [a problem of] millet, so as to eliminate the obstacle.

此都術也。以空言難曉，故特繫之禾以決之。[11]

This chapter is centered on the interpretation of this set of assertions. Many elements in the passage are important. Let us examine each of them in turn. The commentator seems to feel the need to account for the reason why *The Nine Chapters* dealt with the procedure "Measures in square" via a problem concerning millet. The question makes sense if we consider that some procedures are simply recorded without any mathematical problem, as we will see. The commentator appears to trace the providing of a problem here to the "deliberate" intention of the compilers of the classic, who aimed to enable the reader to "understand" the procedure. What exactly this type of "understanding" means is an issue I have addressed elsewhere.[12] For now, let me emphasize this: the commentator himself does *not* seem to perceive the problem presented in 8.1 as formulating a question whose *solution* the reader must learn from the text. Instead, he links the existence of the problem with an intention of the authors to aid the reader in "understanding." Thus, an interpretation of problem 8.1 as a "practical problem," as suggested by its format, is at odds with how a very early reader construed problems in *The Nine Chapters*.

The crucial element for us to discern here is this: the commentary makes clear that in this reader's view there was an alternative to a presentation of the procedure in relation to a problem. The alternative for the commentator apparently was to formulate the procedure in what I render as "abstract expressions," literally "empty talk" or "empty expressions" (空言 *kongyan*).[13] Why I choose to translate the binomial expression as "abstract expressions" becomes clear below. For the moment, we need to understand the commentary in this context, and primarily the term *kongyan*, which refers to a crucial alternative. This discloses a phenomenon, which, from the commentator's viewpoint, embodies abstraction, that is, to begin with, abstraction as a commentator's category.

b. Abstraction from a Wider Perspective: Elements of a History of the Term Kongyan

The interpretation of the passage, in no small part because of its deployment of the phrase *kongyan*, has a relevance that far exceeds the scope of the mathematical classic under discussion. For by using the expression *kongyan* to designate a form of writing in mathematics, the commentator draws a parallel between the mathematical classic and the *Annals* (*Chunqiu*, aka *The Spring and Autumn Annals* 春秋), the Classic ascribed to Confucius, writing about the past as editor, compiler, or author.[14] In his postface (Chapter 130) to *Historical Records* (*Shiji* 史記), Sima Qian 司馬遷, writing ca. 90 BCE, characterizes Confucius's chosen method of writing history by reference to the *Annals*, using the term *kongyan*. (Some speculate that Sima refers here to the views of Dong Zhongshu for the expression *kongyan*—but there are differences between the *Shiji* formulation and that found in the received text of the *Chunqiu fanlu*.)[15] Suffice it to say that scholars for millennia have debated the interpretation of *kongyan*, especially in relation to another term that occurs in the same sentence, which Sima Qian's postface ascribes to Confucius, that is, *shi* 事, meaning, "action, event, situation."[16] Because an understanding of both terms is relevant for my analysis, I evoke some interpretations of the sentence from *Historical Records* under focus to point out some of the main options. In the interpretation offered by Li Wai-yee, the statement ascribed to Confucius reads as follows: "I wanted to convey this [*chih* 之, i.e., Confucius's concerns] through *abstract, conceptual language* (*k'ung-yen* 空言, literally, "empty words"), but [to do so] would not be as profound, compelling, and clear as to embody and see [it] through past *events and actions* (*hsing-shih* 行事)" (我欲載之空言, 不如見之於行事之深切著明也).[17] In Li's view, when writing history, the use of *kongyan* thus represents an alternative to *shi*, and *kongyan* has positive connotations, even though it is considered pragmatically inferior to the use of *shi*. Li's interpretation of the sentence contrasts with one proposed by Joachim Gentz to explain the same statement. Gentz attributes more negative connotations to the use of the expression *kongyan*, suggesting that the writing of history to which the statement refers employs both *kongyan* and *shi*. Indeed, Gentz interprets the same lines as follows: "If I had wanted to explain this in detail [only] with empty words, this would not be as deep, incisive, manifest and clear as if I showed it by using past events." In a footnote, Gentz glosses the last word: "literally, completed affairs/tasks."[18] In fact, this difference of interpretation reflects the fact that, as was argued, the expression *kongyan* has been used *at the time*, and in different contexts, with this whole spectrum of meanings.[19]

Another point is worth noticing: whereas the two interpretations just mentioned agree in reading the relation between *kongyan* and *shi* as an opposition between two modes of writing a classic, which are either alternative to one another or conjoined with each other, other scholars have suggested that other forms of relationship were likely meant. For instance, Watson opts for a third interpretation of *kongyan*, in which the expression has positive connotations, and yet an interpretation of the relationship between *kongyan* and *shi* different from those Li and Gentz envision, for Watson translates: "If I wish to set forth my *theoretical judgments*, nothing is as good as illustrating them through the depth and clarity of *actual events*."[20] This interpretation implies that for Watson, *kongyan* can be best formulating using *shi*. We return to the basic issue of the relationship between *kongyan* and *shi* below.

Against this backdrop, the commentary formulated in the context of problem 8.1 in *The Nine Chapters* becomes even more interesting. Clearly, not only the expression *kongyan* but also the whole structure of the sentence is meant to parallel and thus evoke the important statement ascribed to Confucius. The parallel appears even more exact if we consider that the term *shi* was used in the commentaries on *The Nine Chapters* and other mathematical classics to refer to "mathematical problems followed by procedures."[21] Accordingly, when the quotation from the commentary ascribed to Liu Hui draws a contrast between presenting a procedure using *kongyan* or, alternatively, uses a mathematical problem to formulate the procedure (*shi*), it invokes the very terms Confucius purportedly used to describe his way of writing the *Annals* classic. In the commentator's eyes, one could use the same contrast to interpret the choices of the authors of *The Nine Chapters* made with respect to how the mathematical classic should be written. In this regard, how we interpret the commentary following problem 8.1 of *The Nine Chapters* has a relevance far beyond the specific case of mathematics.

Let us therefore return to this statement, then, asking the following questions: How should we interpret *kongyan* in the context of this mathematical commentary? In other words, which connotations does the commentary attributed to Liu Hui ascribe to the expression? And which relationship does the commentator intend between *kongyan* and *shi*? In what follows, I argue an answer to these questions in the specific context of mathematics. Nonetheless, my solution sheds light on the commentator's interpretation of both *kongyan* and the relationship between *kongyan* and *shi* 事 occurring in Confucius's statement. In my conclusion, I return to the more general insights we can gain into these issues after consideration of such concrete and embodied uses of these terms.

As I have emphasized, the commentary on the procedure "Measures in square," in Chapter 8, seems to indicate the commentator might have expected that this procedure would be presented in terms of *kongyan*. Such an expectation is stated explicitly *only* in this one context, even though in *The Nine Chapters* many other procedures are formulated after one or a series of mathematical problems. It is thus tempting to infer that the expectation derives from the specific character that the commentator attaches to the procedure "Measures in square." As the commentary makes clear in the opening statement of the quotation, it is "a *universal procedure*," more literally a "dominant procedure" or a "leading procedure."

Arguably, for the commentator a "universal procedure" requires a formulation using *kongyan*. Significantly, there is only one other procedure in *The Nine Chapters* that the commentary ascribed to Liu Hui describes as a "universal procedure" (都術). In the second case, which occurs in Chapter 2, however, the exegete does not add a comment explicating how the mathematical classic describes the procedure. In particular, he does not emphasize any lack of *kongyan* in that context. One might thus assume that in his eyes this second "universal procedure" is precisely expressed using *kongyan*, in conformity with what his commentary under consideration has led us to assume. Understanding the difference between the formulations in the two contexts may allow us to offer an interpretation for *kongyan*, and its relationship with *shi*, from the commentator's perspective. Let us turn, then, to that procedure, presented at the beginning of Chapter 2, titled "Unhusked and husked grains" 粟米.

II. A Form of Abstraction as Actors' Category in *The Nine Chapters*

The single other procedure that the commentator describes as "universal procedure" is what we call the "rule of three."[22] The rule allows us to solve the following type of problem: given two numerical values expressing the equivalence between a thing A and a thing B, where the given quantity of A has the same value as the given quantity of B, when we are presented with another quantity of the thing A, what is the corresponding equivalent quantity of B?

The name given in *The Nine Chapters* to the meta-operation[23] carried out by the procedure that the commentary calls a "universal procedure" suggests that the authors granted it a fundamental role. In effect, this meta-operation is called "SUPPOSE" 今有, that is, its name consists of the two characters that, in general, begin the statement of a mathematical problem in the classic. Let us read the statement of the procedure before we explain its statement and how it works, indicating the places where commentary and subcommentary are inserted, without translating them:

(**2.0**) Suppose 今有 [. . . commentary]

Procedure: one multiplies, by the quantity of what one has, the *lü* of what one seeks, what makes the dividend; one takes the *lü* of what one has as divisor. [. . . commentary] dividing the dividend by the divisor gives the result.

[. . . commentary] 術曰：以所有數乘所求率爲實。以所有率爲法 [. . . commentary and subcommentary] 實如法而一。"[24]

a. Examining Features of a "Universal Procedure" in The Nine Chapters

The terms designating the operands of the operations, here multiplication and division, are abstract in our sense of the term, that is, according to an observers' category. To push this insight further, we note that the structure of the terminology used to formulate the procedure draws two oppositions. First, the terminology opposes "what one has [already]" to "what one seeks." With the notations introduced above, the former expression ("what one has") refers to any quantity of *A*, while the latter expression qualifies any quantity of the thing *B*, because *B*, as the topic of the question asked, is the thing whose quantity one seeks to determine. Second, the terminology opposes actual "quantities" (*shu*) to *lü* 率. In the passage under consideration, *lü* designates the two values expressing the two equivalent amounts of, respectively, *A* and *B* that were given at the beginning. The term *lü* designates these values by referring to a key property that they share: they can be either both multiplied or both divided by the same number, and the results of these multiplications and divisions will retain the capacity of stating the equivalence between the two things represented.[25] Indeed, if a quantity *Q* of the thing *A* is equivalent to a quantity *Q'* of the thing *B*, then twice the quantity *Q* of *A* is equivalent to twice the quantity *Q'* of *B*, and, similarly, half the quantity *Q* of *A* is equivalent to half the quantity *Q'* of *B*. This property holds true with full generality, meaning, in every conceivable multiplication or division that can be adduced. By contrast, *shu* 數 indicates an actual *quantity* of something, and if one looks for a quantity of something else equivalent to it, one cannot modify the former, or else the result will not be the same. The terms *shu* and *lü* thus refer clearly to the *nature* of the values that are used as operands of the operations prescribed by the procedure quoted above, when the meta-operation indicated by the term "suppose" is put into play. The actions corresponding to the procedure thus seem perfectly clear. And yet I argue there is an essential nuance that would be easily missed if this text were given alone, without additional elements.

In fact, the text of the procedure quoted above is placed at the beginning of Chapter 2, outside the framework of any problem. (This is why I refer to it as 2.0.) Did the commentator have this in mind when, for the procedure placed after problem 8.1, he opposes a formulation using *kongyan* (which he seems to expect for such procedures) to the formulation that he reads in *The Nine Chapters*? The key fact that enables us to address this question is this: in Chapter 2, the procedure for "suppose" is followed by *dozens* of problems, each with its related procedure, and for each of these problems, the commentary or subcommentary (more often the latter)[26] shows that its procedure of resolution is an instantiation of the first procedure (2.0). Let us quote the first problem of this kind, with the related procedure, to explain this point:

(2.1) SUPPOSE THAT, HAVING ONE *DOU*[27] OF UNHUSKED GRAIN (OF FOXTAIL MILLET), ONE WANTS TO MAKE COARSELY HUSKED GRAIN. ONE ASKS HOW MUCH IT YIELDS.

ANSWER: IT MAKES SIX *SHENG* OF COARSELY HUSKED GRAIN.

PROCEDURE: IF, HAVING UNHUSKED GRAIN, ONE SEEKS COARSELY HUSKED GRAIN, ONE THREE-FOLDS THIS, AND ONE DIVIDES BY FIVE [Chemla: literally, "five then/thus one."]

今有粟一斗，欲爲糲米。問得幾何。

答曰：爲糲米六升。

術曰：以粟求糲米，三之，五而一。[28]

The subcommentary on this problem explains the link between the "universal procedure" and this procedure by showing how the latter derives from the former. The numerical values defining the equivalence between unhusked millet (the thing *A* of our general explanation above) and coarsely husked millet (the thing *B*) have been given earlier in *The Nine Chapters*, and they are used directly without being repeated in this passage. These numbers come from a general table about grains placed before the "universal procedure" at the beginning of Chapter 2. There, unhusked millet is tied to the number 50, whereas coarsely husked grain is tied to 30. The meaning of these numbers is this: 50 units of unhusked millet have the same value as 30 units of coarsely husked grain, with the statement holding true whichever the chosen unit might be, as long as the units are the same in the two related statements. Note that these integral values, which in *The Nine Chapters* express the equivalence between different types of grain, tally with the ratio used in administrative practice at the time. In the Table of *The Nine Chapters*, the values are in fact given without measuring units, for the very good reason that they are meaningful in relation to each other, and not separately. By contrast, in the statement of mathematical problems like 2.1, the quantity of grain whose equivalent amount is sought for (like 1 *dou* of unhusked millet) is expressed with respect to measurement units of capacity, as is the unknown amount sought for. (The answer of problem 2.1 is 6 *sheng*.)

The subcommentary on problem 2.1 brings all these numerical values in relation to the terms occurring in the "universal procedure." In the mathematical problem, as the seventh-century subcommentary explains, one "seeks" "coarsely husked millet," and hence the quantity (of one *dou*) of unhusked millet is the "quantity of what one has." The numbers 50 (for the unhusked grain) and 30 (for the coarsely husked grain) are respectively interpreted as the "*lü* of what one has" and the "*lü* of what one seeks." The procedure attached to problem 2.1 thus derives from the "universal procedure." However, a change was applied to 50 and 30, because the concrete procedure uses 5 and 3 instead.[29]

The subcommentary insists that the transformation of 50 and 30 into, respectively, 5 and 3 stems from the fact that, in the situation of problem 2.1, these numerical values are identified as *lü* 率. Accordingly, they can be jointly divided by 10, and the ensuing numbers 5 and 3 will continue to correctly express the relationship between unhusked millet and coarsely husked grain. The term *lü* refers to the *nature* of 50 and 30 with respect to the situation, and this fact accounts for the correctness of the transformation applied to both numerical values, which changes them into 5 and 3, respectively. These simple explanatory remarks have important consequences.

First, we discover that by the simple designation of two operands in this procedure as *lü* 率, the text of the "universal procedure" in fact *prescribes* a joint transformation of their numerical values. This transformation must be carried out before executing the operations explicitly mentioned in the text of the "universal procedure." Accordingly, the mode of prescription in the "universal procedure" is less simple than it appeared at first sight. Moreover, the fact that the "universal procedure" designates the numbers in question as *lü* 率 not only states but also supplies the *reason* why these numbers can be simultaneously modified, as this derives from the specific nature of these numerical values. Note that these two layers of meaning of the term *lü* are brought to light both in the specific problem 2.1 with the related procedure and in the subcommentary on it. The same analysis holds true for the problems 1–32, of a total of forty-six problems in Chapter 2. This is the nuance that might easily have been overlooked, had *The Nine Chapters* only given the procedure 2.0. We will return to this issue.

b. Interpreting Kongyan in The Nine Chapters

The key point here is that the term *lü* 率 occurs in the "universal procedure" and not in any of the other thirty-two specific procedures. As a result, although the "universal procedure" and each of the specific procedures refer to the *same* computations, they are *not expressed in the same way*. We have reached the crux of the argument with this observation, which prompts a most simple question: why do we have, in fact, for each and every problem in the chapter, *two* procedures prescribing the same operations, the one placed at the beginning of the chapter and common to all problems and a specific one attached to each of the problems in turn? We observe that the textual *dispositif* in *The Nine Chapters*[30] displays a procedure *in a relation of abstraction* with a sequence of other procedures. This the commentary makes explicit. I thus put forward the hypothesis that when the commentary after Problem 8.1 speaks of expressing a procedure in terms of *kongyan*, it refers to a procedure placed in such a specific relation to other procedures, as we find in Chapter 2. This explains why I chose to translate *kongyan* as "abstract expressions."

This conclusion is supported by these facts: in Chapter 8, where the commentary claims that there is no *kongyan*, and puts forward an explanation why this is so, we precisely do *not* have such a *dispositif*. There is no text describing a procedure placed before the statements of all the mathematical problems (no procedure 8.0, if you like). As a result, we do not have this phenomenon whereby two texts of procedure correspond to the operations used to solve each problem. In contrast to the unique Chapter 8, several chapters in *The Nine Chapters* present the same feature as Chapter 2, insofar as these chapters begin with a procedure, and commentary as well as subcommentary show that these opening procedures are tied with the other procedures stipulated for the mathematical problems in the same chapter in exactly the same way as the "universal procedure" just described. By way of consequence, similarly, two texts of procedures refer to the same sequences of operations, but they use different terms.

My interpretation is thus that *kongyan* is an expression used by the commentary to refer to this form of abstraction. It is most important for us to note that these "abstract" procedures are *not* given alone, but systematically in relation to mathematical problems and their specific procedures, namely *with* what I have argued the commentators designate as *shi* 事. In other words, in this way of writing mathematics, as seen in Chapter 2, the composition of a chapter supplies what the commentator deems both *kongyan* and *shi*. It is the lack of *kongyan* in Chapter 8 that the commentary with which we began this chapter attempts to account for.

So far, we have dealt with the commentator's category for abstraction. However, through his remark the commentator does point out a property of the text of *The Nine Chapters*: the existence of double procedures (*shu* 術) solving the same problem and referring to the same operations to do so. One of the two procedures is in a relationship of abstraction with the other, or, for that matter, with several others. This is illustrated by the procedure associated with the operation called "suppose," which similarly underlies all the specific procedures placed after each of the first thirty-two problems in Chapter 2 (and also elsewhere).[31] In effect, the "abstract procedure" occurs most often at the beginning of a chapter. Importantly, whether such "abstract" procedures are formulated in relation to a specific mathematical problem or not does not seem to matter to either the commentator or the subcommentator. What matters is their relationship with the other procedures that follow in the same chapter.[32] As a result, a correlation appears between what was placed at the beginning of a chapter and the remaining contents of the chapter. Let us be clear about what this suggests: that in *The Nine Chapters*, a chapter is not simply a given section in a book, but it conveys in and of itself a mathematical meaning, which is expressed by its structure: specifically, it is divided into two parts, with the first part corresponding to the abstract procedure placed at its beginning and a second part containing mathematical problems and operations "covered" by the abstract procedure. This meaning of a chapter is underscored by the propensity of the editor(s) to title the chapter by the name of the meta-operation carried out by the first procedure. In other terms, the structure of *The Nine Chapters* appears to be closely related to the epistemological value of abstraction, as the commentator understands it.

All the foregoing remarks might explain the term chosen by the commentator to refer to such procedures. As we have emphasized, these "universal procedures" are literally "dominant procedures" or "leading procedures."[33] In any case, with these observations, we have moved from a statement of the commentator to a textual property of the mathematical classic. I suggest that this textual feature of *The Nine Chapters* be read as evidence for a category of "abstraction" that we can attribute to the editor(s) of the mathematical classic. In this case, the "category" of abstraction is not a term that we find dwelt upon in *The Nine Chapters*, but a textual property. In relation to this, the commentator clearly intends to ascribe a practice of "abstraction" to the author(s) or the editor(s) of the book, attributing to this term the meaning expounded above.

We have offered a hypothesis with respect to what the commentator designates as *kongyan*. What remains is to identify the features that procedures must possess before they are perceived to fall into this category, that is, the features possessed by procedures placed in a relationship of abstraction with respect to others. Let us consider the set of these procedures. Is there anything that is distinctive to their subset?

To begin with, one point is clear: not all of the procedures in this set qualify as "abstract" in a modern mathematical sense of the term. By contrast with the text of the procedure given for the meta-operation "suppose," which appears to be "abstract" in both an actors' and observers' sense, other procedures in our set of "abstract procedures" still refer to specific situations. This is the case for the procedure "Reducing the width" (*shao guang* 少廣), for example. As in our example from Chapter 2, "Reducing the width" is placed at the beginning of chapter 4, outside the framework of any specific problem. Moreover, "Reducing the width" has a relation of abstraction with respect to the first eleven procedures of the chapter. Nonetheless, the text of "Reducing the width" refers to specific numerical values and the situation of rectangular cropland, in other words, using components that today few would consider to be abstract.[34]

Conversely, some procedures that could be considered abstract in a modern mathematical sense are not "abstract" in the sense identified through the structure of *The Nine Chapters*. One can mention, as examples, the procedures for "Gathering parts" (*He fen* 合分), that is, adding fractions together, or "Multiplying parts" (*Cheng fen* 乘分). This difference between modern expectations and the commentator's justifies a posteriori the attempt to recover actors' categories in relation to "abstraction." Only on this basis can we hope to understand what was at stake for actors when they practiced abstraction in (one of) the mathematical sense(s) we have identified that they impute to the operation.

c. The Characteristic Features of Abstract Procedures

What, then, can we identify as common to the texts of procedures gathered in the set of "abstract procedures" in *The Nine Chapters*? In the case of the procedure for the meta-operation "suppose," we have mentioned a feature that seems to give an essential clue. The text of the "universal procedure" referred to quantities in a situation, using terminology that showed a complex structure. These terms helped practitioners distinguish between different types of numerical values in a situation in which the procedure could be applied. Moreover, the text of the procedure prescribed operations using the term *lü* 率. This term specifically stated a property of the numerical values that accounts for their possible transformations. Most importantly, all these terms and expressions introduced in the "universal procedure" occur when the commentator and the subcommentator establish the correctness of a procedure by showing that the procedure amounts to the use of the "rule of three." Put another way, it is with the terms introduced in the "abstract" procedure that the commentators formulate the reasons why the other procedures are correct. The "abstract" procedure thus provides essential conceptual tools to understand the "derived" procedures.

More generally, in the set of abstract procedures identified, all the texts of procedures make use of specific terms to designate the operands or to prescribe the operations, with these terms pointing to the reasons why the procedure is correct. Accordingly, they often occur in the proofs of the correctness of procedures offered by the commentator and the subcommentator for other "derived" procedures. Among these terms, let us mention, in addition to *lü*, *tong* 通 ("to make communicate"), *tong* 同 ("to equalize"), *cui* 衰 ("coefficient for weighing in function of the degree"), *ji fen* 積分 ("parts of the product"), and so on. The passages that describe these procedures thus more generally introduce concepts with which to understand the meaning of the operations to be carried out in procedures that, as the commentator and the subcommentator establish, all derive from the "abstract" procedures. In addition, interestingly enough, commentary and subcommentary on these "abstract" procedures also aim at establishing their correctness. They regularly do so by borrowing the context of a concrete problem, sometimes using a problem that follows in the same chapter, and sometimes a problem that was invented for the purpose of proving. By such means, the commentator and the subcommentator make explicit the meaning of their operations in this context.

If we now go back to revisit the commentary on problem 8.1 and the procedure "Measures in square," it seems the commentator considers that in this case, despite the property of the first procedure as "universal," the dissociation between an "abstract procedure" and the following problems would not have worked in this way. Interestingly enough, the commentary for this chapter (as in the other chapters) has all the procedures deriving from the first. However, in contrast with procedures found in all the other chapters of *The Nine Chapters*, Chapter 8 alone relates the successive procedures for problems 8.2 to 8.18 with 8.1 by the unique formula: "One follows 'Measures in square'" 如方程. I have argued elsewhere that key terms used in the formulation of the first procedure exhibit the property characteristic of the abstract procedures.[35] However, in the case of the first procedure in Chapter 8, the text in *The Nine Chapters* plays with their meaning, using these terms with a concrete meaning linked to millet as well as with an abstract meaning.[36]

In conclusion, *The Nine Chapters* regularly gives two texts of procedures referring to the same operations for the solution of a problem, one to which one might apply the term *kongyan*, introduced in a commentary, and the other to which both the commentator and the subcommentator apply the term *shi*. Why, one might ask, should *The Nine Chapters* give two procedures? It appears that the one that is placed directly in relation to a problem prescribes operations in a direct way. The other procedure, which is usually placed earlier in the book, deploys a more "abstract" language to denote them, thereby highlighting the reasons for the correctness of the operations. In the example developed above, the "abstract" procedure thus highlights the origin of the numbers 3 and 5 that the derived procedure puts into play. Clearly, the choice made in *The Nine Chapters* for writing down mathematics is to make use of both types of formulation simultaneously.

III. The Form of Abstraction Identified and the Excavated Manuscripts

To recapitulate: so far we have offered an interpretation for the form of abstraction to which the commentator on *The Nine Chapters* referred when using the expression *kongyan*. We have noted the term seems to have qualified specific procedures given in relation to several other procedures that derive from them, as the commentator and the subcommentator highlighted. In view of this, I have suggested that *The Nine Chapters*' supplying of two procedures for the same problem—one from which the other, or indeed, several others derived—could also be seen as a form of abstraction, in the practice of the actors who contributed to the production of the classic. Having established this, I next argue that this practice of abstraction is not specific to the text of *The Nine Chapters*. In fact, we can identify this practice in a mathematical manuscript that has been excavated from a Western Han tomb. However, it seems that the phenomenon is *not* attested in the second mathematical manuscript that is dated to Qin but lacks provenance. Hence, this practice appears to help us perceive a difference between manuscripts that at first sight look very much alike.

As was mentioned earlier, the Zhangjiashan 張家山 *Writings on Mathematical Procedures* was found in a Western Han tomb in present-day Hubei province. The archaeologists who excavated the tomb (number 247) during the winter of 1983–4 determined it had been sealed in about 186 BCE.[37] The document, written on 190 slips, is divided into sixty-nine sections, which bear headings, and it records three main components: numerical tables, mathematical problems, and procedures (called *shu* 術, the same term used in *The Nine Chapters*). Clearly, the organization of this Zhangjiashan material does not seem to follow the same uniform principles as in *The Nine Chapters*. Moreover, it does not have "chapters."[38] However, we can identify the same type of "abstraction" in the Zhangjiashan manuscript as was described above for *The Nine Chapters*. For instance, the text recorded in the section on slips 74 and 75 consists of a procedure given outside the framework of any problem and carrying out the meta-operation called "Determining the standard price on the basis of the *shi* [a measurement unit]" (*Shi lü*). This name, which likewise constitutes the heading of the section in slips 74 and 75, occurs as the meta-operation to be carried out in the mathematical problem recorded in the section that follows the heading "Selling salt" (*Gu yan* 賈鹽), which Peng Hao places right after the previous one. In this second section, which features on slips 76 and 77, the problem is then followed by a procedure.[39]

Two features of these four strips (construed as two sections) are essential for my purposes. First, the procedure attached to the problem and the procedure given for the meta-operation "Determining the standard price on the basis of the *shi*" exhibits precisely the same relationship described above between "abstract" and "derived" procedures in *The Nine Chapters*. In other words, in both the Zhangjiashan manuscript and *The Nine Chapters* classic, to deal with the same mathematical problem, the document records two texts of procedures that refer to the same operations, one more "abstract" and one "derived." Second, the text of the procedure carrying out the meta-operation "Determining the standard price on the basis of the *shi*" and recorded in the section bearing that title (strips 74–75) presents the same characteristics as what we have described above for "abstract" procedures: it prescribes operations using terms that explicate the reasons why the operations should be carried out.[40] (Meanwhile note that this text is not formulated abstractly in a modern mathematical sense because it refers to cash and to concrete measurement units).

The same phenomenon occurs in several other sections of the Zhangjiashan manuscript but, interestingly enough, apparently not in the unprovenanced mathematical manuscript titled *Mathematics* (deemed by its editors to be Qin in date). To date, I have not been able to identify in it a set of two procedures that could be described as "abstract" and "derived" in the sense I have established above. This gives us a perspective from which we can perceive a difference between the two manuscripts. Though they are often related in the secondary literature, the two manuscripts do not seem to reflect the same type of mathematical practice. More research is required to fully appreciate the significance of this difference.

Let us return to the meta-operation "Determining the standard price on the basis of the *shi*," as seen in the excavated Zhangjiashan *Writings on Mathematical Procedures*. In an earlier work,[41] I was able to establish several points, which can now be reinterpreted in a new light. First, I showed that the procedure for "Determining the standard price on the basis of the *shi*" prescribed the same operations as the procedure corresponding to the meta-operation "Directly determining the standard price" (*Jing lü* 經率) in *The Nine Chapters*. However, their formulations differ. Using observers' categories at the time, I showed that the latter was clearly "abstracted" from the former. The text of the procedure carrying out the meta-operation "Directly determining the standard price" uses a set of terms referring to the operations that points out the reasons for carrying out the operations in a way different from what we find in the Zhangjiashan *Writings on Mathematical Procedures*. Accordingly, what I described as "abstraction" from an observers' viewpoint can be reformulated in actors' categories as follows: the abstract procedure corresponding to "derived procedures" keeps being reworked, precisely with respect to how it refers to the reasons for the correctness of the operations. Simply put, the more abstract procedure in *The Nine Chapters* formulates the reasons for the correctness of the same specific procedure in another way than the abstract procedure in the Zhangjiashan *Writings on Mathematical Procedures*. Coincidentally, in *The Nine Chapters* the text of the procedure for the meta-operation "Directly determining the standard price" is not formulated abstractly in a modern mathematical sense of the word. By contrast, "abstraction" in the sense that I have attributed to early actors is an operation that was likely the object of further reflection during the intervening years between the time of the production of *Writings on Mathematical Procedures* and the later date when *The Nine Chapters* came together, and we can find evidence for this reflection in how the higher-level procedures are reformulated.

Second, in *The Nine Chapters* the procedure corresponding to the meta-operation "Directly determining the standard price" is also placed in Chapter 2 but not at the beginning of the chapter. Actually, Li Chunfeng's subcommentary establishes the fact that this procedure derives from the "universal procedure," which carries out the meta-operation "suppose" in the same way that it does for the procedure following problem 2.1. Thus, Li's subcommentary underscores the same relationship between the two derived procedures and the "universal procedure." In other words, the "abstract" procedure for "Directly determining the standard price" became itself the basis for further abstraction, and, in this process, it became in turn a "derived" procedure with respect to the "abstract" procedure for the meta-operation "suppose."

Both conclusions reveal other facets of this mathematical work on a form of abstraction that, I claim, the structure of *The Nine Chapters* indicates and the commentary on the problem 8.1 refers to. Interestingly enough, this work reveals an intimate relationship between *Writings on Mathematical Procedures* and *The Nine Chapters* from the viewpoint of actors' operations. This feature awaits further research.

IV. Conclusion

This inquiry into abstraction began with a statement in which the commentator formulated a remark contrasting two ways of writing down a "universal procedure," or a "leading procedure." The commentator used the expression *kongyan* 空言, which I translate as "abstract expressions," so as to refer to one type of formulation. I then analyzed what I have suggested interpreting as the deployment of this type of formulation in relation to the meta-operation "suppose," as recorded in *The Nine Chapters*. I have noted that the commentary contrasts this formulation with another type (identified in connection with problem 8.1 examined above), that is, using a specific problem and a procedure attached to it. Interestingly enough, both Liu's commentary and Li's subcommentary consistently use the term *shi* 事 to designate this type of mathematical problem with its own procedure. As I have shown, the commentator on *The Nine Chapters* points out a contrast between two types of formulations, which he ties to a statement that Han sources ascribed to Confucius's decision about writing history in the *Annals*.

In my view, the commentator's remark highlights a structural feature of *The Nine Chapters*, which represents a reflection of a practice of "abstraction." I have suggested that this structural feature might be given the status of an actors' category, even though in the main text of *The Nine Chapters* itself, no specific term refers to it. Indeed, for a significant number of problems, the sequence of operations to be used to solve them is prescribed by means of two procedures in the classic. Specifically, the text for one of these procedures is placed directly in relation to the problem, whereas the text of the other is placed before the problem, often at the beginning of a chapter.[42] The commentary and subcommentary further establish the correlation between the text of the more "abstract" procedure and the reasons why the operations of the specific procedure are put into play. Thus, one of these procedures is shown to be in a relationship of abstraction with the other (and in fact, as we have seen, with several others). As I have repeatedly emphasized, the practice of abstraction evidenced differs from the one we, as moderns, might expect to find. Moreover, the textual structure relating "abstraction" aims to highlight certain concepts, both with respect to objects dealt with and the operations applied to them, that shed light on the meaning of the lower-level procedures. The commentary and subcommentary make this meaning explicit. Conversely, the proof of the correctness of the higher-level procedures makes use of the context of a mathematical problem. For instance, the procedure for the meta-operation "suppose" is established as correct by using an interpretation of its operations of multiplication and division that is formulated in the context of the situation of problem 2.1. In sum, the analysis developed in this chapter establishes that, for the commentator and subcommentator, a concern with one form of abstraction determined the specific way that *The Nine Chapters* was composed. In the exegetes' view, therefore, abstraction thus enjoyed an epistemological value in mathematics already in Han times (a view confirmed by the Zhangjiashan slips). This conclusion provides an interpretation of how the commentator understood that technical writing in mathematics had been achieved in the case of *The Nine Chapters*.[43] Note that my first emphasis concerns mathematical practice, and the second, how abstraction figures in actual technical writing.

If my analysis is correct, it yields an insight into how the commentator who wrote the commentary following problem 8.1 interpreted Confucius's statement about his formulation of knowledge about the past, from which the commentator borrowed his sense of two key terms, *kongyan* and *shi*. As we have seen, for this commentator, *The Nine Chapters* (which is comparable in this respect to the Zhangjiashan *Writings on Mathematical Procedures*) would present mathematical knowledge in light of *both kongyan* and a sequence of *shi*. Both of these are conjoined in what I have called a textual *dispositif*. The choice to deploy *kongyan* and *shi*, the formulations of procedures that they designate, and the role that the commentary attributed to them—all these suggest how the exegete who wrote the key passage on which this chapter centers understood mathematical knowledge should be formulated. It seems to me that the commentator, like Gentz, read the statement ascribed to Confucius to mean that any proper exposition of knowledge required the deployment of both *kongyan* and *shi*. Certainly, in *The Nine Chapters* commentator's understanding, *kongyan* is not an alternative to *shi*, but a necessary complement to it.[44] We have seen the form that, in his view, this conjunction took in *The Nine Chapters*. Moreover, for the commentator, as for Li Wai-yee, *kongyan* is characterized by its use of "abstract, conceptual language." What such language entails, we can see in the context of mathematics, where this language achieves a form of generality that derives from grasping general concepts at the level of reasoning about correctness. Clearly, in this context, *kongyan* carries no derogatory connotation, and so it cannot mean "empty talk." Finally, if my hypothesis proves to be correct, through the comparison between the field of mathematics and that of early conceptual history, the commentator's use of the *kongyan-shi* dichotomy would inevitably imply a view of how *kongyan* and *shi* relate to each other.[45] His interpretation, which his comparison reveals, combines features that occur in both Joachim Gentz's and Li Wai-yee's translations, without being identical to either.

Somewhat surprisingly, the commentator constructed a parallel between the composition of the mathematical classic *The Nine Chapters* and the textual practice ascribed to Confucius in the *Annals* traditions in early Western Han sources. In modern historiography, *The Nine Chapters* was often deemed a "practical" book. Indeed, a superficial reading invites the conclusion that its formulations are straightforward and that it makes no second-order comments on mathematical knowledge and practice. Moreover, the outlines of its problems evoke concrete situations to which officials active in administrations had to attend. However, the foregoing analysis of the commentator's reading calls into question that we should limit ourselves to that simplistic picture. In particular, our analysis shows that ancient readers perceived features in the classic that attached it to ancient sages' way of approaching knowledge. For these readers, both the wording and the structure of *The Nine Chapters* bespoke precisely a much greater interest in abstraction than we, modern readers, would admit, unless we strive, as I have done in this chapter, to understand abstraction in actors' terms.

This conclusion has an import that goes beyond the case study that I presented. The majority of Chinese mathematical texts that were written up to the sixteenth century took the shape of a set of problems and procedures solving them. How should we interpret this genre of technical literature? The foregoing analysis has shown that interpreting these two types of textual components (problems and procedures) in a modern way is what has led most modern historians to assert that these works were practically oriented, meant as exercises or reference books to solve problems that officials encountered in their daily practice. Drawing on an interpretation of this kind, modern historians have accordingly argued that these works do not manifest any interest in abstraction. We have seen that such a view is contradicted by how ancient commentators on a Han classic interpreted this text. In their eyes, the classic put abstraction into play in a specific way. Moreover, for them, this practice of abstraction brought this way of writing up mathematics in relation with the sage-kings' larger administrative arts. This conclusion thus invites us to return more broadly to Chinese mathematical works that consist merely of problems and procedures, and to question systematically the mode of reading we should apply to them. Interpreting them in actors' terms might lead to a revision of our ideas about the history of mathematics in China.

Notes

1. On Edouard Biot, see Han and Duan, "Bi Ao dui Zhongguo tianxiang jilu de yanjiu ji qi dui xifang tianwen de gongxian"; Chemla, "L'histoire des sciences." On Alexander Wylie, see Libbrecht, *Chinese Mathematics*; Han, "Chuanjiaoshi Weilieyali zaihua de kexue huodong"; Chen, "Weilieyali de zhong xi daishuxue bijiao ji qi shixue yuanyuan"; Chen, "Shifting to Sinological Tradition"; Schneider and

Chemla, "The Reception of Wylie's 1852 *Jottings*." The latter publications refer to a more ample bibliography on Wylie's activities in China. The research has received funding from the European Research Council under the European Union's Seventh Framework Programme (FP7/2007–2013)/ERC Grant agreement n. 269804. I have pleasure in extending my thanks to Michael Friedrich for the help he provided me in getting access to some publications. Michael Nylan, Mark Csikszentmihalyi and Vanessa Davies have provided deep and extensive editorial advice, which has improved this chapter tremendously and for which I feel most grateful. I thank Michael and Mark for the invitation to write this contribution and for most helpful discussions. All remaining problems are my sole responsibility.

2. See Chemla, "Abstraction as a Value," for an outline of these discussions in Europe.

3. Park, *Africa, Asia, and the History of Philosophy*, 90–94 in particular, shows how philosophers like Kant held the view that only "the Greeks" were capable of abstraction and that "the others," including "the Chinese," could only think *in concreto*. I thank Mark Csikszentmihalyi for having drawn my attention to this book.

4. Chemla, "Generality above Abstraction." Mathematical writings from early imperial China regarding these practitioners' activities provide various kinds of evidence attesting these facts.

5. I have offered conclusions in this respect in Chemla, "Documenting a Process of Abstraction." In the conclusion of this chapter, I return to them.

6. The date of completion of this book is still debated. For reasons published elsewhere, I take the book to have been completed in the first century CE. In the last years, many critical editions and translations of *The Nine Chapters* and its oldest known commentary and subcommentary have appeared. See, for instance, Li, *Jiuzhang suanshu daodu yu yizhu*; Li, *Jiuzhang suanshu jiaozheng*; Shen, *Jiuzhang suanshu daodu*. Unless otherwise specified, the reader will find the critical edition on which I rely here in Chemla and Guo, *Les Neuf Chapitres*. Moreover, Guo and I offer a translation of *The Nine Chapters* and the commentary and subcommentary into French (on the commentary and subcommentary, see below, esp. note 11). My views on these texts certainly reflect my joint work with Professor Guo over twenty years, for which I thank him.

7. This document is the first mathematical manuscript excavated from a tomb in China. It was found during the winter of 1983–84. Peng, *Zhangjiashan hanjian "Suanshu shu" zhushi*, describes the context of the tomb and provides the first annotated edition of the document. *Kankan Sansûsho* [*The Han Bamboo Strips from Zhangjiashan* Writings on Mathematical Procedures] gives a new critical edition, with translations into Japanese and modern Chinese. Two English translations of Writings on Mathematical Procedures have already appeared: Cullen, *The Suan shu shu* 筭數書 *"Writings on Reckoning,"* and Dauben, "算數書. Suan Shu Shu." I have recently begun a project with Daniel Morgan on this document, and our joint work has shaped my views.

8. The editors date this manuscript, bought on the antiquities' market (and hence impossible to date with certainty), to the Qin period (221–210 BC) and suggest it could not have been completed later than 212 BCE. See Xiao, "Yuelu shuyuan cang Qin jian 'Shu' yanjiu," 1, 16; Zhu and Chen, *Yuelu shuyuan cang Qin jian (er)*, foreword.

9. Chemla and Guo, *Les Neuf Chapitres*, 616–17. In what follows, I translate the text of *The Nine Chapters* itself using capital letters to distinguish it from the text of the commentary and subcommentary (see below). Throughout my chapter, I refer to a problem in *The Nine Chapters* by giving first the number of the chapter in which it was placed, and then the number defining the position it had in the chapter, as in this problem, labeled "8.1."

10. Several critical editions of the anthology have appeared recently: Qian, *Suanjing shishu. Qian Baocong jiaodian*; Guo and Liu, *Suanjing shishu. Guo Shuchun, Liu Dun dianjiao*. About the use of the anthology, with two other mathematical books, immediately after the completion of the editions produced by Li Chunfeng and company in 656, and for mathematical education in the context of the College of Mathematics (*Suanxue* 算學), see Siu and Volkov, "Official Curriculum"; Volkov, "Mathematics Education."

11. See Chemla and Guo, *Les Neuf Chapitres*, 616–17, emphasis is mine. Wagner, "Doubts Concerning the Attribution," has expressed doubts regarding the actual author of parts of the commentary attributed to Liu Hui. He has argued that parts of the commentary that ancient editions featured and ascribed to Liu Hui had in fact originally been composed by Li Chunfeng et al. for their subcommentary, and were subsequently included in Liu Hui's commentary by mistake. Wagner was followed by other scholars. I have summarized their doubts and explained why I shared them in Chemla and Guo, *Les Neuf Chapitres*, 472–73. More recently, I have gathered hints allowing us to describe Li Chunfeng's editorial work in Chemla, "Ancient Writings." In fact, I have doubts regarding whether this quotation from the commentary must really be attributed to Liu Hui, despite the ancient ascriptions. My reasons for considering that it was probably originally part of the subcommentary include the position of the piece of commentary in which it occurs, the terminology this piece of commentary uses, the kind of mathematical understanding to which it refers, the ideas it presents, and in particular the use of the term *lü* 率 (see below). These doubts do not affect the argument I develop here.

12. See Chemla, "On Mathematical Problems," where I show that the commentary and the subcommentary both make use of mathematical problems to make explicit the "meaning 意" of the operations in a mathematical procedure. Commentary, like subcommentary, thereby establishes a relationship between the procedure and fundamental operations that also underlie other procedures in *The Nine Chapters*. Chemla, "On Mathematical Problems," shows that this embodies the "understanding" to which the passage under discussion refers.

13. Elements of a history of this notion are given in the following section.

14. On this issue, see Nylan, *The Five "Confucian" Classics*, chapter 6, in particular p. 254, and related note 2 in the online endnotes.

15. *Shiji* 130.3285–3322. The sentence discussed below is on p. 3297. A translation of the whole postface is given in Watson, *Ssu-Ma Ch'ien*, 42–69. The sentence under discussion here is translated on p. 51. See below for further interpretation. For a general introduction to Sima Qian's *Historical Records*, see Hulsewé, "Shih chi." On Dong Zhongshu, see Loewe, *Dong Zhongshu*. For the statement about *kongyan* ascribed to Dong Zhongshu, see Su, *Chunqiu fanlu yizheng*, 159, and for a translation, Queen and Major, *Luxuriant Gems*, 182. Editors' note: Much recent scholarship, building upon Japanese scholarship, queries the attribution of the *Chunqiu fanlu* to Dong Zhongshu.

16. See the most recent and most detailed discussion in Van Ess, "Die leeren Worte des Konfuzius." In this extremely useful article, Van Ess examines the various interpretations that have been given to this statement in the past and at the present day, and analyzes the problems they raise from a grammatical viewpoint as well as from the viewpoint of how they fit with the contexts in which this statement was quoted. The present chapter adopts a new viewpoint, because it looks at this issue from the perspective of specialized literature.

17. Li, "The Idea of Authority," 353 (mod.).

18. The original German reads, "Wenn ich dies [bloss] in *leeren Worter* darlegen wollte, dann wäre dies nicht so tiefgründig, scharf, offensichtlich und klar, als wenn ich es an vergangenen *Ereignissen* zeige." Gentz adds in a footnote: "wortlich, durchgeführten *Angelegenheiten*." See Gentz, *Das Gongyang zhuan*, 545 (my emphasis), in the context of a development devoted to Sima Qian's postface (541–50). See also footnote 8, where Gentz discusses various uses and interpretation of *kongyan*.

19. See the nuanced presentation of scholars' diverging uses of the expression *kongyan* at the time and later in China in Watson, *Ssu-Ma Ch'ien*, 87–89. Recent discussions of the topic include Li, "The Idea of Authority"; Van Ess, "Die leeren Worte des Konfuzius."

20. Watson, *Ssu-Ma Ch'ien*, 51, my emphasis. See his endnote 67, where he accounts for this interpretation.

21. See the entry "*shi* réalisation/situation" in the glossary of technical terms used in ancient Chinese mathematical writings that I have compiled in Chemla and Guo, *Les Neuf Chapitres*, 982–83.

22. As above, in all early editions, this part of the commentary is ascribed to Liu Hui. However, for exactly the same reasons, I have doubts about the actual author of this section as well as about the author of the previous statement that I have quoted. For many reasons, which I will develop elsewhere, I believe that both were written by the group of scholars working under Li Chunfeng's supervision, and thus represent pieces from the subcommentary. As seen above in note 11, these

reasons include the position of the piece of commentary; the terminology it uses; the references it contains; the ideas it presents, in particular the use of the term *lü* 率; and the parallels with other sections of the subcommentary and statements ascribed to Li Chunfeng elsewhere. In the case of the rule of three, presented at the beginning of Chapter 2 of *The Nine Chapters*, it is interesting and relevant to note that Li Chunfeng and his colleagues sign their names to most of the commentary (which thus is a subcommentary) in Chapter 2, in contrast with most of the other chapters.

23. For the sake of clarity, I introduce the term meta-operation, by contrast with the term operation, to refer to the following distinction. By definition, a procedure fulfills a task, for instance, the task of computing the quantity of *B* equivalent to a given quantity of *A*. This procedure corresponds to the execution of an operation. However, any such procedure, to fulfill this task, also brings into play various operations, as in this case, multiplication and division. Each of these operations can be in turn executed by another procedure. To distinguish between the operation carried out by a procedure and the operations put into play by a procedure, I use the opposition between "meta-operation" versus "operation." It is important to note that this distinction is entirely determined by context, because an operation that is a "meta-operation" can be used as an operation for an even higher-level "meta-operation." In particular, we will see that the term "universal procedure" seems to refer to a "meta-operation" of the highest level.

24. Chemla and Guo, *Les Neuf Chapitres*, 222–25.

25. Li, "'Jiuzhang suanshu' zhong de bilü lilun," offers the first systematic treatment of the concept of *lü*. His work has been followed by numerous others.

26. In fact, in a majority of these cases, it is the subcommentary, which early editions ascribe to the group working under Li Chunfeng's supervision.

27. *Dou* 斗 and *sheng* 升 are measurement units of capacity. This system of units is decimal, the *dou* being equal to 10 *sheng*.

28. Chemla and Guo, *Les Neuf Chapitres*, 224–25.

29. In other words, 5 is to 3 as 50 is to 30.

30. By the expression "textual *dispositif*," I mean the fact of having two procedures, one at the beginning of the chapter and then another one after each and every problem, with this type of relation between their texts.

31. For an account of the other procedures beyond problems 1 to 32 in Chapter 2, I refer the reader to Chemla, "Documenting a Process of Abstraction." In fact, the procedure "suppose" underlies the procedures for problems 1 to 32 in one way, whereas it underlies the procedures for problems 33 to 46 in another way. Chemla, "Generality above Abstraction," has established that a problem with the specific procedure attached to it was read as a paradigm, that is, as a general statement. One might have wondered which part abstraction could have played if generality was expressed mainly in this way. The fact that an "abstract procedure" relates to several other procedures attached to different problems yields one part of the solution to this issue. More follows.

32. This remark accounts for the fact that the text of the procedure corresponding to the Pythagorean "rule" (to avoid the term theorem, because it is not an actors' category in *The Nine Chapters*) in Chapter 9 occurs after a set of three problems. However, the commentator and subcommentator do show how all the procedures in the same chapter relate to the figure of a right-angled triangle and this procedure. It is in this way that they account for the correctness of these procedures.

33. In the "Monograph on Pitch-pipes and the Calendar" (*Lüli zhi* 律曆志) of the *Suishu (History of the Sui)*, Li Chunfeng inserts a statement about the structure of mathematical knowledge, showing that the procedure for the meta-operation "suppose" might have played an even more "leading" role; conceivably, in Li's view, it might even embody a still higher type of abstraction, as noted in Chemla, *Classic and Commentary*. I return to this topic below.

34. This case is analyzed in detail in Chemla, "Abstraction as a Value."

35. See Chemla, "On Mathematical Problems."

36. See my endnotes on the text of the procedure in Chemla and Guo, *Les Neuf Chapitres*, 861–6nn1–35.

37. Peng Hao, *Zhangjiashan Hanjian "Suanshu shu" zhushi*, 1, 4.

38. Slips 120, 121, 123, 125, and 181 are damaged, but these lacunae are too localized to challenge the conclusion that the manuscript lacks explicit chapters.

39. The two sections are edited in Peng, *Zhangjiashan Hanjian "Suanshu shu" zhushi*, 73–75. I do not want to enter in any mathematical detail here. For greater detail on the meaning of the texts and the relationship between the two sections, see Chemla, "Documenting a Process of Abstraction."

40. I have described in detail this feature and other specificities of the texts of these procedures in Chemla, "Describing Texts for Algorithms."

41. Chemla, "Documenting a Process of Abstraction."

42. This is true for Chapters 1 (for the interpretation, see Chemla, "Résonances entre Démonstration et Procédure"), 2, 3 (with a difficulty for the final part), 5 (for the argument, also see Chemla, "Résonances entre Démonstration et Procédure"), 6 (with an argument that I have developed in my introduction to Chapter 6 in Chemla and Guo, *Les Neuf Chapitres*), and 7–9.

43. The expression *kongyan* is used a second time by a commentator on *The Nine Chapters*. This is in the context of the commentary on problem 8.18, which is attributed to Liu Hui. This attribution is, however, also problematic for several reasons. (I will return to this issue in another publication.) The stakes do not relate to this chapter, and I hence do not dwell on it. Readers may email me if they wish to learn more about this point. In the commentary on problem 8.18, the expression *kongyan* describes how the commentator will present a new procedure that he has developed. The text that follows has two main parts, one in which one might recognize an "abstract procedure," and another one that presents detailed calculations for problem 8.18. In this context, the commentator refers to the detailed calculations by the term *li* 例 (here, "paradigm") and accounts for his choice of this mode of

presentation. This implies that this model for writing technical prose also guided how a commentator composed his own text. For the text of the commentary, and for the entry on *li* in the glossary, see Chemla and Guo, *Les Neuf Chapitres*, 652–59, 953, respectively. The commentary on the measurement of the circle attributed to Liu Hui also once employs the term *kong* with a similar meaning, when Liu explains why he foregoes an abstract formulation and chooses to present a procedure in a detailed way. Reasons of the same kind are put forward, as shown in the text and translation given in Chemla and Guo, *Les Neuf Chapitres*, 179–80.

44. After all, because the commentator immediately reacts when he sees the only case when we have one without the other, this reinforces the conclusion that, for him, they must complement each other. The commentator also explains reasons why *The Nine Chapters* deviated from this model, in terms that echo Confucius's statement on the epistemological virtues of the *shi* 事.

45. It is interesting to recall here that, if the piece of commentary on which this chapter centers is to be attributed to Li Chunfeng, it is not so surprising that he makes reference to statements about the writing of history, given his appointment to the Bureau of History in early Tang. While working in that capacity, Li contributed many monographs for some of official histories in preparation at the time. See McMullen, *State and Scholars in T'ang China*, 169, 335, endnote 37.

Asian-language Bibliography

Chen Zhihui 陳志輝. "Weilieyali de Zhong-Xi daishuxue bijiao ji qi shixue yuanyuan" 偉烈亞力的中西代數學比較及其史學淵源 [Alexander Wylie's Comparisons of Algebra between China and Europe, and Its Historiographical Origins]. *Ziran kexueshi yanjiu* 自然科学史研究 36, no. 4 (2017): 502–18.

Guo Shuchun 郭書春, and Liu Dun 劉鈍. *Suanjing shishu. Guo Shuchun, Liu Dun dianjiao* 算經十書。郭書春, 劉鈍 點校 [*Ten Mathematical Classics. Punctuated Critical Edition by Guo Shuchun and Liu Dun*]. 2 vols. 1998. Reprint, Taipei: Jiuzhang chubanshe, 2001.

Han Qi 韓琦. "Chuanjiaoshi weilieyali zaihua de kexue huodong" 傳教士偉烈亞力在華的科學活動 [*The Missionary Alexander Wylie and His Scientific Activities in China*]. *Ziran bianzhengfa tongxun* 自然辯證法通訊 [*Journal of Dialectics of Nature*] 20, no. 2 (1998): 57–70.

Han Qi 韓琦, and Duan Yibing 段異兵. "Bi Ao dui Zhongguo tianxiang jilu de yanjiu ji qi dui xifang tianwen de gongxian" 毕奥對中國天象記錄的研究及其對西方天文學的貢獻 [*Edouard Biot's Researches on Chinese Records of Celestial Phenomena and His Contribution to Western Astronomy*]. *Zhongguo keji shiliao* 中國科技史料 [*Chinese Documents for the History of Science and Technology*] 18, no. 1 (1997): 80–87.

Kankan Sansûsho 漢簡『算數書』 [*The Han Bamboo Strips from Zhangjiashan* Writings on Mathematical Procedures]. Edited by Chôka zan kankan Sansûsho kenkyûkai Research Group on the Han Bamboo Strips from Zhangjiashan *Writings on Mathematical Procedures* 張家山漢簡『算數書』研究会編. Kyoto: Hôyû shoten, 2006.

Li Jimin 李繼閔. *Jiuzhang suanshu daodu yu yizhu* 九章算術導讀與譯註 [*Guidebook and Annotated Translation of* The Nine Chapters on Mathematical Procedures]. Xi'an: Shaanxi renmin jiaoyu chubanshe, 1998.

———. *Jiuzhang suanshu jiaozheng* 九章算術校證 [*Critical Edition of* The Nine Chapters on Mathematical Procedures]. Xi'an: Shaanxi kexue jishu chubanshe, 1993.

———. " 'Jiuzhang suanshu' zhong de bilü lilun" 九章算術中的比率理論 [*The Theory of Ratios in* The Nine Chapters on Mathematical Procedures]. In *"Jiuzhang suanshu" yu Liu Hui* 九章算術與劉徽 [The Nine Chapters on Mathematical Procedures and Liu Hui], edited by Wu Wenjun 吳文俊, 190–209. Beijing: Beijing Shifan daxue chubanshe, 1982.

Peng Hao 彭浩. *Zhangjiashan Hanjian "Suanshu shu" zhushi* 張家山漢簡《算數書》注釋 [*Commentary on* Writings on Mathematical Procedures, *a Document on Bamboo Strips Dating from the Han and Discovered at Zhangjiashan*]. Beijing: Kexue chubanshe [Science Press], 2001.

Qian Baocong 錢寶琮. *Suanjing shishu. Qian Baocong jiaodian* 算經十書. 錢寶琮校點 [*Critical Punctuated Edition of* The Ten Classics of Mathematics]. 2 vols. Beijing: Zhonghua shuju, 1963.

Shen Kangshen 沈康身. *Jiuzhang suanshu daodu* 九章算術導讀 [*Guide for the Reading of* The Nine Chapters]. Hankou: Hubei jiaoyu chubanshe, 1997.

Shiji 史記 [*Writings by the Senior Archivist*, aka *Historical Records*]. Sima Qian 司馬遷 (?145–?86 BCE). 10 vols. Beijing: Zhonghua shuju, 1959. Reprint, 1982.

Su Yu 蘇輿. *Chunqiu fanlu yizheng* 春秋繁露義證 [*Explication of the Meaning of the* Chunqiu fanlu]. Beijing: Zhonghua shuju, 1914. Reprint, 1992.

Xiao Can 蕭燦. "Yuelu shuyuan cang Qin jian 'Shu' yanjiu" 嶽麓書院藏秦簡《數》研究 [*Research on the Qin Strips* Mathematics *Kept at the Academy Yuelu*]. PhD diss. Academy Yuelu 嶽麓書院, Hunan University 湖南大學, Changsha, 2010.

Zhu Hanmin 朱漢民, and Chen Songchang 陳松長, eds. *Yuelu shuyuan cang Qin jian (er)* 嶽麓書院藏秦簡（貳）[*Qin Bamboo Slips Kept at the Yuelu Academy* (2)]. Shanghai: Shanghai cishu chubanshe, 2011.

Western-language Bibliography

Chemla, Karine. "Abstraction as a Value in the Historiography of Mathematics in Ancient Greece and China: A Historical Approach to Comparative History of Mathematics." In *Ancient Greece and China Compared*, edited by Geoffrey

Lloyd and Jingyi Jenny Zhao, in collaboration with Dong Qiaosheng, 290–325. Cambridge: Cambridge University Press, 2017. [2018].

———. "Ancient Writings, Modern Conceptions of Authorship: Reflections on Some Historical Processes That Shaped the Oldest Extant Mathematical Sources from Ancient China." In *Writing Science: Medical and Mathematical Authorship in Ancient Greece*, edited by Markus Asper, 63–82. Berlin: de Gruyter, 2013.

———. *Classic and Commentary: An Outlook Based on Mathematical Sources*. Vol. 344, *Preprint/Max-Planck-Institut für Wissenschaftsgeschichte*. Berlin: Max-Planck-Institut für Wissenschaftsgeschichte, 2008. http://www.mpiwg-berlin.mpg.de/Preprints/P344.PDF.

———. "Describing Texts for Algorithms: How They Prescribe Operations and Integrate Cases. Reflections Based on Ancient Chinese Mathematical Sources." In *Texts, Textual Acts and the History of Science*, edited by Karine Chemla and Jacques Virbel, 317–84. Dordrecht: Springer, 2015.

———. "Documenting a Process of Abstraction in the Mathematics of Ancient China." In *Studies in Chinese Language and Culture: Festschrift in Honor of Christoph Harbsmeier on the Occasion of his 60th Birthday*, edited by Christoph Anderl and Halvor Eifring, 169–94. Oslo: Hermes Academic Publishing and Bookshop A/S, 2006.

———. "Generality above Abstraction: The General Expressed in Terms of the Paradigmatic in Mathematics in Ancient China." *Science in Context* 16, no. 3 (2003): 413–58.

———. "L'histoire des sciences dans la sinologie des débuts du XIXe siècle: Les Biot père et fils." In *Jean-Pierre Abel-Rémusat et ses successeurs: Deux cents ans de sinologie française en France et en Chine* 雷慕沙及其繼承者: 紀念法國漢學兩百週年學術研討會, edited by Pierre-Etienne Will and Michel Zink, 411–37. Paris, 2014. https://halshs.archives-ouvertes.fr/halshs-01509318/document. Paris: Académie des Inscriptions et Belles-Lettres & Collège de France, 2020.

———. "On Mathematical Problems as Historically Determined Artifacts: Reflections Inspired by Sources from Ancient China." *Historia Mathematica* 36, no. 3 (2009): 213–46.

———. "Résonances entre Démonstration et Procédure: Remarques sur le Commentaire de Liu Hui (IIIe siècle) aux *Neuf Chapitres sur les Procédures Mathématiques* (1er Siècle)." In *Regards Obliques sur l'Argumentation en Chine. Extrême-Orient, Extrême-Occident. 14*, edited by Karine Chemla, 91–129. Saint-Denis: Presses Universitaires de Vincennes, 1992.

Chemla, Karine, and Guo Shuchun. *Les Neuf Chapitres: Le Classique Mathématique de la Chine Ancienne et ses Commentaires*. Paris: Dunod, 2004.

Chen Zhihui 陳志輝. "Shifting to Sinological Tradition and Science: Alexander Wylie's Study on the History of Astral Sciences in China, with a Comparison with His Historiography of Mathematics." In *Writing Histories of Ancient Mathematics: Reflecting on Past Practices and Opening the Future, 18th–21st*

Centuries, edited by Karine Chemla, Agathe Keller, and Christine Proust. Dordrecht: Springer, forthcoming.

Cullen, Christopher. *The Suan shu shu* 筭數書 *"Writings on Reckoning": A Translation of a Chinese Mathematical Collection of the Second Century BC, with Explanatory Commentary*. Edited by Christopher Cullen. Needham Research Institute Working Papers 1. Cambridge: Needham Research Institute, 2004.

Dauben, Joseph W. "算數書. Suan Shu Shu (A Book on Numbers and Computations). English Translation with Commentary." *Archive for History of Exact Sciences* 62 (2008): 91–178.

Gentz, Joachim. *Das Gongyang zhuan: Auslegung und Kanonisierung der Frühlings- und Herbstannalen (Chunqiu)*. Wiesbaden: Harrassowitz, 2001.

Hulsewé, Anthony F. P. "Shih chi 史記." In *Early Chinese Texts: A Bibliographical Guide*, edited by Michael Loewe, 405–14. Berkeley: The Society for the Study of Early China and the Institute of East Asian Studies, University of California, Berkeley, 1993.

Libbrecht, Ulrich. *Chinese Mathematics in the Thirteenth Century: The Shu-shu Chiu-Chang of Ch'in Chiu-Shao*. 1973. Reprint, New York: Dover, 2006.

Li, Wai-Yee. "The Idea of Authority in the *Shih chi* (Records of the Historian)." *Harvard Journal of Asiatic Studies* 54, no. 2 (1994): 345–405.

Loewe, Michael. *Dong Zhongshu: A "Confucian" Heritage and the Chunqiu Fanlu*. Leiden: Brill, 2011.

McMullen, David. *State and Scholars in T'ang China*. Cambridge Studies in Chinese History, Literature, and Institutions. Cambridge: Cambridge University Press, 1988.

Nylan, Michael. *The Five "Confucian" Classics*. New Haven: Yale University Press, 2001.

Park, Peter K. J. *Africa, Asia, and the History of Philosophy: Racism in the Formation of the Philosophical Canon*. Albany: State University of New York Press, 2013.

Queen, Sarah A., and John S. Major, eds., trans. *Luxuriant Gems of the Spring and Autumn Annals, Attributed to Dong Zhongshu*. New York: Columbia University Press, 2016.

Schneider, Martina, and Karine Chemla. "The Reception of Wylie's 1852 *Jottings* in 19th Century Europe." In *Writing Histories of Ancient Mathematics–Reflecting on Past Practices and Opening the Future, 18th–21st Centuries*, edited by Karine Chemla, Agathe Keller, and Christine Proust. Dordrecht: Springer, forthcoming.

Siu, Man-Keung, and Alexei Volkov. "Official Curriculum in Traditional Chinese Mathematics: How Did Candidates Pass the Examinations?" *Historia Scientiarum* 9 (1999): 85–99.

Van Ess, Hans. "Die leeren Worte des Konfuzius." In *Han-Zeit: Festschrift für Hans Stumpfeldt aus Anlaß seines 65. Geburtstages*, edited by Michael Friedrich, Reinhard Emmerich, and Hans van Ess, 147–67. Wiesbaden: Harrassowitz, 2006.

Volkov, Alexei. "Mathematics Education in East- and Southeast Asia." In *Handbook on the History of Mathematics Education*, edited by Alexander Karp and Gert Schubring, 55–72, 79–82. New York: Springer, 2014.

Wagner, Donald Blackmore. "Doubts Concerning the Attribution of Liu Hui's Commentary on 'Chiu-chang suan-shu.'" *Acta Orientalia. Societates Orientales Danica Fennica Norvegica Svecica* 39 (1978): 199–212.

Watson, Burton. *Ssu-Ma Ch'ien: Grand Historian of China*. New York: Columbia University Press, 1958.

8

Commentarial Episodes in Early Chinese Medicine

An Experiment in Decentering the Standard Histories

Miranda Brown 董慕達

This chapter explores the transmission of medical knowledge. Toward this end, it asks how medical knowledge circulated. To what extent was it transmitted through commentaries on authoritative medical texts and passed down from master to disciples? Or were there other routes by which explications on those texts changed hands?

While a dedicated study on this subject has yet to be published in English, scholars have already advanced at least one theory about early Chinese medical commentaries. Years ago, Nathan Sivin and Yamada Keiji proposed that medical commentaries emerged from a system of master-disciple initiation, a system that encompassed all forms of book learning, including the Classics.[1] Under this system, the master-healer transmitted not only canonical texts on medicine to his novices, but also oral explanations of those authoritative works, allowing his followers to subsequently commit those explanations to memory or to writing and pass them down to their disciples. Sivin and Yamada were largely informed by the picture of medical practice found in the *Records of the Historian* (*Shiji* 史記) by Sima Qian. In the "Biography of the Granary Master" (*Canggong liezhuan* 倉公列傳), the court physician and former Qi 齊 official Chunyu Yi 淳于意 (180–154 BCE) described his initiation into the healing arts. Chunyu claimed that a reclusive master, Yang Qing 陽慶, had transmitted his old classics of the various curative arts along with Yang's explanations of the meanings of those works (*jie* 解).[2]

Sivin and Yamada also drew their views from the *Yellow Emperor's Inner Classic of Medicine* (*Huangdi neijing* 黃帝內經), first referenced in the section on the technical arts in the "Treatise on Classics and Writings" (*Yiwen zhi* 藝文志) of the *Han History* (*Hanshu* 漢書).[3] Several chapters of the *Yellow Emperor's Inner Classic* depict mythical medical authorities answering questions from novice healers. In addition, these passages explain concepts found in existing works of medicine, which are referred to as classics (*jing* 經). Such passages suggest that master-disciple exchanges drove the composition of commentaries on authoritative medical texts.[4]

Sivin and Yamada are not alone. Scholars have long privileged the picture of the healing arts preserved in the standard history. It is evident in Michael Loewe's seminal study of the career of Chunyu Yi, which proceeds from a detailed analysis of the Han figure's biography in the the *Historical Records* (*Shiji* 史記).[5] In *Pulse Diagnosis in Early Chinese Medicine,* Elisabeth Hsu offers a thorough analysis of the contents of that chapter, arguing that much of it offers a firsthand account of the second-century healing arts.[6] In *The Art of Medicine in Early China* (2015), I too relied heavily on the standard histories, particularly the "Treatises on Classics and Writings," to trace the creation of a self-conscious medical tradition in the first century CE.

But what happens if we momentarily bracket the picture of the technical arts in the standard histories and instead reimagine Chinese medicine literally from the ground up? Fortunately, this is neither a theoretical question nor a novel proposition. Thanks to a century of archaeological findings, historians are now poised to reconstruct the transmission of Chinese medical knowledge from surviving manuscripts. How, then, would our current picture of "medicine" change if we were to take those manuscripts as our starting point? Would this picture challenge our understanding of both the practice and transmission of the technical arts?

This chapter has several goals. More narrowly, it investigates the different contexts in which medical commentaries circulated in early China. It probes whether commentaries circulated solely in the context of pedagogical exchange. Modern scholars have long assumed that the modes of transmission depicted in those accounts were representative of literate traditions of early Chinese medicine. By focusing on the documents discovered near a military colony in the northwest, this chapter complicates that picture. More broadly, this paper demonstrates what can happen when we "de-center" the standard histories' narrative for the technical arts by reference to archaeologically recovered documents. In this regard, this study joins a growing body of literature that seeks to revise the current understanding of ancient China through detailed analysis of the excavated corpus.⁷

To these ends, this exploration of early Chinese medical commentaries focuses on the case of the medical manuscripts discovered near Wuwei 武威 (Gansu Province), dating to the first century CE. It proposes that the explanatory material in the manuscripts should be considered commentary, based on its *family resemblance* to other works typically regarded as glosses on medical classics. This section moreover shows that by scrutinizing the structure of the Wuwei manuscripts, it is possible to detect hidden commentary added after the formation of the base text. Having made a case for a hidden medical commentary from the Han dynasty, I situate the Wuwei manuscripts within their broader archaeological contexts. I suggest that explanations of medical texts moved through official circuits, as well as through the more familiar context of master-disciple initiation. In the former context, such explanations were read by imperial officials, who encountered the healing arts as part of their administrative duties. This chapter thus reveals how medical knowledge circulated within multiple contexts in early China.

I. Definitions and Contexts

Before plunging into the analysis, it is necessary to say something about commentaries in the early Chinese context, a topic whose meaning and import scholars continue to debate. Some sinologists distinguish commentaries from other forms of intertextual practice, including *zhuan* 傳 (traditions), *shuo* 說 (discourses), and *jie* 解 (explanations). To take one example, Timothy Chan claims these works were not "true" commentaries, as they interpreted and elaborated on authoritative texts to form new compositions that sometimes even circulated independently of the base text.⁸ By contrast, true commentaries, which explain the meaning of authoritative texts, typically took the form of line-by-line glosses (*zhangju* 章句, literally, "chapter and verse"). According to Chan, such glosses *explain* the meaning of the classics.

Chan's definition of commentaries—as line-by-line glosses that merely explain rather than interpret or elaborate—hardly represents the consensus view in the field, since his definition is both overdetermined and excessively restrictive. As we will see below, commentaries, even those that shared the same format, were hardly unified in their goals: some just glossed, others explained and freely expanded on the meaning of texts, whereas still others attempted to restore a text to its original form. By identifying commentaries with interlineal remarks, this definition further neglects a wide range of explanatory materials discovered in excavated manuscripts from the pre-Qin and early imperial periods. Such explanatory materials, however, did not take the form of interlineal remarks.

Given the pitfalls of Chan's definition, I follow Charles Sanft and Hermann-Josef Roellicke in adopting a more liberal understanding of commentaries, one that encompasses a wide range of remarks that gloss or illustrate other texts. In some cases, the exegetical matter circulated with the base text, and, over time, crept into it. Roellicke refers to this submerged matter as "hidden commentaries."[9]

This exploration of early Chinese medical commentaries focuses on the case of the medical manuscripts discovered near Wuwei. First, a few words about the Wuwei manuscripts themselves: their location, dates, and owner. Unlike the Mawangdui 馬王堆 find (ca. 168 BCE), the Wuwei manuscript cache, discovered the year before Mawangdui, in 1972, has received little scholarly attention until recently, when Yang Yong and I (2017) published the first English translation and study of these sources. The name derives from the manuscripts' place of discovery, in a tomb located in present-day Hantanpo 旱灘坡 hamlet, ten kilometers from Wuwei City, in today's Gansu province. Unfortunately, the identity of the early Eastern Han tomb occupant, who died sometime in the mid- to late first century CE, is lost to us. Based on other clues such as the style of the tomb and the presence of grave goods, we may surmise that the occupant was neither a young nor a very poor man.[10] Most likely, he was in life an official, with a rather extensive interest in medicine. For the tomb occupant's manuscripts not only explained taboos related to acupuncture but also set forth the contents and preparation of several dozen formulas for a wide variety of ills and afflictions: cold damage, chronic diarrhea, discomfort and lumps in the abdomen, pain in the eyes, wounds, and impotence (an abiding preoccupation of Chinese medical authors).[11]

II. The Wuwei Manuscripts: A Case for a Hidden Commentary

We must now make a case that the Wuwei manuscripts contain commentaries. While portions of these manuscripts look different from interlineal glosses, I argue that they nevertheless should be treated as commentary based on their similarity to medieval exegesis of the *Yellow Emperor's Inner Classic* and the extraneous quality of such explanatory matter.

Two formulas in the Wuwei manuscripts present hidden commentarial episodes, both of which concern afflictions that plagued men. To give readers a full sense of the larger context of these episodes, the formulas and commentarial matter appear below in their entirety. An analysis of their structure and a justification for labeling them as commentaries follows below.

Excerpt 1

Main Text	Commentary
白水矦所奏治男子有七疾方 A formula to treat the Seven Afflictions of men memorialized by the Lord of White Waters.	
	何謂七疾？一曰陰寒，二曰陰痿，三曰苦衰，四曰精失，五曰精少，六曰橐下養濕，☒不卒，名曰七疾。令人陰☐小，橐下養濕，☐之，黃汁出。☒遠行，小便時難。溺☐赤黃泔白，☐便赤膿，餘酒☒苦痛。膝脛寒，手足熱且煩，臥不安牀，涓目泣出☒下常痛，溫_(溫)下溜，旁急，時蘇☐☐☐☐☐陰☒有病如此，名為少傷。何巳☐☐☐尚☐☒伏下☒焉，已汙☐孫☐內餳，除☒其坐則應中☒，人不見☐☐☐驚☐☐酒大樂，久坐不起，☐便不☒，有病如此，終古毋子。治之方 What is meant by the Seven Afflictions? The first is called coldness in the privates; the second is called impotence; the third is called "bitter decline" [i.e., exact meaning unclear]; the fourth is called the loss of essence. The fifth is called small amounts of essence. The sixth is called itchiness and moistness at the base of the scrotum. [Indeterminate number of missing characters] . . . no conclusion [i.e., in ejaculating?]. These are called the Seven Afflictions. This condition causes the privates [illegible character] small the base of the scrotum to be itchy and moist . . . [1 illegible character and 1 untranslatable character], a yellow discharge comes out . . . [Indeterminate number of missing characters and 2 untranslatable

continued on the next page

Excerpt 1 (continued)

Main Text	Commentary
	characters] . . . Sometimes, there is difficulty in urinating. Urine . . . [illegible character] red and yellow milky white. When [illegible character] red pus . . . [Indeterminate number of missing characters and 4 untranslatable characters] . . . The knees and shins are cold, and the feet and hands are hot and irritated. One does not get rest from sleep. The eyes are watery with tears coming out . . . [Indeterminate number of missing characters]. . . . there will constantly be pain below. Slowly, water will flow, but the bladder will [leak] frequently. At times, there will be relief . . . [5 illegible characters, one untranslatable character, and an indeterminate number of missing characters] . . . In cases of illness like these, the name is 'small damage.' [Many untranslatable and illegible characters, as well as an indeterminate number of missing characters.] . . . taking great pleasure in alcohol. Sitting for periods of time without rising . . . [1 illegible character, 2 untranslatable characters, and an indeterminate number of missing characters] . . . in cases of illness like this, the person is ultimately childless. **A formula to treat [these conditions]:**
活樓根十分，天雄五分，牛膝四分，續斷四分，□□五分，菖蒲二分，凡六物皆并冶合和，以方寸匕一，為後飯，愈。久病者，卅日平復，百日毋疾苦。建威耿將軍方，良，禁，千金不傳也 Roots of *Trichosanthes*, 10 *fen*. Chinese or Szechuan aconite, 5 *fen*. *Achyranthes*, 4 *fen*. Japanese teasel, 4 *fen*. [2 illegible characters], 5 *fen*.	

Commentarial Episodes in Early Chinese Medicine | 345

Main Text	Commentary
Sweet flag or calamus, 2 *fen*. Pulverize, combine, and mix all six of these substances, taking them in a square-inch spoon after food. The medicine will cure even those who have been long ill. In thirty days, they will recover; in one hundred days, they will be without suffering. A formula of General Geng of Jianwei. Efficacious. Proscribed. Do not transmit even for 1000 coins.*	

*Gansusheng bowuguan, *Wuwei Handai yi jian*, strips 84A–84B; slightly modified from transcription and translation by Yang and Brown, "The Wuwei Medical Manuscripts," 292–93.

Excerpt 2

Main text	Commentary
●治東海、白水矦所奏方，治男子有七疾及七傷。 A formula to treat [illnesses] memorialized by the lords of Donghai and White Waters (?): In treating male ailments, there are Seven Afflictions and Seven Injuries.	何謂七傷？一曰陰寒；二曰陰痿；三曰陰衰；四曰橐下濕而養，黃汁出，辛恿；五曰小便有餘；六曰莖中恿，如林狀；七曰精自出，空居獨怒，臨事不起（起）死玉門中，意常欲得婦人。日甚者，更而菖輕重，時腹中恿，下弱旁光，此病名曰［內傷］。 What are the Seven Injuries? The first refers to the privates being cold; the second is impotence; the third refers to decline in the privates; the fourth to moistness and itchiness at the base of the scrotum with yellow discharge coming out with extreme pain.

continued on next page

Excerpt 2 (continued)

Main text	Commentary
□□桔梗十分，牛膝、續［斷］、［方］風、遠志、杜仲、赤石脂、山茱臾、柏實各四分，月從容、天雄、署與、蚖☒。□□五物皆并治□，☒ Balloon flower, 10 *fen*. Four *fen* each of: *Achyranthes*; Japanese teasel; fangfeng; Chinese *senega*; *Eucommia*; halloysite; Asiatic cornelian cherry; and arborvitae. As for Mongolian broomrape; Chinese or Szechuan aconite; yam; *Cnidium* . . . [Indeterminate number of missing characters and 2 illegible characters] . . . all five of the substances, pulverize [1 illegible character and an indeterminate number of missing characters] . . .*	The fifth refers to excessive urination; the sixth refers to pain in the penis like *lin* [i.e., an ailment causing the urine to be yellow-red in color]. The seventh refers to spontaneous ejaculation and the ill living in isolation but nevertheless getting excited. With respect to matters [of the bedroom], there is no erection. During an erection, there is "death" [i.e., lost or premature erection] within the "gates of life" [i.e., the vagina]. Thoughts are constantly directed at obtaining women, which increase by day [5 untranslatable characters] . . . Sometimes, there is pain in the abdomen; below there is urine leaking from the bladder. This illness is called "internal injury." [2 illegible characters] . . .

*Gansusheng bowuguan, *Wuwei Handai yi jian*, strips 85A–85B; translated by Yang and Brown, "The Wuwei Medical Manuscripts," 294–95.

8.1. A photograph of wood slips 84 (obverse and reverse) from the Wuwei manuscripts. Source: Gansusheng bowuguan 甘肅省省博物館館 ed., *Wuwei Handai yi jian* 武威漢代醫簡 (Beijing: Zhonghua shuju, 1975).

The foregoing examples admittedly do *not* look much like commentary. To begin with, there is the format of the text. Figure 8.1 displays a photograph of the obverse and reverse of wood strip 84 (corresponding to the first excerpt). As a casual glance reveals, all of the characters, including the text labeled above as commentary, are of identical size and occupy the same position on the line as the base text. In addition, other devices found in contemporary manuscripts—for example, the large bullet points seen in the Xuanquanzhi 懸泉置 manuscript of the *Edict of Monthly Ordinances* (ca. 5 CE), used to delineate main text from commentary, are missing.[12] In other words, there is no way to distinguish main text from commentary just by looking.

The layout of the Wuwei manuscripts supplies a contrast to that of medieval commentaries on the *Yellow Emperor's Inner Classic*—arguably the two earliest surviving interlineal glosses of what was by then an authoritative medical canon.[13] There are several reasons to compare the Wuwei manuscripts with these commentaries as opposed to, say, the *Classic of Difficulties* (*Nanjing* 難經), also known as the *Yellow Emperor's Classic of Eighty-One Difficulties* (*Huangdi bashiyi nanjing* 難經黃帝八十一). While the *Difficulties* is generally regarded as being earlier than the medieval commentaries (dated to around the first century CE), scholars disagree about whether it should be regarded as a commentary on the *Yellow Emperor's Inner Classic*. The *Difficulties* not only circulated as a stand-alone work; it also bore an uncertain relationship to the *Yellow Emperor's Inner Classic*, for much of the *Difficulties* dealt with general issues found within the medical corpus, as opposed to glosses on specific passages in the *Yellow Emperor's Inner Classic*.[14]

Figure 8.2 contains a photograph of one excerpt from a seventh-century manuscript recovered from the Dunhuang caves. The manuscript is possibly the earliest surviving physical copy of the interlineal commentary on a canonical medical text. The main text draws from a section of the *Basic Questions* (*Suwen* 素問) edition of the *Yellow Emperor's Inner Classic* ("Treatise on the Three Sectors and Nine Positions" 三部九候論). Based on observed taboos, the cataloguers believe the commentary to predate the mid-seventh century (though this dating is far from definitive, as taboos were often violated in handwritten manuscripts). The identity of the commentator, unfortunately, was not recorded. The commentary is sparse: it merely elucidates the main text, instructing readers, for example, to peel botanical agents such as licorice or specifying the location of acupoints such as *qimen* 期門. What catches the eye above, however, is the layout of the manuscript, which literally elevates the main, canonical text from its commentaries. In so doing, this format rigorously distinguished the remarks of the original author from that of the commentator.

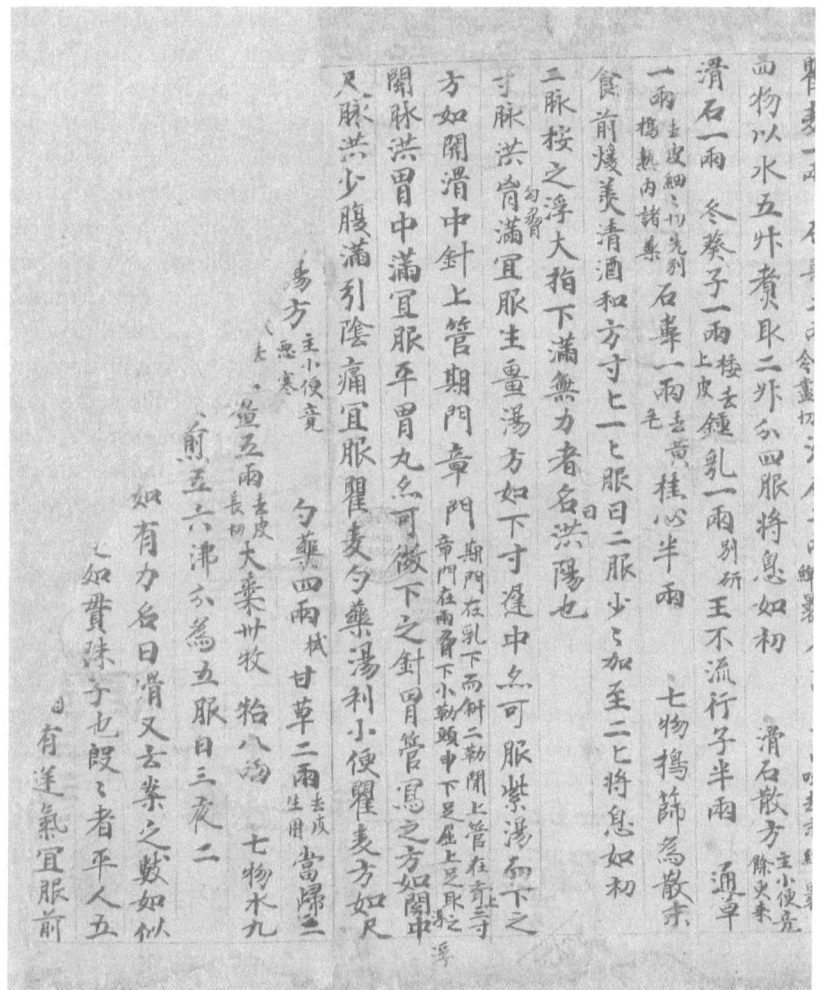

8.2. An example of an interlineal commentary from a manuscript from the Dunhuang site. PelliotChinois_III: 3287. Source: http://idp.bl.uk/database/oo_scroll_h.a4d?uid=572031089;recnum=60567;index=4

Although the layout of the Wuwei manuscripts does not closely resemble that of medieval commentaries, portions of the Han manuscripts under examination functioned precisely like their later, interlineal counterparts. The Wuwei commentaries explain the meaning of words presented in the text. The main text names the illness treated by the formula, specifically, the Seven Afflictions (*qiji* 七疾) and Seven Injuries (*qishang* 七傷) of men. It then supplied the ingredients and methods of preparation for treating those illnesses. The commentaries, sandwiched between the name of the formula and the treatment, then poses the question in each case: "What is meant by . . . ?" It then proceeds to explain the meaning of the term in question, namely, the Seven Afflictions and the Seven Injuries. The commentaries then list in numerical order each of the illnesses: "The first is called coldness in the privates; the second is called impotence; the third is called 'bitter decline'; the fourth is called the loss of essence . . ." In a few instances, the commentary goes beyond naming the illnesses and expands on the physical discomfort or dysfunction associated with it. For example, the second excerpt describes one of the Seven Injuries in the following way: "The seventh refers to spontaneous ejaculation and the ill living in isolation but nevertheless getting excited. With respect to matters [of the bedroom], there is no erection. . . ." As these remarks reveal, the commentary does more than gloss difficult terms: it adds greater texture to the reader's understanding of the illness by supplying the range of symptoms associated with this broad class of male afflictions.

In this regard, the Wuwei commentaries are quite similar to the authoritative commentary on the *Yellow Emperor's Classic* by Wang Bing 王冰 (fl. 762 CE) from the Tang dynasty. This commentary was preserved in the edition promoted by the Northern Song government in the eleventh century, and it provides a fuller example than the sparsely commented Dunhuang manuscript.[15] A translation of the main text, the commentary, and the subcommentary appear below. The left column displays the canonical text as it appeared in the eleventh-century print version. The middle column, which represents the main commentary, contains the remarks of Wang Bing. The far right reproduces the subcommentary by Lin Yi 林億 and his colleagues, the eleventh-century Song officials responsible for producing the received, print version of the text.

As the eleventh-century commentaries reveal, most of the remarks by Wang Bing are like those from Wuwei in that they draw out the meaning of the text. To be sure, Wang Bing occasionally took some interpretative license with his remarks. The original passage described the treatment of an illness called blood withering or drying (*xueku* 血枯), which presumably affected women. Like the Wuwei commentators, Wang Bing's observations elaborated on the ideas presented in the base text, explaining the relationship between amenorrhea and dysfunction in the liver and kidneys. In addition, Wang took this opportunity to explore an evidently more pressing subject, one not explicitly referenced in the original: the root of impotence and sterility in men. In other places, however, Wang Bing focused on explaining the meaning of terms, fleshing out the symptoms of illness, and justifying the use of particular medical substances in the formulas found in the base text. Just as the Wuwei commentaries explicated the meaning of the Seven Afflictions and Seven Injuries, Wang Bing glossed an ambiguous but crucial phrase that appeared in the *Yellow Emperor's Inner Classic*: *houfan* 後飯. By the Tang dynasty, commentators were divided on whether *houfan* meant that the medicine was administered before *or* after a meal.[16] In addition, like the Wuwei manuscripts, Wang also elaborated on references in the original text, supplying greater specificity and context to the brief description of the treatment prescribed in the original. For instance, they accounted for the choice of cuttlefish in terms of its pharmacological properties, cross-referencing later pharmacological classics.

In this regard, the Wuwei commentaries, insofar as they explain and elaborate on concepts in the base text, are closer to Wang Bing's exegesis than the later Song-dynasty subcommentary. Wang Bing thought he should do more than just gloss terms. He clearly saw his task as extending to elaborating on the discussion of the original text, to explain concepts more fully, and to make novel connections. By contrast, Lin Yi and his colleagues in their subcommentaries endeavored to peel away the interpretative layers imposed by later commentators. Their mission was to turn back the clock and return the text to its original form, as they imagined it had existed in antiquity. For example, Lin Yi issued corrections to Wang Bing's version of the text, noting places in which Wang had amended the text incorrectly. Needless to say, such corrections speak not only to Lin's outsized ego. They also reveal efforts to unveil the implicit meanings of the text or to supply greater context, and also to restore it to its original, presumably "pristine" form: to reproduce the original characters of the text as they purportedly appeared in antiquity and to remove any infelicities inadvertently introduced through the process of transmission.

Main text	Tang Commentary (by Wang Bing)	Song Subcommentary (by Lin Yi et al.)
岐伯曰：病名血枯，此得之年少時，有所大脫血。若醉入房，中氣竭，肝傷，故月事衰少不來也。		

continued on next page

Main text	Tang Commentary (by Wang Bing)	Song Sub-commentary (by Lin Yi et al.)
Qibo said, "The name of the illness is 'blood withering.' If this is gotten in youth, there will be those who will experience a serious case of hemorrhaging. If one has sexual relations while inebriated, the internal *qi* will be spent and the liver damaged. Consequently, the menses will decrease or will not come at all."		
	出血多者謂之脫血。漏下鼻衄嘔吐出血皆同焉。夫醉則血脈盛。血脈盛則內熱。因而入房髓液皆下。故腎中氣竭也。肝藏血以少大脫血。故肝傷也。然於丈夫則精液衰乏。女子則月事衰少而不來。 Those who bleed a lot are referred to as [suffering from] hemorrhaging of blood (*tuoxue* 脫血). Leaking below, nasal hemorrhaging, and vomiting are all the same with this regard. If one is inebriated, then the blood vessels overflow; and if the blood vessels overflow, then the internal parts will be hot. For this reason, if one has sexual relations, the bone marrow will all leak out. Consequently, the *qi* within the kidneys will be depleted. As a result, the	

Commentarial Episodes in Early Chinese Medicine | 353

Main text	Tang Commentary (by Wang Bing)	Song Sub-commentary (by Lin Yi et al.)
	blood stored within the liver viscera, to a greater or lesser degree, will hemorrhage, and so the liver will suffer damage as a consequence. When this is the case, men will suffer from a lack of semen, and women will experience decreased menses or periods that do not come.	
帝曰：治之奈何？復以何術？ The [Yellow] Emperor asked, "How is this condition treated? What techniques are used for recovery?"		
		新校正云按別本一作傷中 The *New Annotated and Corrected Edition* observes: In one different edition, the characters are rendered "damages the interior."
岐伯曰：以四烏鰂骨，一蘆茹，二物併合之，丸以雀卵，大小如豆，以五丸為後飯，飲以鮑魚汁，利腸中，及傷肝也		

continued on next page

Main text	Tang Commentary (by Wang Bing)	Song Sub-commentary (by Lin Yi et al.)
Qibo said, "Take four cuttlefish bones and one root of a reed, combining both of these two substances. Make them into a pellet with the egg of a sparrow, the size of a bean. Take five pellets before eating (?) (*houfan* 後飯), drinking them with preserved fish sauce. This will improve the stomach and reach the damaged liver."		
	飯後：藥先謂之後飯。按古本草經云烏鰂魚骨蘆茹等並不治血枯。然經法用之是攻其所生所起爾。夫醉勞力以入房則腎中精氣耗竭。月事衰少不至則中有惡血淹留。精氣耗竭則陰萎不起而無精。惡血淹留則血痺著中而不散。 *Houfan* 飯後: the medicine coming first is what is referred to as *houfan*. In the *Old Version Pharmacopeia*, cuttlefish bone and the roots of reed are not for treating blood withering. This being the case, the methods found in the classic still employ these substances to attack the sources of the illness. If one is inebriated and exerts one's power by having sexual relations, the refined *qi* within the kidney will be	

Main text	Tang Commentary (by Wang Bing)	Song Sub-commentary (by Lin Yi et al.)
	spent. If the menses decrease or do not come at all, then there is noxious blood retained for a long time within the body. If the refined *qi* is spent, there will be impotence and no semen. If the noxious blood is retained for long, there will be paralyzed blood stuck inside that will not disperse.	
	故先玆四藥用入方焉。古本草經曰烏鰂魚骨味鹹冷平無毒。主治女子血閉。蘆茹味辛寒平有小毒。主散惡血。雀卵味甘溫平無毒。主治男子陰萎不起強之令熱多精有子。鮑魚味辛臭溫平無毒。主治瘀血痺。在四支不散者尋文會意方義如此而處治之也。 Hence, prioritize and augment the four medicinal substances and use them to make a penetrating formula for the condition. *The Old Version Pharmacopeia* observes, "Cuttlefish. Flavor: salty, chilled, level, without poison. Controls and treats congestion (blockages?) of the blood in women. *Lüru*. Flavor: bitter, cold, level, slightly poisonous. Controls and disperses noxious blood. Sparrow eggs. Flavor: sweet, warm, level, without poison. Controls and treats impotence for men. Strengthens (the erection), causing it to be hot and there to be an abundance of semen and offspring. Preserved fish sauce. Flavor: bitter, pungent, warm, level, without poison. Controls and treats stasis of the blood and paralysis in the four limbs that do not disperse. I have looked through texts, bringing together the methods and meanings, which are similar in terms of their treatment and remedying of this illness.	

continued on next page

Main text	Tang Commentary (by Wang Bing)	Song Sub-commentary (by Lin Yi et al.)
		新校正云。按甲乙經及太素蘆茹作藺茹。詳。王注。性味乃藺茹當。改蘆作藺。又按本草烏鰂魚骨冷作微温。雀卵甘作酸與王注異 The *New Annotated and Corrected Edition* says: In the *AB Classic of the Yellow Emperor* and *Great Simplicity*, lüru 蘆茹 is rendered lüru 藺茹 (root of chrysanthemum). This is clear. Wang's annotation says that its characteristics and flavor match that of lüru 藺茹 (root of chrysanthemum). He amended the character for lü 蘆 as lü 藺. It is also noted in the *Pharmacopeia* that cuttlefish is chilled and faintly warm; also it has sparrow's egg as sweet, when it should be sour, which differs from Wang's annotation.*

*Chongguang buzhu Huangdi neijing suwen, 11/28a–28b.

The family resemblance between the Wuwei explanatory matter and Wang Bing's commentary is not the sole clue to the type of the explanatory material. To begin with, a similar, though not verbatim, explanation of the "Seven Injuries" appears in a later medieval compilation, the *Treatise on the Origins of the Various Illnesses* (*Zhubing yuanhou lun* 諸病源候論) by Chao Yuanfang 巢元方 (fl. 605–616 CE).[17] To be sure, one may wonder whether the parallels simply reflect the fact that the Wuwei manuscripts served as the basis of Chao's account. Although that possibility cannot be entirely ruled out, there are still other signs that the explanatory material became part of the formula through some hasty—or even sloppy—cutting and pasting. Take the first example, which contained a superfluous sentence, stuck in the middle of the formula (marked in bold in Table 1): "A formula to treat [these conditions]." The sentence is superfluous because the title of the formula was already given at the beginning, "A formula to treat the Seven Afflictions of men submitted by the Lord of White Waters. . . ." The second reference to the formula title, appearing in the middle of the formula, suggests that the compiler had lifted the sentence, along with the preceding material, from another source, inserting it into the middle of the formula. Had he been more careful and removed the second reference to the title, the explanatory material would have fit more snugly into the text of the formula.

The recurring structure of the formulas in the Wuwei manuscripts, finally, indicates that the explanatory material was not originally part of the text. All told, the manuscript comprised several dozen formulas, only two of which contained explications of any of the key terms. The rest of the formulas, or virtually all of them, were identical in structure. They began by informing readers of the titles of the formulas, then moved on to list the substances and the portions to be used, explain the methods of preparation, offer comments regarding the efficacy of the formulas, and so forth. More tellingly, the structure of the doubled-formulas was anomalous when considered against the corpus as a whole. Once the explanatory matter is removed, however, the doubled-formulas no longer deviate from the structure of the other formulas in the manuscripts, a strong indication that the commentarial matter was inserted later.

III. Social Context

What do the examples from the Wuwei manuscripts reveal about the larger context of medical commentaries in early China? Did commentaries circulate solely for pedagogical purposes, or were there other contexts in which clarifications of existing medical texts changed hands? Certainly, we have evidence to support the received wisdom, because anecdotes exist that indicate that students sometimes recorded their teacher's explanations of old texts. But students were not the only ones passing along explanations of existing works. As suggested below, officials, including someone like the occupant of the Wuwei tomb, routinely copied formulas and, in some cases, copied or added elucidations of the formulas' contents.

Any view of the genesis of medical commentaries depends on the available sources. We can make a case for the scholastic context based on materials ostensibly from early China ("ostensibly" because we lack ironclad dates for many works transmitted through the ages, the *Classic of Difficulties* being a good case in point). The problem of dating notwithstanding, the sources suggest medical novices—sometimes nameless, always faceless, figures—actively wrestling with the meaning of earlier medical treatises. Such novices asked their teachers questions pedantic enough to suggest the dialogues between masters and pupils. The *Classic of the Pulse* (*Maijing* 脈經), compiled by the Jin 晉 court physician of the mid-third century CE, supplies the best evidence for the scholastic context for commentaries generated in master-disciple conversations, for it not only furnishes numerous examples of textual exegesis, but also refers explicitly to the "explanations of the master" (*shiyan* 師言).[18] Such references thereby lend credence to the hypothesis advanced by Yamada and Sivin: the commentarial tradition was linked to the exchanges between master healers and their initiates.

The Wuwei manuscripts, however, offer a different perspective on many received sources, particularly dynastic histories such as Sima Qian's *Historial Records* and the *Yellow Emperor's Inner Classic*. Unlike the other early texts, the Wuwei manuscripts make no reference to masters whatsoever, indicating that commentaries were hardly just for the instruction of novices. To be sure, we find prohibitions ("Proscribed. Do not transmit" [*jin bu chuan* 禁不傳]), remarks one might interpret as surviving signs of an initiation ceremony, wherein a master swears his novice to secrecy before transmitting the contents of his formulas. Yet such remarks—especially in the absence of other corroborating evidence—are hardly smoking guns. As Donald Harper proposes, the prohibitions could have been tacked on later to enhance the perceived value of the formulas, much like phrases as "previously tested" (*yan* 驗) or "excellent" (*liang* 良).[19]

Instead, the Wuwei manuscripts hint that medical commentaries circulated in a bureaucratic context. As argued in the *Art of Medicine in Early*

China, the Han imperial administration developed a rudimentary system of health care, something evident from references to various staffs comprising medical officers, officials charged with distributing medicaments, court healers, and physicians working in the counties and frontiers. This probably took place as early as the Qin dynasty, but was certainly in place by the first century BCE. Imperial administrators, furthermore, tracked illnesses afflicting members of frontier armies, officials, conscript laborers, and even horses (concern for the last reflected the military value of animals, rather than a real love of pets). Toward this end, officials also kept sick logs noting the identity of the ill, the nature of the affliction, and the period of incapacitation; they wrote up records of medical consultation, jotting down the name of not only the sick person, but also the physician, as well as a brief description of the treatment. Such descriptions often included information such as whether acupuncture or moxibustion had been administered (and at how many locations) and the number of doses taken and the outcomes of treatment. Officials also kept an archive of formulas on hand. We find directions for the making and administering of formulas at virtually all administrative sites, mixed in with other official documents—tax records, criminal cases, civil suits, and legal statutes. The fact that health care was part of the imperial administration's oversight functions need not surprise us. The outbreaks of epidemics potentially threatened the success of military campaigns (particularly in the frontiers), contributing to the cost of public works projects and disrupting the social peace.[20]

The contents of the Wuwei manuscripts overlap with those of documents found at other administrative sites, for example, in mentioning the same pharmacological substances: Szechuan aconite, Szechuan peppercorn, cassia twigs, balloon flowers, and so forth. In addition, the Wuwei manuscripts emphasized precise measurement, something rather unusual for early personal collections but standard in administrative documents, which also used identical units of measurements (for example, the *fen* 分, *liang* 兩, *sheng* 升, and so forth).[21] Most strikingly, a copy of one of the formulas for chronic coughs, found in the Wuwei manuscripts, was discovered at Liushazhui 流沙墜, a former administrative site in the Northwest. The duplication of formulas indicates that information flowed freely between local bureaus and the personal collection of the Wuwei tomb occupant, which would make sense had the occupant been an official.[22]

The repetitive format of the Wuwei manuscripts also points to the official's hand, in my view. As mentioned above, the formulas were uniform in their structure—so uniform, in fact, that the manuscripts make for a less-than-thrilling read. For a sense of this monotonous format, consider the titles of the formulas, all of which begin exactly the same way (*zhi* 治 . . . *fang* 方):

●治久欬上氣, 喉中如百蟲鳴狀, 卅歲以上方 (strip 3)
A formula to treat those over thirty years of age with a condition where there is a persistent cough, *qi* rising, and something in the throat that sounds like the crying of a hundred insects [wheezing?]:

治傷寒遂風方 (strip 6)
A formula to treat cold damage disorders and to expel wind:

The repetitive structure of the Wuwei manuscripts was not an inevitable, or even a common feature (or defect) of medical compositions in early China (although we do find it in other technical works, such as the mathematical classics reviewed by Karine Chemla in this volume). Instead, it set the manuscripts apart from other early works about healing. Take the *Treatments for the Fifty Two Illnesses* (*Wushi'er bingfang* 五十二病方), recovered from a noble tomb and arguably the most famous excavated work of pharmacology from early China. The formulas came in differing formats. The ruled structure of the Wuwei manuscript is only rivaled by the formulas unearthed from sites administered by officers from the court—for example, those uncovered from the Liye 里耶 site, dating to the Qin dynasty, which also exhibited an invariable legal structure.[23]

The similarities make sense given the larger context. Much like their modern counterparts, Qin and Han officials were creatures of form. They used forms when communicating in their official capacity: to report the results of criminal procedures, to document the use of torture during interrogations, to relay census data, to communicate with superiors, to make requests of leave, to submit difficult cases for judgment, to explain the circumstances of suspected homicides, to track the time required for parcels to be sent from one location to another, and so forth. While they often made for repetitive, even clunky-sounding documents, forms were also indispensable as aids to communication. Forms helped the early bureaucrat, who sometimes was barely literate, to "filter" reality by creating a standard metric. They rendered legible, even readily accessible, what otherwise would have been a host of unwieldy particulars. Given their usefulness, it makes sense that court officers sometimes took their forms with them into the hereafter.[24]

IV. Conclusions

The Wuwei manuscripts represent one of the earliest attested commentaries, even if it is not a dedicated commentary in the tradition of the interlineal remarks. For the commentator is, above all, a teacher of the text, and here we see the compiler explaining preexisting texts. Perhaps more revealing, the manuscript came out of the broader context of the imperial administration, one in which medicine was commonly practiced. In all likelihood, its compiler was some kind of functionary, making it likely that he had encountered the healing arts in the context of governance. This situation stands apart from the better-known context of initiation, in which knowledge of medicine was transmitted from teacher to novice. Within the administration, by contrast, formulas and other curative techniques—like other kinds of technical knowledge vital to the functioning of the imperial administration, such as legal codes, tax collection, and flood control—represented common knowledge.[25]

In closing, I would like to suggest directions for future research, ones that might also profit potentially from a look from the "ground up." While conducting the research for this chapter, I was surprised by the paucity of commentaries on matters relating to healing. To be sure, it is unclear whether the lack of attested commentaries reflects preservation bias. But it was striking that, apart from the small handful of commentarial episodes discussed above, few explanations or illustrations of existing medical texts survive. For example, interlineal commentaries on medical classics postdate antiquity by many centuries. Such commentaries appear, for example, in a number of the Dunhuang manuscripts and the Japanese manuscript editions of the *Cold Damage Disorders* (*Shanghan lun* 傷寒論), the latter of which derived from Tang-dynasty editions.[26]

The belated arrival of an established, rather than sporadic, practice of writing medical commentary demands explanation. By the time medical commentaries apparently became an established practice in the medieval period, scholars had long been composing this type of commentary for other kinds of texts. Such scholars glossed the meaning of terms, explained the rationales behind word choice, and expounded upon trivial, if not tedious, details such as the fabric of mourning garments. For a millennium by then commentaries had offered a space for exploring the meaning of the canons and classics.

The mystery deepens when we also consider the fact that ancient medical authors were often classical scholars, the very individuals most immersed in the commentarial traditions. Take Huangfu Mi 皇甫謐 (215–282 CE), a scholastic par excellence, described by one historian as nothing less than the leading mind of the fourth century.[27] Historian, poet, hermit, drug addict, and classical scholar, Huangfu not only wrote authoritative essays about cold-food powder, a heavy metal concoction; he also compiled editions of old texts that he attributed to ancient sages. Such efforts culminated in the work for which he is now best known, the *AB Classic of Needling and Moxibustion* (*Zhenjiu jiayi jing* 鍼灸甲乙經).[28]

What is more, the conditions should have been ripe for commentaries in late antique China. Medical controversies raged, about how to administer cold-food powder, to take one example. Presumably someone could have glossed or amended an ancient work to make it say the "right" thing. Furthermore, beginning in late antiquity, bellyaching about the poor state of old medical texts became a common refrain, with scholars caviling about the vagaries of textual transmission and the abstruseness of old classics. In response, elite men, beneficiaries of a classical education, took it upon themselves to rearrange old works and create new editions. In some cases, they even started from scratch by composing entirely new treatises.[29] So the question must be asked, Why not, under these circumstances, write a commentary?

An answer to this question, however, may require taking another look at recovered manuscripts. As we have already seen, archaeologists have discovered old texts not just from early imperial tombs, but also from medieval sites in the northwest. It is quite common for us to find overlapping content between different sites and manuscript collections. This is common not only with elite manuscripts like those excavated from tombs and abandoned administrative sites. Medical authors, for example, often worked in the same vein as Huangfu Mi: they found it useful to collect, rearrange, and repurpose existing textual matter, rather than to gloss or emend. For example, the formula attributed to General Geng, discussed above, found its way into a seventh-century anthology. To answer this question, we will need to know much more about how manuscripts on the ground circulated, who copied and expanded on them over the centuries, and why. But that is for future study.

Notes

1. Sivin, "Text and Experience"; Yamada, *Zhongguo gudai yixue de xingcheng*, 414–36.

2. *Shiji* 105.2794, 2815–16.

3. *Hanshu* 30/1776.

4. Sivin, "Text and Experience," 184–86; Harper, *Early Chinese Medical Literature*, 68–69.

5. Loewe, "The Physician Chunyu Yi and His Historical Background."

6. Hsu, *Pulse Diagnosis in Early Chinese Medicine*.

7. For recent examples of this approach, see Lai, *Excavating the Afterlife*; Yates and Barbieri-Low, *Law, State, and Society*; Harper and Kalinowski, *Books of Fate*; Cook, *Ancestors, Kings, and the Dao*; Sanft, *Literate Community in Early China*.

8. Chan, "Jing/Zhuan Structure."

9. Sanft, "Edict of Monthly Ordinances," 139; Roellicke, "Hidden Commentary."

10. Yang and Brown, "The Wuwei Medical Manuscripts," 242–43.

11. Xie, "Han Bamboo and Wooden Medical Records," 79–83.

12. Sanft, "Edict of Monthly Ordinances."

13. Fan, "The Period of Division," 89; Brown, *The Art of Medicine*, 142.

14. See, for example, the *Nanjing benyi xinjie* (no. 77), *xia* (no. 22), 60a–60b. The passage discusses the concept of prevention in a number of Chinese texts and did not quote from any specific passage in the *Yellow Emperor* corpus. For a discussion of the *Nanjing* in relation to the *Yellow Emperor's Inner Classic*, see Unschuld, *Nan-ching: The Classic of Difficult Issues*, 3.

15. On the efforts of the Song bureau to promote health care through the compilation and editing of ancient medical classics, see Goldschmidt, *The Evolution of Chinese Medicine*; and Sivin, *Healthcare in Eleventh-Century China*.

16. See Yang and Brown, "The Wuwei Medical Manuscripts," 255–57.

17. Chao, *Zhubing yuanhou lun*, 3/88.

18. Wang, *Maijing*, 7/16.138, 8/8.157, 9/2.179, 9/5.185.

19. Harper, *Early Chinese Medical Literature*, 66.

20. Brown, *The Art of Medicine*, 63–88; Brown, "'Medicine' in Early China," 472–74.

21. Yang and Brown, "The Wuwei Medical Manuscripts," 254–55.

22. See Yamada, *Shinhatsugen Chūgoku kagakushi shiryō no kenkyū*, 407.

23. Yang and Brown, "The Wuwei Medical Manuscripts," 246–47.

24. For examples of forms in the Qin corpus, see McLeod and Yates, "Forms of Ch'in Law," 131–41.

25. Brown, *The Art of Medicine*, 63–88; Brown, "'Medicine' in Early China."

26. See Zhang, *Shanghanlun banben daquan*, 61 (Kangping 康平 edition); for photographs, see Chen and Ouyang, "Zhongjing xuwen yingxi houren tuozuo yu Sun Simiao zhi hou, Wang Bing zhi qian," 28–29.

27. DeClercq, *Writing Against the State*, 159–205.

28. Brown, *The Art of Medicine*, 110–29; Yu, *Yu Jiaxi wenshi lunji*, 166–209.

29. On the later point, see Ge, *Zhouhou beiji fang*, *xu*, 1a–b.

Asian-language Bibliography

Chao Yuanfang 巢元方. *Zhubing yuanhou lun* 諸病源候論. Third ed. Beijing: Renmin weisheng, 1996.

Chen Senhe 陳森和, and Ouyang Yu'e 歐陽玉娥. "Zhongjing xuwen yingxi houren tuozuo yu Sun Simiao zhi hou, Wang Bing zhi qian" 仲景序文應係後人託作於孫思邈之後,王冰之前. *Zhongyiyao yanjiu luncong* 中醫藥研究論叢 13, no. 1 (2010): 25–42.

Chongguang buzhu Huangdi neijing suwen 重廣補注黃帝內經素問 (Sibu congkan 四部叢刊 edition). Shanghai: Shangwu yinshuguan, 1937–1938.

Gansusheng bowuguan 甘肅省博物館, ed. *Wuwei Handai yi jian* 武威漢代醫簡. Beijing: Zhonghua shuju, 1975.

Ge Hong 葛洪 (284–364). *Zhouhou beijifang*. 肘後備急方. In *Zhongguo yixuedacheng, san pian, si* 中國醫學大成.三編; 4. Changsha: Yuelu, 1994.

Hanshu 漢書. Compiled by Ban Gu 班固 (32–92). Annotated by Yan Shigu 顏師古 (581–645). 12 vols. Beijing: Zhonghua shuju, 1962. Reprint, 1996.

Nanjing benyi xinjie 難經本義. Edited by Siku quanshu. Taipei: Taiwan shangwu, 1983.

Shiji 史記 [*Writings by the Senior Archivist*, aka *Historical Records*]. Compiled by Sima Qian 司馬遷 (?145–?86 BCE). 10 vols. Beijing: Zhonghua shuju, 1959. Reprint, 1982.

Wang Shuhe 王叔和. *Maijing* 脈經. Hong Kong: Shangwu yinshuguan, 1961.

Yamada, Keiji 山田慶兒. *Shinhatsugen Chūgoku kagakushi shiryō no kenkyū* 新發現中國科學史資料の研究. Kyoto: Kyōto Daigaku Jinbun Kagaku Kenkyūjo, 1985.

——. *Zhongguo gudai yixue de xingcheng* 中國古代醫學的形成. Translated by Liao Yuqun 寥育群 and Li Jianmin 李建民. Taipei: Dongda tushu gongsi, 2003.

Yu Jiaxi 余嘉錫. *Yu Jiaxi wenshi lunji* 余嘉錫文史論集. Changsha: Yuelu shushe, 1997.

Zhang Ji 張機. *Shanghanlun banben daquan* 傷寒論論版本大全. Shanghai: Xueyuan chubanshe, 2001.

Western-language Bibliography

Brown, Miranda. "'Medicine' in Early China." In *Routledge Handbook of Early Chinese History*, edited by Paul R. Goldin, 465–78. London: Routledge, 2018.

——. *The Art of Medicine in Early China: The Ancient and Medieval Origins of a Modern Archive*. New York: Cambridge University Press, 2015.

Chan, Timothy W. "The Jing/Zhuan Structure of the *Chuci* Anthology: A New Approach to the Authorship of Some of the Poems." *T'oung Pao* 84 (1998): 293–327.

Cook, Constance A. *Ancestors, Kings, and the Dao*. Cambridge, MA: Harvard University Asia Center, 2017.

Declercq, Dominik. *Writing Against the State: Political Rhetorics in Third and Fourth Century China*. Leiden: Brill, 1998.

Fan Kawai. "The Period of Division and the Tang Period." In *Chinese Medicine and Healing: An Illustrated History*, edited by T. J. Hinrichs and Linda L. Barnes, 65–97. Cambridge: Harvard University Press, 2013.

Goldschmidt, Asaf. *The Evolution of Chinese Medicine: Song Dynasty 960–1200*. London: Routledge, 2008.

Harper, Donald. *Early Chinese Medical Literature: The Mawangdui Medical Manuscripts*. London: Kegan Paul International, 1998.

Harper, Donald, and Marc Kalinowski, eds. *Books of Fate and Popular Culture in Early China: The Daybook Manuscripts of the Warring States, Qin, and Han*. Leiden; Boston: Brill, 2017.

Hsu, Elisabeth. *Pulse Diagnosis in Early Chinese Medicine: The Telling Touch*. Cambridge: Cambridge University Press, 2010.

Lai, Guolong. *Excavating the Afterlife the Archaeology of Early Chinese Religion*. Seattle: University of Washington Press, 2015.

Loewe, Michael. "The Physician Chunyu Yi and His Historical Background." In *En suivant la voie royale*, edited by Jacques Gernet, Marc Kalinowski, and Jean-Pierre Diény, 297–313. Paris: École Française d'Extrême Orient, 1997.

McLeod, Katrina C. D., and Robin D. S. Yates. "Forms of Ch'in Law: An Annotated Translation of the Feng-chen shih." *Harvard Journal of Asiatic Studies* 41, no. 1 (1981): 111–63.

Roellicke, Hermann-Josef. "Hidden Commentary in Pre-Canonical Chinese Literature." *Bochumer Jahrbuch zur Ostasienforschung* 19 (1995): 15–24.

Sanft, Charles. "Edict of Monthly Ordinances for the Four Seasons in Fifty Articles from 5 C.E.: Introduction to the Wall Inscription Discovered at Xuanquanzhi, with Annotated Translation [Zhaoshu sishi yueling wushitiao, the Sole Known Example of a Han Edict in Wall Inscription Form, That Was Recovered from the Site of Xuanquanzhi, a Han-Era Outpost, Located Near Dunhuang, Gansu Province]." *Early China* 32 (2008–2009): 125–208.

———. *Literate Community in Early Imperial China: The Northwestern Frontier in Han Times*. Albany: State University of New York Press, 2020.

Sivin, Nathan. *Healthcare in Eleventh-Century China*. Cham, Switzerland: Springer, 2015.

———. "Text and Experience in Classical Chinese Medicine." In *Knowledge and the Scholarly Medical Traditions*, edited by Don Bates, 177–204. Cambridge: Cambridge University Press, 1995.

Unschuld, Paul U. *Nan-ching: The Classic of Difficult Issues*. Oakland, CA: University of California Press, 1986.

Xie, Guihua. "Han Bamboo and Wooden Medical Records Discovered in Military Sites from the North-Western Frontier Regions." In *Medieval Chinese Medicine:*

The Dunhuang Medical Manuscripts, edited by Vivienne Lo and Christopher Cullen, 77–106. London: Routledge Curzon, 2005.

Yang Yong, and Miranda Brown. "The Wuwei Medical Manuscripts: A Translation and Study." *Early China* 40 (2017): 241–301.

Yates, Robin D. S., and Anthony J. Barbieri-Low. *Law, State, and Society in Early Imperial China: A Study with Critical Edition and Translation of the Legal Texts from Zhangjiashan Tomb no. 247*. 2 vols. Leiden: Brill, 2015.

9

Narratives of Decline and Fragmentation, and the *Hanshu* Bibliographic Taxonomies of Technical Arts

MARK CSIKSZENTMIHALYI 齊思敏 AND ZHENG YIFAN 鄭伊凡

Bookstores the world over segregate fiction from nonfiction.[1] Despite the truism that "truth is stranger than fiction," the ubiquity of similar distinctions in modern life reflects and inscribes a central belief that people can distinguish what is objective or real from what is subjective or pretend. Whether or not this belief is true (or whether or not we may *know* that it is true), the categories that a bookstore uses to arrange its books fuse ancient and modern taxonomies that preserve key distinctions arising from the history of ideas about how people gain knowledge, how that knowledge was written down, and how that writing was transmitted to the present.

The Eastern Han (25–220 CE) "Yiwen zhi" 藝文志 (Treatise on Classics and Writings) stands apart from bibliographical materials from the ancient world, insofar as it is complete, as far as we know, and so preserves its unique taxonomy of categories. It lists the titles of more than 596 works and divides those works into six sections and thirty-eight subsections. The titles and brief descriptions attached to those titles have been a tremendous resource for information on the early history of the book in East Asia. This chapter, however, concentrates less on the titles included in the "Yiwen zhi" than on its taxonomies, scrutinizing them for what they may reveal about Han views of knowledge, writing, and transmission—the very issues raised in connection with our observations about the way bookstores are organized. As it turns out, these views in *Hanshu* are closely tied to a set of narratives about the decline and fragmentation of an earlier, more unified body of knowledge.

Today, some might examine a bibliography compiled more than two thousand years ago because of a wish to retrieve information about the distant past. The organization and structure of the "Yiwen zhi" reveals that its creators were similarly enthralled by their relationship to their past. In particular, the arrangement that Han bibliographers adopted meshed closely with two stories that they told themselves about that past. The first narrative, one of decline and fragmentation, explains the first two categories of both Liu Xin's 劉歆 catalogue *Seven Summaries* (Qilüe 七略)[2] and the subsequent "Yiwen zhi." The narrative relates how the Classics, and the past practices that they were intended to preserve, had degraded and divided into multiple textual and interpretive transmissions, to become the so-called "hundred schools."[3] This tale of decline and fragmentation may be found across a number of pre-imperial texts. Yet for many Western Han writers, both of these levels of text—the Classics and the writings associated with these different modes of expertise—and the entire range of cultural practices from the pristine ones originating from the sages of antiquity to their corrupted forms, no longer constituted the full set of writings and procedures relevant to administering an empire. A second, concurrent narrative developed based on the view that the ineffable Way (*dao* 道) was knowable through its constituent techniques (*shu* 術), a view that meshed with the reality of the new empire's absorption and standardization of diverse regional cultural practices. The famous discussion of Sima Tan 司馬談 and Sima Qian 司馬遷 now known as the "Essential Points of the Six Kinds of Expertise" (Lun liujia yaozhi 論六家要旨) reflects this view, in that the sixth kind of expertise, actually named *dao* or *daode* 道德, subsumes the other five. Han writers at times adapted this view into narratives of decline by arguing that access to the pristine Way through the technical arts had also become fragmented. This fragmentation was a result of social decline, and the technical arts were dispersed via specialization—divided among different archaic government offices—or via geography, with the transmission and interpretation of the Classics in particular associated with the states of the most celebrated exponents of their associated practices, Zou 鄒 and Lu 魯.

The complex taxonomy of the "Yiwen zhi," we argue, is consistent with these two narratives. As mentioned above, the first two categories of the "Yiwen zhi"—the Classics and the "Many Masters"—tell the story of how classical practices degraded, a story repeated in other taxonomies that likely date from the Western Han period, such as the final chapter of the

Zhuangzi 莊子, now titled "Tianxia" 天下. In addition, by comparing the "Yiwen zhi" to the taxonomy of techniques that aid governing described in the final chapter of the *Shiji* 史記, presented around 86 BCE, some two centuries prior to the "Yiwen zhi," we can see the relationship between the fragmentation narrative around the Classics and a similar one about different kinds of technical arts valued in the Han. Finally, turning to the details of the taxonomy, we demonstrate the connections between the other categories of the "Yiwen zhi" and parts of the literature of the different experts, showing how the historical origins of the different technical arts were seen as lesser, fragmentary perspectives on the ancient "techniques of the Way." In this way, we hope to show how the categories of the "Yiwen zhi" can indeed tell us quite a bit about Han views of knowledge, writing, and transmission.

I. Lost in Transmission: A Diminishing Classical Legacy

The "Yiwen zhi" catalogue is not simply a synthesis of prior typologies or records of lineages of textual transmissions. Its general preface contextualizes its structure through a narrative of the fragmentation of an authentic textual transmission, caused by the degradation of the classical legacy following the death of Kongzi (traditionally 551–479 BCE) and his disciples. Specifically, "the *Annals* classic fragmented into five, the *Odes* classic was fragmented into four, the *Changes* classic had multiple expert transmissions."[4] The nature or cause of this fragmentation is not specified, but the preface begins with an acknowledgement that Kongzi's subtle teachings (*weiyan* 微言) were lost when he died, and his direct teachings degraded when his "seventy disciples died, so his complete understanding [of the Classics] was broken off."[5] The legacy of the Classics is not entirely broken off, because there remain the texts that were handed down, but the process of transmission had introduced divergent editions and interpretations.

The second part of the "Yiwen zhi" preface's narrative continues the story of decline by describing two recent historical tendencies that affected transmission. First, the degradation of sources from antiquity was supposedly accelerated by the politically motivated destruction of texts by the Qin hegemons, carried out to "keep the masses ignorant." When the Han dynasty replaced the Qin, it staged a "great collection of the fragmentary records, and re-opened the path to presenting manuscripts in tribute." By way of extolling the Han dynastic rulers, the "Yiwen zhi" continues by saying that during the reign of Emperor Wu 武 (r. 141–87 BCE), "writings had missing sections and lost slips, rituals had decayed and music had declined."⁶ In response, the emperor declared himself troubled by the situation and set up institutional changes, such as those catalyzed by directives to gather and preserve texts by copying their manuscripts and storing the collated versions in the Mifu 祕府 (Palace Archive).

Following the reign of Emperor Wu, Emperors Cheng and Ai continued to address the problem through a process that culminated in the creation of the "Yiwen zhi" itself. The prefatory remarks continue:

> When it came to the time of Emperor Cheng (r. 33–7 BCE), because of the loss of manuscripts, he sent Imperial Messenger Chen Nong to search for manuscripts among the people of the realm. He issued an edict directing Counsellor of the Palace Liu Xiang to collate the Classics and their affiliated works, the masters, and the poetry and rhymeprose (fu 賦); Colonel of Infantry Ren Hong to collate the military texts; Director of Astronomy Yin Xian to collate "Computational arts and techniques"; and Physician in Attendance Li Zhuguo to collate the "Methods and skills." As each text was finished, Liu Xiang would assemble a list of chapter numbers and titles, extracting their general intent, all of which he recorded and submitted. When Liu Xiang died, Emperor Ai sent Liu Xiang's son, the Palace Attendant Commandant of Imperial Carriages Liu Xin, to complete his father's work. Collecting the manuscripts together, Liu Xin submitted the *Seven Summaries*, comprising the "General Summary," "Summary of the Six Attainments [i.e., Classics]," the "Summary of the Masters," the "Summary of the Poetry and Rhymeprose," the "Summary of the Military Texts," the "Summary of Technical and Computational Arts," and the "Summary of Methods and Skills." In the following, we have extracted the essentials in order to lay out the records.⁷

> 至成帝時，以書頗散亡，使謁者陳農求遺書於天下。詔光祿大夫劉向校經傳諸子詩賦，步兵校尉任宏校兵書，太史令尹咸校數術，侍醫李柱國校方技。每一書已，向輒條其篇目，撮其指意，錄而奏之。會向卒，哀帝復使向子侍中奉車都尉歆卒父業。歆於是總群書而奏其七略，故有輯略，有六藝略，有諸子略，有詩賦略，有兵書略，有術數略，有方技略。今刪其要，以備篇籍。

The organization of the *Seven Summaries*, then, in part purportedly derives from the divisions between different court bureaus. For the purposes of this discussion, the key distinction is between the Five Classics and masterworks under Liu Xiang's charge, and the three other categories connected to other kinds of texts. Liu Xiang's official brief as *Guanglu daifu* 光祿大夫 or Counsellor of the Palace placed him at the intersection of communication between the imperial clan and the Nine Ministers of State. Michael Loewe describes Liu's orientation to the legacy of the past by saying that he "deplored the loss of the ancient virtues, reiterated the need to take account of the lessons of the past and protested at extravagance."[8] Liu Xiang had deployed classical literature in the composition of his official communications, just as the compilers of the military, astronomical, and medical summaries had evinced mastery of their subjects in the course of duty. While this is the process that created the *Seven Summaries*, it is important to note that, almost a century after Liu Xiang's death, Ban Gu 班固 (32–92) rationalized the categories of the "Yiwen zhi" rather differently.

After this general preface, the listings in the individual sections and subsections of the *Hanshu* "Yiwen zhi" are followed by brief summaries. The "Liuyi lüe" 六藝略 summaries provide descriptions of the state of the transmission of the Classics in the Western Han, as well as three sections for the paraclassics *Lunyu* 論語, *Xiaojing* 孝經, and "Xiaoxue" 小學.[9] These remarks note the division of the study of the Five Classics into different *xueguan* 學官—a phrase that may refer to an Academician in charge of, or the recognition of, an "official academy" (teaching an interpretative tradition of a classical text).[10] Through the official recognition of particular transmissions and readings of the Classics, official bureaus of the first century BCE preserved fragments of the Five Classics (in somewhat the same way that the *Hanshu* would derive particular types of technical arts from particular offices in a distant, halcyon age, as below). By Ban Gu's time, each of the Five Classics (omitting the sixth, the *Classic of Music* or *Yuejing* 樂經 entry, which does not discuss *xueguan*) had three or four official transmissions:

- *Changes*: There were four *xueguan* associated with Shi 施 (Shi Chou 讎), Meng 孟 (Meng Xi 喜), Liangqiu 梁丘 (Liangqiu He 賀), and Jing 京 (Jing Fang 房).

- *Documents*: There were three *xueguan* for Ouyang 歐陽 (Ouyang Gao 高), and Xiaohou the Elder and Younger 大小夏侯 (Xiahou Sheng 勝 and Jian 建).

- *Odes*: The *Shiji* sharply distinguishes the explanatory glosses (*xungu* 訓故) of Shen Gong 申公 of Lu, from the transmissions (*zhuan* 傳) of Yuan Gu 轅固 of Qi and Han Sheng 韓生 (Han Ying 韓嬰) of Yan, all three of which each had their own *xueguan*.

- *Record of Rites*: There were three *xueguan* associated with Dai De 戴德, his nephew Dai Sheng 戴聖, and Qing Pu 慶普.

- *Annals*: Four transmissions are identified, but only two (those of the *Gongyang* 公羊, ascribed to Gongyang Gao 高 of Qi, and *Guliang* 穀梁) become *xueguan*. The teachings of Zou 鄒 and Jia 夾 did not because Zou's transmission lacked teachers and Jia's lacked a written record.[11]

The Five Classics were taught in these court-recognized transmissions, which do not exhaust the competing written and oral traditions associated with the Five Classics. At once a de facto official recognition of the earlier fragmented state of classical learning, this institutional recognition of the particular editions and interpretations is arguably an attempt to reintegrate the traditions of classical learning.

By the late Western Han, however, other texts besides the Classics were subject to similar processes. The summary makes similar comments about transmissions included in the "Many Masters" summary (*Zhuzi lüe* 諸子略), which is made up of the six kinds of expertise outlined by Sima Tan and his son Qian in chapter 130 of the *Shiji*, as well as the "Zongheng" 縱橫 (Horizontal and Vertical [alliances]), "Za" 雜 (Miscellaneous), "Nong" 農 (Agricultural), and "Xiaoshuo" 小說 (Minor Narratives). The summaries to these subsections describe how people "struggled over the distinction between authentic and false works associated with the Warring States Horizontal and Vertical alliances," while "the words of the many masters became confused and disordered."[12] The *Hanshu* "Yiwen zhi" bibliographic treatise mentions specific ways that the classical traditions of learning have become confused: not only are there divergent regional transmissions and multiple interpretive traditions, but also imitations that are difficult to distinguish from authentic texts. The process of fragmentation did not hold out much hope for understanding the past, and this was offered by the compilers of the treatise as the justification for the manuscript collection and remediation.

As described in the treatise itself, the underlying information loss occurs on two levels. On one level, manuscripts in their material aspect are literally degrading: "writings had missing sections and lost slips." On a less material level, it is the ritual and musical forms described in the Five Classics that were also fading from memory, and with them the ability to make sense of the surviving writings, which had all sustained damage. Clearly, these two levels are connected. The consistent narrative of the fragmentation and decline relating to the first three "summaries" serves simultaneously as justification for the creation of the catalogue and also for the need to gather and preserve a wide range of works. This is because the summaries cast these sections in a part/whole relationship with the entire classical legacy from the pre-Qin period. The practices they preserve were generally not, according to the Five Classics, ones in which the sages engaged in remote antiquity, but the practices themselves were efficacious because of their reliance on the same underlying regime of truth.

The *Hanshu* "Yiwen zhi" is hardly alone in embedding a taxonomy of texts and related practices in a justification based on degrading information. Two early taxonomic works deeply concerned with the transmission of practices are the final chapter of the *Zhuangzi*, "The World" (Tianxia 天下), composed as late as the third or second century BCE, and the "Essential Points of the Six Kinds of Expertise" included in the final chapter of the *Shiji* 史記 compiled at the end of the second century BCE.[13] As much as or more than the *Hanshu*, these works are deeply concerned with situating new "technical arts" of governance in relation to the Dao 道, usually translated as "Way," here indicating the substratum of truth that underlies equally the techniques of the past and those of the present. Both chapters integrate the traditions of the Five Classics into the broader universe of technical arts, anticipating several key tropes in the *Hanshu* "Yiwen zhi."

II. The "Tianxia" 天下 and the Techniques of the Way

The start of the imperial period saw the wide circulation of technical genres based on the practices related to *fang* 方 (formulae), *shu* 術 (technical arts), *shu* 數 (computational arts), and *fa* 法 (patterns or methods) and codifying fields like divination, physiognomy, and hemerology. Their vital importance in some social contexts has been confirmed by their significant representation in excavated tombs across today's China. While the origins of these genres and the titles of individual works were sometimes linked to semi-divine figures from the past, including the Yellow Emperor (Huangdi 黃帝), Great Unity (Taiyi 太一), and the Divine Farmer (Shen Nong 神農), today it remains unclear how these early imperial genres relate to the textual legacy of the Zhou period.

In this chapter, we use the phrase "technical arts" to refer strictly to practical techniques that were preserved in genres such as those defined above. The English-language phrase "technical arts" comes from *techne* (τέχνη), once associated with a wide range of skills or crafts or arts, and sometimes contrasted with *episteme* (ἐπιστήμη).[14] Following on the distinction between *techne* and *episteme*, derivative terms like techniques, technical arts, and technology have been at the heart of a set of interwoven habits of distinguishing abstract principles from concrete knowledge, of separating things made by people from things that were not, and of differentiating artistic productions from mechanistic ones. Applying the English phrase to these Chinese practices justifiably raises comparative questions about whether such traditions grow out of a culturally bound set of concerns or reflect more universal patterns that develop when complex societies mark off "techniques" and "technologies" from other aspects of culture. Here, however, applying the phrase should not be taken as an assertion of categorical universality, but rather as simply a recognition that similar kinds of practices were classed as *fang* 方 or *shu* 術, at times in contrast with methods based on the Classics, in China; and as *techne*, at times in contrast with *episteme*, in Europe.

Many of the texts in China associated with the technical arts are additionally distinguished by their connection with ancient legendary rulers or mythical figures of yore, but they were not usually the same figures connected with the Classics. The propensity to identify and valorize texts by connecting them with the sage-rulers of the distant past is criticized in a passage from the *Huainanzi* 淮南子 (compiled by 139 BCE): "ordinary people today often respect the ancient but look askance at the modern, and so those who aspire to the Way necessarily attribute their ideas to Shen Nong and Huangdi, in order to have them enter the conversation."[15] While classical works were likely to be associated with Kongzi, the Duke of Zhou 周公, or the sage-kings of antiquity, the practices of the technical arts texts were often associated with primeval figures even earlier than Yao, Shun, and Yu, and hence to culture heroes. Gu Jiegang's 顧頡剛 (1893–1980) essay "HuangLao zhi yan" 黃老之言 makes a similar point.[16] Gu argues that this was an attempt by Han writers to attribute their HuangLao writings on good governance to even earlier figures before Yao, Shun, and Yu, and hence culture heroes even more venerable than these so-called "Confucian" sage-kings. Slightly modifying Gu Jiegang's point, the fact that pre-Zhou period sages are associated with the invention or exemplary practice of such technical arts as agriculture, astronomy, medicine, or military methods rather than with the Five Classics, largely conceived of as the legacy of the Western Zhou period (1050–771 BCE), may also represent a projection of the Han social location of the works associated with the professions onto earlier, more "primitive" societies. The distinction between the "Classics" and "writings" in the title of the "Yiwen zhi" itself plays on the same binary distinction, and a related contrast is critical to the way that narratives of the decline and fragmentation of classical knowledge described the rise of technical arts.

The "In the Realm" (Tianxia) chapter of the *Zhuangzi* plainly distinguishes between the Five Classics as product of the sages before the Eastern Zhou, and other traditions associated usually with later pre-imperial masters. Setting aside for the moment the rich history of debate about the authenticity and role of the chapter within the *Zhuangzi* corpus,[17] the chapter structurally falls into two parts. The first component consists of a short introduction plus a narrative describing the fragmentation of the unitary Way, as understood by the sages, which led to the rise of the discrete "masters," each of whom taught one specific aspect of the Way, only to have their teachings further fragmented over time. A key formal feature of its taxonomy of masters is the repetition of the phrase "some of the ancient techniques of the Way are preserved in their teachings" (*gu zhi daoshu you zai yu shi zhe* 古之道術有在於是者) to open the presentation of five of the six early masters. The second component is composed of a set of lengthy descriptions of the methods and legacies specific to each of the major early masters.

This unique structure is just one obvious way the "Tianxia" differs from the rest of the *Zhuangzi*. Formally, the masters' analyses have little in common with the speculative expositions and fantastic stories that make up the rest of the *Zhuangzi*. In terms of content, the mode of description and lists of evaluations, and the inclusion of one such evaluation for Master Zhuang himself, portrayed as yet another of the masters, are arguably also at odds stylistically with the rest of the *Zhuangzi*. Differences such as these led some Ming and Qing scholars to identify the chapter as a postface, or a relocated preface, for the entire *Zhuangzi*.[18] This view might conceivably explain the perceived formal differences between the "Tianxia" chapter and the rest of the *Zhuangzi*, with the postface effectively taking a step back to explicate the entire work.

A critical approach to the descriptions given for the different masters in the "Tianxia" chapter reveals problems with this traditional view, however.

The structural feature in the chapter's second part, which links it to the preceding narrative introduction, is the identification of sets of masters (in twos and threes, except for Master Zhuang himself) as maintaining and transmitting particular aspects of "the ancient techniques of the Way" (*gu zhi daoshu* 古之道術). In total, five sets of names are marked off in this way:

1. Mo Di 墨翟 and Qin Guli 禽滑釐
2. Song Xing 宋鈃 and Yin Wen 尹文
3. Peng Meng 彭蒙, Tian Pian 田駢 and Shen Dao 慎到
4. Guan Yin 關尹 and Lao Dan 老聃
5. Zhuang Zhou 莊周

The above list of five sets of masters leaves out a final section describing Hui Shi 惠施, who is *not* identified as continuing "the ancient techniques of the Way." Because the description for Hui Shi, in both form and content, differs from those given for the five sets of masters that precede it, the modern scholar Jiang Boqian 蔣伯潛 suggests that this paragraph was once an independent text, which was incorporated later into the *Zhuangzi*, perhaps because Hui Shi is partnered with the persona Zhuang Zhou elsewhere in the *Zhuangzi*.[19] Whether or not the Hui Shi section was originally part of the one-chapter Huizi 惠子 (Master Hui) listed in the *Hanshu* "Yiwen zhi,"[20] Jiang's suggestion that the Hui Shi section was not part of the original taxonomy of the "Tianxia" chapter is based on sound formal analysis.

Beyond the problem of the final section on Hui Shi, others have argued that the assessments of individual masters are inconsistent with the rest of the *Zhuangzi*. Liu Xiaogan 劉笑敢 notes that the "Tianxia" chapter shares direct parallels with a stratum of *Zhuangzi* chapters that does not disparage virtues like *ren* 仁 and *yi* 義.[21] Liu notes several ways that this stratum is unique relative to the rest of the *Zhuangzi*:

> In general, this kind of text is relatively positive toward Ru and Fa [here, "Confucian" and "Legalist"] and absorbs and accepts aspects of both, compared to the kinds of text in both the inner, and the outer/miscellaneous sections of the work. Further, this kind of text emphasizes the hierarchical relation between ruler and minister . . . in clear contrast to the so-called Inner Chap-

ters [conventionally, the first seven chapters of the *Zhuangzi*]. Lastly, this kind of essay clearly promotes a vision of the ruler following *wuwei* while the ministers act, which is rather unique in the *Zhuangzi*.[22]

> 总的看来，这一类文章对儒法两家比较宽容，并有所吸收和肯定. 这与内篇及外杂篇的其他两类文章也造成对照. 再次，这类文章强调上下尊卑的君臣关系，. . . 与内篇也是明显不同的. 最后，这类文章明确提出了君无为而臣有为的观点，这在《庄子》书中，是比较特殊的。

Liu's analysis underscores Jiang's point that the "Tianxia" chapter differs from the Inner Chapters of the *Zhuangzi*. As we argue below, Liu's description arguably fits not just the "Tianxia" but also the hybrid taxonomy offered by the *Shiji*, as we will see. Liu's view that the chapter shows explicit philosophical inconsistencies with all but a handful of other *Zhuangzi* chapters makes it even harder to accept the position that "Tianxia" chapter was intended as a postface to the entire *Zhuangzi*.

Such debates about the nature and place of the "Tianxia" 天下 chapter in the *Zhuangzi* as a whole have long constituted the prism through which the chapter's content traditionally has been refracted. Here, however, they distract from the story the chapter tells about the fragmentation of knowledge and the relationship between the sagely Classics, the fragmentary techniques of the "Many Experts," and the debased knowledge of the present—a story told in slightly different ways in the "Tianxia" narrative, Sima Qian's "Essential Points," and Ban Gu's bibliographic treatise, the "Yiwen zhi." In this context, let us examine the contemporary scholar Fang Yong's 方勇 assessment of Zhuangzi's "Tianxia" chapter as an early work providing a scholastic taxonomy, which should be read against other works of that genre:

> These are doubtless continuations and developments of this form of academic classification based on a narrative style of tracing the historical origins of each to one of the ancient "techniques of the Way." When [the *Shiji*] "Essential Points of the Six Kinds of Expertise" and *Seven Summaries* are critiquing each "expertise," they talk about both their advantages and their shortcomings, a feature that should be seen as their development of the unique critical spirit of this chapter.[23]

这无疑是对本篇试行学派分类，并追溯各派的历史起源到古代的某一"道术"这一叙述方式的继续与发展。而《论六家要指》、《七略》在评述各家时，都既谈其优点，又谈其缺点，这应当视为是对本篇那种独特批判精神的发扬光大。

Fang Yong writes that the "Tianxia" account of the way in which the technical arts emerged from a fragmented Way became the template for critically important Han period taxonomies found in the *Shiji* and *Hanshu*. While we are not as confident as Fang Yong is about the chronological order of the three works, we do feel that he is absolutely right about the core similarities.

The introductory section of "Tianxia" is perhaps the clearest and most elaborate attempt to link the Five Classics to other textual genres. As noted above, the last chapter of the *Zhuangzi* is unique in some senses, and it may even represent a point of view in the text that is "relatively positive toward Ru and Fa." This is certainly true of the introduction, which foregrounds the value of the classical legacy, even while underscoring its location in a specific place and time. The "Tianxia" begins by contrasting the present with the past, laying out three stages of progressive fragmentation of the Way:

> "There are many people in the realm who administer (*zhi*) methods and techniques, and all believe that none could improve on the efficacy of their own. But where are the ones that the ancients called the 'techniques of the Way?'"
>
> I say: "They are everywhere" [literally, there is no place where they are not].
>
> If you ask, "From where do the spirits descend? From where does their percipience emanate?"
>
> I say: "The Sages possess what gave birth to them, the Kings possess what matured them, and all of them originated from Unity."[24]
>
> 天下之治方術者多矣，皆以其有為不可加矣。古之所謂道術者，果惡乎在？曰：「無乎不在。」曰：「神何由降？明何由出？」「聖有所生，王有所成，皆原於一。」[25]

This opening passage describes the current fragmentation of knowledge, one in which people mistakenly believe that their branch of knowledge is superior to all others. It ends with the claim that all branches of knowledge, past and present, derive from a primal unity. The chapter does not explicitly recognize the potential to go back and tap into that unity, but its survey of the various methods and techniques in circulation implicitly holds out the hope that in learning these the unity may be glimpsed. While the unitary Way is the original source of knowledge, the exceptional sages and rulers of the past discovered and developed that knowledge, creating further techniques as they adapted to the challenges of their time.

Following this brief section comes a prose description of several different kinds of exceptional person. It describes five exemplary kinds: *tianren* 天人 (heavenly person), *shenren* 神人 (spirit person), *zhiren* 至人 (ultimate person), *shengren* 聖人 (sage), and *junzi* 君子 (gentleman). The first members of this list have a simpler relationship to pure forms of knowledge: "The heavenly person is not separate from his or her ancestral source (*zong* 宗)," while the latter members are credited with mastery of bodies of practical knowledge: "The one who takes benevolence as kindness, righteousness as principle, ritual propriety as action, music as harmony, and who is warmly compassionate and humane we call the gentleman."[26] Beyond the fact that the chapter presents these classical virtues and roles in a positive light, the first three exemplary kinds are not mentioned again, which relegates them to a perhaps irrecoverable past, connected more directly to a time when knowledge and understanding were one.

Instead, once that primal unity—cosmic and political—was no longer available, even the sages and rulers had to rely on multiple sources of partial knowledge. The cardinal example in the "Tianxia" introduction is the study of the Classics in the states of Zou and Lu, the homes of Mengzi and Kongzi. The study of the Classics is a means to master multiple kinds of knowledge through one of the Six Classics, each of which addresses a different aspect of positive behavior. The passage reads:

> Many of the men in service and high-ranking officials from Zou and Lu could understand the *Odes* and *Documents* (Shangshu 尚書), and the rites and music. The *Odes* was used to guide their intentions, and the *Documents* to guide them in affairs. The rites were used to guide them in actions, and music to guide them in harmony. The *Changes* was used to guide [alternations of] *yin* and *yang*, and the *Annals* to guide differentiation of titles and status hierarchies. When their regular processes were dispersed among the people of the realm and established in the Central States, from time to time some of the scholarly traditions of myriad kinds of expertise commended and used them for guidance.[27]

其在於《詩》、《書》、《禮》、《樂》者,鄒、魯之士、搢紳先生多能明之。《詩》以道志,《書》以道事,《禮》以道行,《樂》以道和,《易》以道陰陽,《春秋》以道名分。其數散於天下而設於中國者,百家之學時或稱而道之。²⁸

Here the Classics and their contents are not valorized because they are connected with Kongzi directly, but rather the scholars of Zou and Lu included classicists determined to preserve the legacy of the past, pursuing its unitary knowledge through the multiple kinds of knowledge preserved in the Classics. Those in Zou and Lu used the Six Classics (at one point, four) to guide them in the different aspects of their lives.²⁹

Yet this was not the only stage of decline described in the quotation, because the methods of the Classics were "dispersed among the people of the realm and established in the Central States." No longer restricted to Zou and Lu, the classicists' expertise spread to other regions and was somehow incorporated in the contested perspectives of the "Hundred Schools." Here it is important to note that all five sets of figures in the second part of the "Tianxia" *Zhuangzi* chapter belong to this stage of the process. However, as in the introductory part of the chapter, the partial knowledge and understanding of these masters underwent even further decline over time. For example, the *Zhuangzi* chapter describes the misunderstanding by the various followers of Mozi of key elements of his teachings: they "all chanted the Mohist canons, but each one's interpretation diverged from the next, and each called the others 'splinter' Mohists."³⁰ The fragmentation of the unitary Way into the traditions represented by the many masters, it turns out, was repeated over and over again with the transmissions of the masters' teachings to successive generations of disciples.

As described in the "Tianxia" narrative, the final stage of decline is one in which all attempts at comprehensive knowledge have failed, and (as foreshadowed in the chapter's first line) no one anywhere remembers that there ever was more than partial knowledge. This is the state of affairs in the present day, according to the text:

> But the people of the world are greatly disordered, the worthies and sages are misunderstood, the way and its virtues are not unified. Most people of the world are proud of themselves for having attained but a single aspect of them. This may be compared with having one's ears, eyes, nose, and mouth each sense something different without being able to communicate with

one another. So too with the Many Experts' many skills, which each have their strength and a time when they are useful, but are neither comprehensive nor complete, making them scholastic "one-trick ponies."

天下大亂，賢聖不明，道德不一，天下多得一察焉以自好 。譬如耳、目、鼻、口，皆有所明，不能相通。猶百家眾技也，皆有所長，時有所用。雖然，不該不遍，一曲之士也。³¹

The transition described here, also described in the case of the later Mohists, is a move from a differentiated scholastic and professional set of experts to contending individuals who no longer can communicate with one another in meaningful ways. The devolution of knowledge in the chapter now has four stages:

1. Unitary knowledge and understanding of the Way (*dao* 道)

2. Partial preservation of the Classics (*jing* 經) in Zou and Lu

3. Extensive practices of the masters and experts (*zhuzi baijia* 諸子百家)

4. Individuals' mastery of skills (*ji* 技) derived from a debased understanding

The present is like the period of the "Hundred Schools," absent the "mind" and "will" that tries to connect the partial understandings of the world to the derivation of the skills. At present, each practice instead is taken by its practitioner to be true in itself, and the potential for the unified and comprehensive knowledge and understanding celebrated by the generalist (where the chapter began) is gone. The motif of fragmentation in the "Tianxia" narrative is echoed in a number of Western Han works.³² As Fang Yong notes, it shares quite a few features with the *Shiji*, a work that has had an outsized influence on the taxonomies later used as the basis of the "Yiwen zhi" bibliographical categories.

III. The Simas 司馬氏 and Expertise in the Way

The final chapter of *Shiji* (chapter 130) contains a number of different elements, some of which are autobiographical, and is often called the

"Personal Narrative of the Senior Director of the Archives" (Taishigong zixu 太史公自序).³³ While not as descriptively detailed as the *Zhuangzi* "Tianxia" chapter concerning the different kinds of knowledge it discusses, this *Shiji* 史記 chapter (sometimes dubbed a postface) "Personal Narrative" more directly sets out to make sense of diverse textual genres and types of practice of which the Senior Director of the Archives was aware, using a fragmentation narrative about the past. A central element of the chapter is the description of what is conventionally translated as the "Six Schools" (*liu jia* 六家)—more accurately, the "Six Kinds of Expertise"—of *yin* and *yang* 陰陽, Ru 儒, Mo 墨, Ming 名 (Names), Fa 法 (Law), and Daode 道德 (The Way and its Virtue).³⁴ The chapter begins with the clan history of the Simas, then turns to an essay by a Senior Director of the Archives (Sima Tan or Sima Qian, according to early sources), on the "Essential Points of the Six Kinds of Expertise." The six branches fit with the third stage of the "Tianxia" narrative of fragmentation, in that each one is a partial set of texts and practices for "working on behalf of those who govern" (*ci wu wei zhi zhe ye* 此務為治者也). Of these six branches, the last one represents the culmination of the previous five, embodying more profound insights:

> Experts in the Way cause people to concentrate their essential *qi* and spirit, so their every move is in accord with the formless, and thus they sufficiently supply the myriad creatures. In making techniques, they rely on the great succession of *yin* and *yang qi*, select the good from Ru and Mo, assemble the essentials of Ming and Fa, shift in accord with what is timely, and transform in response to external things.³⁵

> 道家使人精神專一, 動合無形, 贍足萬物。其為術也, 因陰陽之大順, 采儒墨之善, 撮名法之要, 與時遷移, 應物變化。

While the experts in the Way are like the previous five kinds of experts, in having a specialization, paradoxically that "specialization" incorporates all the other more particular insights from the previous five groups, and their "making techniques" (*weishu* 為術) reflects a discriminating and deliberative process that adapts the various forms of expertise of the five other groups to each circumstance. The stress on adapting various kinds of expertise to changing situations is a hallmark of the sage in several chapters of the *Huainanzi* 淮南子, including the "Boundless Discourses" (Fan Lun 氾論),

of which John Major and others write that the sage "must be prepared to abandon old policies when they become obsolete, innovate when faced with new challenges, and adopt diverse approaches as the circumstances warrant."³⁶ Elsewhere in the "Personal Narrative" we are told that Sima Tan "studied the *Changes* with Yang He" 受易於楊何 and "studied discussions of the Way with Master Huang" 習道論於黃子.³⁷ Yang He, Sima Tan, Sima Qian, and Liu An's presentation of the *Huainanzi* all were in Chang'an during the 130s BCE, when the taxonomy based on the "Six Kinds of Expertise" was being developed, and when the Way was being widely deployed as a conceptual tool to describe how diverse practices work across times and cultures. Experts in the Way, like Master Huang and his student Sima Tan, held out the possibility that a certain type of person could draw from the partial forms of knowledge in such a way as to recapture the supreme adaptability of the sages of the past, whose mastery of their circumstances is recorded in the Classics. In contrast to the *Zhuangzi* "Tianxia" chapter, then, the *Shiji* explicitly holds out the promise of moving from fragmentation back to a more comprehensive sort of knowledge and understanding on which practice can be built.³⁸

Where the *Shiji* and the "Tianxia" are very similar is in their discussions about the relationship between the techniques embodied in the Classics and the technical methods associated with the other kinds of expertise, such as the calendrical and astronomical methods associated with the experts in *yin* and *yang*. For the Ru, "the Six Classics are the template" for the practice, even if a complete understanding of the Six Classics is no longer possible: "For generations no person has been able to fully comprehend their objects of study, and in a single lifetime no person is able to fully research the rites."³⁹ Recall that in the *Zhuangzi* "Tianxia" chapter, the Classics are linked to traditions of practice, and the mastery of each text (or body of practice) connected to a different aspect of good behavior. In the *Shiji* taxonomy, a very similar list appears:

> For these reasons, the rites are used to regulate people, the music is used to facilitate harmony, the *Documents* is used to guide them in affairs, the *Odes* is used to develop one's intentions, the *Changes* to guide transformations, and the *Annals* to guide them in righteousness.⁴⁰
>
> 是故禮以節人，樂以發和，《書》以道事，《詩》以達意，《易》以道化，《春秋》以道義。

This passage about the Six Classics has multiple points of similarity with the passage in the *Zhuangzi* "Tianxia" chapter examined above, establishing that the link between the two chapters is stronger than simply a formal resemblance. Certainly, both the *Zhuangzi* and *Shiji* texts see the techniques of the Classics as ancient and therefore incompletely transmitted because of fragmentation and decline. However, the elevation of the expertise in the Way indicates that the techniques of the Classics are neither privileged nor sui generis in the context of the broader universe of the technical arts.

The way the taxonomies found in the last chapters of the *Zhuangzi* and *Shiji* integrate classical learning and other technical arts into an overarching framework serves as a precedent, facilitating Ban Gu's integration of the bibliographical project begun by Liu Xiang and Liu Xin into the *Hanshu*. The structure of the *Hanshu* "Yiwen zhi," as we shall see, draws on earlier narratives of fragmentation and decline, but Ban Gu adds an abiding concern with how the machinery of the state is connected to the many kinds of expertise.

IV. Technical Knowledge as the End Stage of Classical Knowledge in the "Yiwen zhi"

The historical stages outlined above for the *Zhuangzi* and the *Shiji* form a backdrop and provide context for the rhetorical framework of the *Hanshu* bibliographic catalogue "Yiwen zhi." The *Hanshu* chapter, as already noted, had its genesis in the project commissioned by Emperor Cheng. While we do not have an explanation for why each particular official was chosen to undertake his own part of the task, except for those posts defined by specialized knowledge, the method of relying on experts in government offices fits neatly with a worldview fostered among members of the governing elite, where the specialization of knowledge is a given. More to the point, the rhetorical frame in which the lists of holdings are embedded—a brief summary essay for each category—tells a story (possibly fictive) that elaborates on the familiar taxonomies in important ways.

The taxonomy used to organize the works listed in the *Hanshu* "Yiwen zhi" essentially begins from a binary in the title's contrast between *yi* 藝 ("Classics") and *wen* 文 ("Writings"). Above, we have seen examples of authors who classified both the classical and technical arts under the epistemic authority of an overarching "Way." This same assumption is reflected in the structure of the part of the catalogue that follows the "Classics" section. Below, we look at how the summaries and organization of three other sections of the treatise—"Military writings" (*bingshu* 兵書), "Computational and Technical arts" (*shushu* 數術), and "Methods and skills" (*fangji* 方技)—are organized in a way that echoes the narratives of decline that link the "Classics" to the "Masters" sections. In addition, the rhetorical framing of these sections returns to the notion that particular ancient aspects of statecraft had become debased into contemporary skills used to promote personal welfare.

Recall that centuries before Ban Gu, some Western Han writings had already theorized the relationship between the Way and its constituent techniques as that of a whole to its parts. The *Xinshu* 新書, attributed to the statesman Jia Yi 賈誼 (200–168 BCE), has a chapter dedicated to this relationship: "Daoshu" 道術 (The Way and its techniques). In it, Jia Yi responds to a question about what reality the word *dao* 道 refers to by saying:

> The Way refers to what one follows to make contact with external things. Its beginning we call "the empty or undifferentiated," its secondary manifestations we call "techniques." The undifferentiated refers to its essential subtlety; it is easy and plain, before it has a particular application. Techniques refer to what one follows to regulate external things via alternations between movement and stillness. All of these are the Way.[41]
>
> 道者，所從接物也。其本者謂之虛，其末者謂之術。虛者，言其精微也，平素而無設施也。術也者，所從制物也，動靜之數也。凡此皆道也。

In the cosmogonic sequence, the Way is undifferentiated, but, as soon as it makes contact with external things, the Way expresses itself through well-defined techniques for regulating them.

An echo of this theoretical construct is seen in Ban Gu's description of the category of calendrical techniques in the "Calendrics and Registers" (Lipu 曆譜 subsection) of the "Computational and Technical Arts" section of the *Hanshu* bibliographic treatise, as below. Ban's description makes use of the narrative of decline we have seen, imputing an original account of the decentralization of knowledge. By this account, the basis of all later techniques of this kind were the innovations by the sage-kings, but once the unitary Way became fragmented, the broader administrative techniques of an official sphere were whittled down into individual techniques of only limited relevance to assorted individuals on their own. The early history is described first:

"Calendars and chronologies" order the positions of the four seasons, correctly align the nodes of the solstices and equinoxes, and correlate the asterisms of the sun, moon and five visible planets. For this reason, it was necessary for the sage-kings to align their calendars and chronologies in order to fix the regulations pertaining to proper colors of clothing under the Triple Concordance calendar, and also to thoroughly observe the conjunctions of the sun, moon and visible five planets. Techniques concerned with worries about bad fortune and adversity, or happiness due to good fortune and prosperity, all derive from these. These were the techniques the sages used to understood allotment. Were the sages not the most skilled among the people of the world, who else could have provided the technique?

曆譜者, 序四時之位, 正分至之節, 會日月五星之辰, 故聖王必正曆數, 以定三統服色之制, 又以探知五星日月之會。凶阨之患, 吉隆之喜, 其術皆出焉。此聖人知命之術也, 非天下之至材, 其孰與焉! [42]

The ancients were responsible for developing the systems for the calendar, and the section begins by listing texts associated with the mythical culture heroes Yellow Emperor and Zhuan Xu 顓頊. The sages' methods of understanding allotment later developed into much more general practices addressing contemporary "worries about bad fortune and adversity, or happiness due to good fortune and prosperity," likely indicating hemerology and other widely used mantic procedures relying on astro-calendrical computations.

The "*Yin* and *yang*" subsection of the "Masters" section of the survey may have contained works associated with these techniques of the sage-kings, as the summary to that section explains the masters copied sages like Yao who used *yin* and *yang* "to arrange and make images of the sun, moon, stars and asterisms."[43] Nonetheless, the decline of the Way after the time of the sage-kings inevitably led to a transition from those antique techniques used by the early rulers for reorganizing the agricultural calendar to broader and more individual application. To address personal anxieties, the techniques of the sages were adapted to matters of narrower scope:

> When the Way fell into chaos, troubles that arose from less perceptive persons who were compelled by their desires to understand the heavens, who diminished the great to make it

the smaller, and reduced the distance to make it nearer. This is the reason why "the Way and its techniques" were diminished and became difficult to know, and were used to examine the reality behind warm and cold, or life and death."[44]

道之亂也, 患出於小人而強欲知天道者, 壞大以為小, 削遠以為近, 是以道術破碎而難知也。以考寒暑殺生之實。

The psychological needs of people during the long stage of gradual distancing from the once unitary Way led to the "diminishing" of the techniques derived from the sage-kings. Troubles were addressed by techniques tethered to concerns about individual health and life span. Mapping the decline onto a model of two historical stages, the compiler is drawing a distinction between the *yin-* and *yang-*based techniques of statecraft used by the sage rulers of the past, during the time the Way prevailed, and the derivative techniques that later became popular when the Way fell into chaos.

Two other sets of technical texts derived from the *yin* and *yang* techniques of the past are found in the "Military *yin* and *yang*" (Bing *yinyang* 兵陰陽) subsection, within the "Military texts" section, and the "Five Phases" (*Wuxing* 五行) subsection in the "Computational and Technical Arts" section of the treatise. The former subsection is composed of eleven texts, and the military methods are contained in the last two, *Master Ding* (*Dingzi* 丁子) and *King Xiang* (*Xiang wang* 項王), whose compilation cannot predate the battles that marked the transition from Qin to Han.[45] The latter subsection begins with six titles that all contain the phrase *yinyang* 陰陽, starting with *Great Unity's yin and yang* (*Taiyi yinyang* 泰一陰陽) and *Yellow Emperor's yin and yang* (*Huangdi yinyang* 黃帝陰陽). Then it moves on to several texts devoted to uses of the five visible planets in military contexts, starting from the *Shen Nong and Da You wuxing* 神農大幽五行.[46] The taxonomy sorts by families of technical arts, and within those family categories arranges works chronologically, which reinforces the hierarchy between the reigns of the sage-kings and the present diminished age.

To describe these more recent, applied *yin* and *yang* techniques, Ban Gu adopts the term *xiaoshu* 小數 or "lesser computational arts." In his discussion of the "Masters" section appearance of *yin* and *yang* techniques, Ban borrows the term when he contrasts their origin in the office of the sage-king Yao's calendrical expert Xi He 羲和 with those of later times. Ban says, "When it came to the time that those with limitations used them, then they rigidly applied taboos and prohibitions, mired in lesser computational arts, abandoning human affairs and relying on the demons and spirits."⁴⁷ For the technical texts based on *yin* and *yang* in the "Five Phases" subsection of the "Computational and Technical Arts" section, the *Hanshu* characterizes a similar kind of change: "However, the experts in lesser computational arts applied them to good and bad fortune, and they became so popular in the ages, that each was steeped in more disorder than the last."⁴⁸ In this telling of the relationship between the different *yin* and *yang* technical works, the domain of the techniques of the "Masters" section texts was social and concerned with human affairs, while that of the lesser "Computational and Technical Arts" texts was self-interested and preoccupied with seeking blessings from the spirits. The normative subtext of the *Hanshu* narrative of decline is familiar, but its attention to personal psychological needs such as allaying anxiety, as well as its projection of a binary between the statist and personal applications backward into mythic time, are both new relative to the two works previously examined.

Bearing out the complaint that the decline of the computational arts from the time of Xi He 羲和 was characterized by a move from "human affairs" to those of the "demons and spirits," the rise of unofficial specialists in methods of the spirit transcendence is central to the summary of the "Spirit Immortality" (Shenxian 神僊) subsection of the "Recipes and Arts" section of the *Hanshu* bibliographic treatise. Ban's discussion of it draws a similar picture of the relationship between recent works and their ancient counterparts. As Li Ling 李零 has pointed out, this textual category does not refer to "spirits" of the dead but rather cultivation via technical practices for healing and longevity in order to become a "spirit" ideally in this life.⁴⁹ Ban Gu describes the way such techniques became debased over time:

> Spirit immortality is the means by which one protects one's legitimate human nature and allotment, and wanders searching through what is external to them. One depends on flowing thoughts and an even mind, in a space where life and death are regarded as the same, with no agitation in one's breast. However, some people took this as their vocation, so that strange, exaggerated, anomalous and overstated writings increased more and more. This is not why the sage kings taught [these practices]. Kongzi said: "Always reclusive and acting atypically, yet becoming a storied figure in later generations—this is not something I would choose to do."⁵⁰

神僊者, 所以保性命之真, 而游求於其外者也。聊以盪意平心, 同死生之域, 而無怵惕於胸中。然而或者專以為務, 則誕欺怪迂之文彌以益多, 非聖王之所以教也。孔子曰:「索隱行怪, 後世有述焉, 吾不為之矣。」

Just as the techniques of *yin* and *yang* became debased because of their application to address personal anxieties, so too the techniques of the Way were debased to advantage specialists who exaggerated the worldly benefits of the techniques of spirit immortality. Kongzi enters the discussion to denounce techniques to secure longevity, wealth, and fame, providing for later generations in the Han an implicit critique of the characters of those who chose not to serve the state, but used the techniques developed for doing so to different ends than those of the sage-kings.

The subsection of the "Masters" section with the closest affinities to the "Spirit Immortality" subsection is that of "Dao." Yet the difference in its level of generality is clear from a comparison of the first three titles of the "Spirit Immortality" subsection—*Way of Fu Xi and Various Masters* (*Fu Xi zazi dao* 宓戲雜子道), *Way of the Ancient Sages and Various Masters* (*Shangsheng zazi dao* 上聖雜子道), and *Essentials of the Way of Various Masters* (*Dao yao zazi* 道要雜子)—with the very last text of the "Dao" in the "Masters" section: *Sayings of the Experts in the Way* (*Daojia yan* 道家言). While the "Dao" subsection of the "Masters" section, like the "Spirit Immortality" subsection, contains many familiar names (e.g., the Yellow Emperor, Laozi, and Zhuangzi), the "Spirit Immortality" subsection ties these legendary figures to particular techniques for individual health and longevity, such as *Guiding and Pulling of Huangdi and Various Masters* (*Huangzi zazi buyin* 黃帝雜子步引),[51] *Massage of Huangdi and Qi Bo* (*Huangzi zazi anmo* 黃帝岐伯按摩), *Zhi and Jun fungi of Huangdi and Various Masters* (*Huangzi zazi zhijun* 黃帝雜子芝菌),[52] and *Smelting of Gold of Taiyi and Various Masters* (*Taiyi zazi huangye* 泰壹雜子黃冶).[53] The particular techniques in the "Spirit Immortality" subsection of the "Dao" section have in common the goal of longevity and immortality, whether by rhythmic movement, mimetic dance, massage, ingesting mushrooms, or medical or alchemical preparations. In this way, the "Dao" subsection (of the "Masters" section) may have been conceived of as a theoretical grounding for the "Spirit Immortality" subsection (of the "Recipes and Arts" section), or perhaps something more along the lines of a root/branch metaphor. The Song writer Sima Guang 司馬光 (1019–1086) described the different foci of the texts in these two subsections in similar terms:

The general thrust of the writings of Laozi and Zhuangzi is a desire to equalize life and death and thereby lighten the burden of staying or departing. However, those who practice "spirit immortality" ingest cinnabar and cultivate refinement in order to seek lightness and rising up, refining plants and stones into gold and silver. As a result, the techniques they practiced [i.e., Laozi and Zhuangzi vs. the "spirit immortality" practitioners] were exactly the opposite of each other. This is why Liu Xin's *Seven Summaries* records the "Experts in the Way" in the "Masters" section and places "Spirit Immortality" among the "Recipes and Arts."[54]

老莊之書，大指欲同死生，輕去就。而為神僊者服餌修鍊，以求輕舉，鍊草石為金銀，其為術正相反。是以劉歆《七略》敘述道家為諸子，神僊為方技。

Sima Guang was writing a full millennium after the bibliographic survey that led to the *Seven Summaries*, so his identification of the key distinctions between the two groups of texts does not necessarily re-create the rationale of the treatise's Han compilers. Nonetheless, his observation points to a key structural feature of the *Seven Summaries* and *Hanshu* bibliographic treatise ("Yiwen zhi").

The summaries of the sections of the *Hanshu* treatise rhetorically frame several subsections of the "Masters" sections as theoretical grounding for later sections that contain more specific or debased methods based on similar techniques. It is as if the *Hanshu* is taking the move from the "preservation of the Classics" to the "extensive practices of the many masters and experts"—stages outlined above in the discussion of the "Tianxia" chapter—as a template for the further move from those "masters' and experts' applications" to "individuals' mastery of skills derived from a debased understanding." The "Yiwen zhi," in common with the previous examples, used multiple iterations of narratives of decline and fragmentations to create its taxonomy, even while it introduced a new description of the cause of this final iteration: a shift in the location of technical arts from officials with public goals to the private exploitation of personal anxieties.

V. Bibliography and Administration

We have described above how narratives of decline and fragmentation provided a *rhetorical explanation* accounting for the organization and structure used in the *Hanshu* bibliographic treatise, but of course we are not asserting that a "decline" in techniques from the time of the sage-kings actually happened. What was behind this rhetoric? A close reading of Ban Gu's adaptation of the earlier bibliographic summaries reveals that this structuring principle was combined with one other, more distinctive rhetorical feature: the projection of the bibliographical taxonomy onto an idealized administrative framework. We have already seen how the *Seven Summaries* traced its histories of expertise to several court bureaus whose practices Ban Gu said influenced the final taxonomy, an administrative map influencing the drawing of a bibliographic one. Along with the narrative of decline and fragmentation, the summaries of the *Hanshu* treatise regularly trace the origins of sections, subsections, or individual texts back to a set of "royal offices" or "royal officers" (*wang guan* 王官). Of the ten masters sections, four of them (the Ru, *yin* and *yang*, Ming, and Zongheng) were directly tied to these legendary or imagined ancient offices or officials.

This *Hanshu* construction of a notional link to the Western Zhou administration connects these masters to another text, the *Rites of Zhou* (*Zhouli* 周禮). For the treatise not only borrowed the names of these ancient offices mentioned in the *Rites of Zhou* (and less systematically in other Classics); it moreover ties the *Rites of Zhou* descriptions of the duties of these ancient offices to the textual traditions it catalogues. A representative entry, the summary of the "Ru" subsection of the "Masters" section, reads:

> The current of the Ru experts originates from the office of Minister over the Masses (*Situ* 司徒), which assisted the ruler in following [alternations of] *yin* and *yang*, and promoting educational measures to transform the people.[55]
>
> 儒家者流，蓋出於司徒之官，助人君順陰陽明教化者也。

In the *Rites of Zhou*, the "Minister over the Masses" section is part of the section of "Offices of Earth" (*di guan* 地官), with duties described as "commanding subordinates to manage the teachings of the state, so that he could assist the king to secure and pacify the states and principalities."[56] The primary duty of the ministerial office is "instruction," then, in both the *Rites of Zhou* and the *Hanshu* bibliographic treatise. A similar connection is present between the *Rites of Zhou* and the *Hanshu* catalog's subsection summaries for two Classics, the *Odes* and the *Annals*, when the author associates these texts with the work of ancient offices in charge of collecting poems (*cai shi zhi guan* 采詩之官), or in charge of keeping the archives, with their divinations and charts (*shi guan* 史官).[57] The connections between ancient offices and the sections of the Classics, Masters, and Military texts is summarized in the following chart:

Table 9.1. Masters and "Ancient Offices" in the "Yiwen zhi"

Section	Derivation	Source Text, apparent
Masters 諸子	Ancient royal offices (*wang guan* 王官)	Ancient royal offices in the *Rites of Zhou* 周禮 and other texts
Ru experts 儒家	Offices of the Minister over the Masses (Situ zhi guan 司徒之官), "assisted the ruler in following the *yin-yang* principles and promoting educational measures to transform the people." 助人君順陰陽明教化者也。	Based on the *Rites of Zhou*, the Situ "commands subordinates to manage the teachings of the state, so that he could assist the king to secure and pacify the states and principalities" 使帥其屬而掌邦教，以佐王擾邦國。 In the second part of the "Yao dian" chapter of the *Documents*, when Shun appointed Qi 契 as the Situ, his description of the duty of this office is "to respectfully extend the notion that the five teachings abide in leniency." 敬敷五教在寬。* The description highlights a continuity in the function of "transformation by teaching" (*jiao hua* 教化).
Experts in the Way 道家	Office of Scribes (Shi guan 史官)	The apparent connection of this post with the Daoist tradition originates from the anecdote that Laozi served as the archivist of Zhou, recorded in *Shiji*: "Laozi . . . was an official in the Zhou archives [of documents.]" 老子者......周守藏室之史也。**
Experts in *yin* and *yang* 陰陽家	Office of the Astronomer (Xi He zhi guan 羲和之官).	The "Yao dian" 堯典 chapter of the *Documents*: "Yao commanded the Xi He to solemnly comply with August Heaven, to calculate and model the regular movements of the sun, moon, planets, and asterisms, and to dutifully confer the seasons to the people." 乃命羲和，欽若昊天，歷象日月星辰，敬授人時。
Experts in Law 法家	Office of Justice (Li guan 理官)	

Shangshu zhengyi 尚書正義 3.44b.
**Shiji* 61. 2139.

Narratives of Decline and Fragmentation | 393

Section	Derivation	Source Text, apparent
Experts in Names 名家	Office of Ritual (Li guan 禮官)	From the *Rites of Zhou*: the Zongbo 宗伯 "commanded subordinates to manage the rituals of the state and assist the ruler in bringing harmony to the states and principalities" 使帥其屬而掌邦禮，以佐王和邦國。
Mohist Experts 墨家	Guardians of the Pure Ancestral Temple (Qingmiao zhi shou 清廟之守)	There is a set of hymns under the subtitle of "Ten Hymns of the Pure Ancestral Temple" 清廟之什 in the *Odes*. Kong Yingda 孔穎達 interprets it as the place where "the Spirit of King Wen dwells, thereupon where sacrifice should take place" 此解文王神之所居稱為清廟之意.* The connection of this post with Mozi is presumably due to the Mozi's positive attitude towards the ghosts and spirits.
Zongheng experts 縱橫家	Offices of Diplomacy, Experts (Xingren zhi guan 行人之官)	According to the *Rites of Zhou*, the Major Diplomacy Experts "manage the ritual around major guests and major visitors as well as in order to treat the lords with affection" 掌大賓之禮及大客之儀，以親諸侯。The Minor Diplomatic Experts "manage the ritual rolls of the guests and visitors from the states and principalities in order to host the emissaries from other parts of the world" 掌邦國賓客之禮籍，以待四方之使者。
Miscellany of Experts 雜家	Offices in charge of debates (Yiguan 議官)	
Experts in Agriculture 農家	Offices for Agricultural Activities (Nong Ji zhi guan) 農稷之官	The Zhou ancestor Qi 弃 was believed to be the earliest Chief Officer of Agriculture, per the *Shiji*, "Yao had heard of him, and appointed Qi as the Officer of Agriculture" 帝堯舜聞之，舉弃為農師。**
Xiaoshuo Experts 小說家	Millet Office (Bai guan 稗官)	

*Maoshi zhengyi 毛詩正義 19.706b.
**Shiji 4.111.

Just as the *Hanshu* bibliographic treatise used the Classics and Masters texts as prototypes for the sections devoted to the technical arts in discrete professions, the treatise identifies technical arts texts as products of particular offices in an ancient, perhaps fictive, administration. Relevant offices are seen in the following chart:

Table 9.2. Technical Arts sections and "Ancient Offices" in the "Yiwen zhi"

Section	Derivation	Probable Source Text
Section on Military Texts (Bingshu lüe 兵書略)	Ancient Post of the Major (Gu Sima zhi zhi) 古司馬之職	From the *Rites of Zhou*: "they commanded subordinates to lead the administration of the kingdom, in order to assist the ruler in settling the fiefs and principalities" 使帥其屬而掌邦政，以佐王平邦國。
Section on Computational and Technical Arts (Shushu lüe 數術略)	Posts associated with the Mingtang 明堂, Xi He 羲和, archivists 史, and diviners 卜之職	This section includes six sub-sections: "Heavenly Patterns" (Tianwen 天文), "Calendrics and Registers" (Lipu 曆譜, "Five Phases" (Wuxing 五行), "Tortoise and Milfoil [Divination]" (Shigui 蓍龜), "Miscellaneous Divinations" (Zazhan 雜占), "Five Phases" (Wuxing 五行), and "Methods based on Forms" (Xingfa 形法). Some of the posts are clearly connected to these sub-sections, for example, Xi He with "Calendrics and Registers,"* and diviners with "Tortoise and Milfoil [Divination]" and "Miscellaneous Divinations."
Section on Methods and Skills (Fangji lüe 方技略)	From one department of the Royal Offices (wangguan zhi yishou 王官之一守)	

*The "Yao dian" 堯典 chapter of the *Shangshu*: "Yao commanded the Xi He to solemnly comply with Highest Heaven, to calculate and model the regular movements of the sun, moon, planets, and asterisms, and, to attentively impose them upon human seasons" 乃命羲和，欽若昊天，歷象日月星辰，敬授人時。 Kong Yingda's 孔穎達 commentary reads: "Since Chong Li's time, Xi and He were the officers of the four seasons of the heaven and earth" 孔傳曰：重黎之後，羲氏、和氏世掌天地四時之官職。

These charts lay out the similar "origin stories" of individual sections in the *Hanshu* bibliographic treatise, a way in which the technical arts were integrated into the intellectual scheme that explained the origins of all authoritative writings. Scholars have long thought that Liu Xin himself fabricated the *Rites of Zhou* as a resource to support Wang Mang's (r. 9–23 CE) reforms and political ambitions.[58] The way that both the *Seven Summaries* and the *Rites of Zhou* inform the prose of the *Hanshu* bibliographic treatise supports the idea that the imaginary of the Western Zhou administrative structure was important to Liu Xin, Ban Gu, or both.[59]

This practice of "tracing back" each section to the ancient royal offices supports the narratives of decline and fragmentation, as described above, because the glorified practices of the courts of the remote past eventually spawned over time writings that applied best to individuals and families, in service or not. The brief treatments constructed for many of the sections of the bibliographic treatise contrasted the administrative use of authoritative writings with their personal and domestic uses.[60] This rhetorical feature of the *Hanshu* bibliographic treatise should also be placed within the context of the roles of Liu Xiang, Liu Xin, and Ban Gu as advocates for a system of institutionalized checks on the ruler's powers and privileges, as well as for the role played by local officials in regulating nonofficial practitioners. While Han reconstructions of the archaic royal offices associated with certain branches of the technical arts were arguably not historically accurate, the insistence that each kind of technique derived from an individual office reflected the view of the summaries that technical practices were developed by the sages within the context of governance before being watered down by later self-interested and "lesser" specialists. The iterative rhetorical construct of decline and fragmentation accounts for the layered aspect of the structure of the many categories in the present *Hanshu* bibliographic treatise, and while the treatise does not explicitly address the possibility of a return to a unitary Way, its connection of technical arts to offices implied that the path to doing so was via the comprehensive governance and centralized control of past ages.

VI. Conclusion

Unquestionably, in the eyes of the Eastern Han compilers of the *Hanshu* bibliographic treatise, the "Masters texts" were seen as the repository of "techniques of the Way," which, in the distant past, had all applied to the domain of statecraft. By contrast, many of the works most concerned with the specialized technical arts were presumed to be derived from those same texts, in a second devolution. We have shown this to be the case for three subsections in the *Hanshu* bibliographic treatise: (1) the "*yin* and *yang*" subsection of the "Masters" section; (2) the "Military *yin* and *yang*" subsection of the "Military" section; and (3) the "Five Phases" subsection of the "Computational and Technical Arts" section. It is equally the case with the "Dao" section within the "Masters" section and the "Spirit Immortality" subsection within the "Recipes and Arts" section.

Consistent with Han narratives of the decline from the Classics to the literature of the diverse period of the "many masters," the second-stage decline from the masters' authoritative writings to more specialized texts concerned with narrower technical arts maps neatly onto the presumed shift posited from good governing techniques supporting the health of the entire body politic to lesser techniques designed to prove advantageous to the single body or family. Centuries after this taxonomic shift, the Tang historian Wei Zheng 魏徵 (580–643) would argue for the logic of his more general "Masters" (*zi* 子) category, which combined the *Hanshu* "Masters" category, with its "Computational and Technical Arts," "Spirit Immortality," and "Recipes and Arts" categories. His reasoning was that the "Masters" were the *teachings* of the sages, while things like "military" and the "medical recipes" were aspects of the *governing* of the sages. In Wei's view, while all good things began with the sages of yore, each ended up in a different domain: "in each case to what they were applied was different" (*suo shi ge yi* 所施各異).[61] Wei's rationale points to the same iterative process that we argue was the theoretical basis for the expansion of the catalogue to include technical arts texts.

To return to our initial observations about bookstores and the proclivity for taxonomies to reflect cultural priorities, it is clear that the importance of technical arts texts in both Western and Eastern Han, coupled with the need to integrate authoritative texts about them into classical rubrics, resulted in a great deal of cultural work being done to frame new taxonomies like the ones we have seen in the last chapters of the *Zhuangzi* and *Shiji* and most clearly, perhaps, in the *Seven Summaries* and *Hanshu* bibliographic treatise. This cultural work, preserved in the connective tissues of the bibliographic treatise that purport to explain the origin and long history of the authoritative techniques that are the subject in each section and subsection, prioritized public use over individual or domestic use while pitting the allegedly selfless practices of the Ancients against recent adaptations of those antique practices by specialists for their own benefit. While technical arts texts themselves had many readers, the *Hanshu* bibliographic treatise arranged them according to hierarchies of value that assigned a higher value to the Classics and the masters corpus, even as it normalized and integrated the technical arts texts into its narrative of the history of the "Way and its techniques."

Notes

1. The authors appreciate the helpful comments of Michael Nylan on an earlier draft of this chapter and helpful discussions with Esther Klein and Sara Barrera-Rubio.

2. The opening lines to the "Yiwen zhi" specify that "Qilüe" was its basis and that its editors extracted (*shan* 刪) its essentials to fill out the written record. There, the seven outlines are specified as "Six Classics" (*Liuyi* 六藝), "Many Masters" (*Zhuzi* 諸子), "Poetry and Rhymeprose" (*Shifu* 詩賦), "Military Writings" (*Bingshu* 兵書), "Technical and Computational Arts" (*Shushu* 術數), and "Methods and Skills" (*fangji* 方技). *Hanshu* 30.1701.

3. While "schools" is still often used to translate *jia* in *zhuzi baijia* 諸子百家, Csikszentmihalyi and Nylan made the case for "experts" being a better translation, especially in the context of the discussion of the six kinds of expertise in *Shiji* 130, and so here we use that translation. See "Constructing Lineages and Inventing Traditions."

4. *Hanshu* 30.1701: 春秋分為五, 詩分為四, 易有數家之傳. Here, and elsewhere, we reserve "Classic" to refer to the Six Classics in the Han, at times grouped with the *Lunyu* 論語 and/or *Xiaojing* 孝經.

5. Ibid. 七十子喪而大義乖. The use of *weiyan* as metalinguistic communication is seen in *Lüshi chunqiu* "Jingyu" 精諭, but during the Han it is also applied to expressing oneself by choosing the appropriate ritual expression or lyric from the *Odes*.

6. Ibid. The three quoted phrases are 以愚黔首; 大收篇籍, 廣開獻書之路; and 書缺簡脫, 禮壞樂崩. The last phrase is basically the same as an edict of Emperor Wu that Liu Xin quotes after explaining the complexity of the Han transmissions of the *Odes*. See *Hanshu* 38.1969. The late Western and Eastern Han story about the Qin is problematic; see Nylan, "Han Views of the Qin Legacy."

7. *Hanshu* 30.1701. For detailed discussions of this project from other perspectives, see Nylan, *Yang Xiong*; and Hunter, "The Yiwen zhi" 藝文志."

8. Loewe, "Liu Xin, Creator and Critic," esp. 300.

9. This category concerns authoritative guides to script, and therefore to the scribal professions. It included works such the *Shi Zhou* 史籀, described in the summary such as "what the official bureaus used to instruct children to write. The form [of the script] differed from the archaic script forms from the walls of the Kong family dwelling" 周時史官教學童書也 與孔氏壁中古文異體; the composite *Cang Jie*

倉頡 (also called the *Cang Jie pian* 倉頡篇), named after the putative inventor of the writing system in the following parts: (a) Fascicles 1–7, a *Cang Jie* 倉頡 compiled by Li Si 李斯 in the Qin; (b) Fascicles 8–13, a *Yuanli* 爰歷 compiled by Zhao Gao 趙高 in the Qin; (c) Fascicles 14–20, a *Wide Learning* 博學 compiled by Huwu Jing 胡毋敬; a *Biography of Cang Jie* 倉頡傳; and two Han works on compiling different accounts of the meanings (*xunzuan* 訓纂) of the graphs in the *Cang Jie* by Yang Xiong 揚雄 (53 BCE–18 CE) and the archaic script *Documents* expert Du Lin 杜林 (d. 47 CE). See the hypothesis of Yao Zhenzong 姚振宗 that the last two were inserted by the compiler of the *Hanshu* (*Hanshu Yiwen zhi tiaoli*, 147).

10. The origins and connotations of the term *xueguan* are still somewhat unclear. Stuart V. Aque's somewhat ambiguous translation as "academy" may have overstated the institutional nature of the term ("Pi Xirui and *Jingxue lishi*," 516n98). At the same time, Cai Liang's reaction in rendering it as "learned officials" perhaps goes too far in divorcing the term from any process of official recognition (*Witchcraft*, 47–53). In "Constructing," Csikszentmihalyi and Nylan translate the term as "local academies in the commanderies and principalities," or officials in charge of these academies (91). Its systematic usage in Western Han texts coincides with Sima Qian's criticism of the connection of mastery of the Classics with attainment of government office, and the *Hanshu*'s use of the term confirms that denial of certain transmissions such a status was at least as important as favoring others.

11. *Hanshu* 30.1704, 1706, 1708, 1710, and 1715: 立於學官, 鄒氏無師, 夾氏未有書. For the *Annals*, see Csikszentmihalyi 2015, 461–62.

12. *Hanshu* 30.1701: 戰國從衡, 真偽分爭 and 諸子之言, 紛然殽亂.

13. Certainly these were not the only possible comparisons, but as we will see, they are the most relevant to the *Hanshu* catalogue. Other extant taxonomic works from early China include the *Xunzi* 荀子 "Fei shi'er zi" 非十二子 and "Jiebi" 解蔽 chapters, *Shizi* 尸子 "Guangze" 光澤 chapter, *Lüshi chunqiu* 呂氏春秋 "Bu'er" 不二 chapter, and *Huainanzi* 淮南子 "Yao lüe" 要略 chapter.

14. Richard D. Parry, in his entry "*Episteme* and *Techne*," discusses the difference between the modern view that theory and practice "seem irreconcilable," and ancient Greek ideas in which the two generally overlap. For Aristotle, *techne* "is itself also *epistêmê* or knowledge because it is a practice grounded in an 'account.'" https://plato.stanford.edu/entries/episteme-techne/ (accessed 8/10/2020). Similarly, as we shall see, there is ultimately no epistemic distinction between different spokes of the wheel of the *Dao*.

15. *Huainanzi*, "Xiuwu xun" 脩務訓: 世俗之人, 多尊古而賤今, 故為道者必託之于神農、黃帝而後能入說.

16. Gu, *Handai xueshu shilue*, 35–44.

17. A summary of some of these views is found in Wang Shumin's (d. 2004) treatment of the chapter. Wang cites Ma Su 馬驌 (1621–1673) saying that the chapter is an authorial preface (自序), and Qian Mu's 錢穆 quotation of the late Ming writer Lu Xixing 陸西星 (1520–1606), in his *Nanhua zhenjing fumo* 南華真

經副墨, who calls the chapter "Zhuangzi's postface" 莊子後序 because it "lists and assesses the sources of ancient and modern 'techniques of the Way' and then continues [the list] with its own" 列敘古今道術淵源, 而以己承也. In the end, however, Wang rejects both views, concluding: "This chapter is *not* the work of Zhuangzi, and ought to be seen as neither his authorial preface nor postface, but rather as the statement of his academic school. Hence the section on the 'techniques of the Way' of Zhuang Zhou elevates Zhuangzi to the place of highest honor, something on which Zhuangzi certainly would not have signed off." 此篇非莊子作, 恬悻, 不當視為莊子自序或後序, 蓋莊于學派所述, 故於莊周道術章, 推尊莊子至極。莊子固未嘗自是者也。See Wang, *Zhuangzi jiaoquan*, 3 v., v. 3, 33.1293.

18. See previous note. Several commentators make an interesting connection between "Tianxia" and the *Mengzi*. Lu Xixing writes, "This has the same import as the last section of the *Mengzi*" 即孟子終篇之義, likely following Lin Xiyi 林希逸 (1193–1271) who, in *Nanhua zhenjing kouyi* 南華真經口義, compares "Tianxia" with the last section of the *Mengzi*, which lists the early inspirational sages and the later men of insight who "saw" and "heard" them. The *Mengzi* says the latter "listen to and thereby understand [prior sages]" 聞而知之, which is similar to the "Tianxia" portrayal of the masters' ability to "listen to the influence [of prior practitioners] and take pleasure in it" 聞其風而說之.

19. Jiang, *Zhuzi tongkao*, 422–23, cf. *Nan Qi shu* 南齊書 24.353. Among other arguments, Jiang maintains that the commentary attributed to the "Hui Shi" chapter of the *Zhuangzi* by Du Bi 杜弼 (491–559) shows that the final section of the chapter devoted to Hui Shi was not part of the Guo Xiang edition of the text but may have originally circulated as an independent text, separate from the *Zhuangzi*.

20. *Hanshu* 30.1736.

21. Liu, *Zhuangzi zhexue ji qi yanbian*, 90–91. Liu notes the parallels between the final section of "Zai you" 在宥 and the chapters "Tianyun" 天運, "Tiandao" 天道, "Keyi" 刻意, "Shanxing" 繕行, and "Tianxia."

22. Liu, *Zhuangzi zhexue ji qi yanbian*, 90.

23. Fang, *Zhuangzi*, 567.

24. The binary of *shenming* 神明 has a wide range of meanings; here the verb "descend" seems to indicate they are being associated with "heaven and earth" (*tiandi* 天地).

25. Wang, *Zhuangzi jiaoquan*, v. 3, 33.1294. Wang points out that the quotation "the Way is everywhere" is found in the "Zhi bei you" 知北遊 chapter of *Zhuangzi*.

26. Ibid. 不離於宗, 謂之天人 . . . 以仁為恩, 以義為理, 以禮為行, 以樂為和, 薰然慈仁, 謂之君子.

27. Note the absence of the *Changes* and *Annals* from the first version of the list of the Classics. The subject of the transmission of knowledge is a continuation from the previous section, which reads: "People of antiquity were complete: they matched the spirits and their percipience, were pure as Heaven and Earth, raised the myriad creatures, were at peace with the people of the realm, their benefits reached

the commoners, their understanding reached both original procedures and followed them out to the finest degree, over the six directions and four regions, across matters of any importance or subtlety, operating omnipresently. Their understanding was of procedures and degrees, and much of it may still be found in the ancient standards and transmitted histories." 古之人其備乎！配神明，醇天地，育萬物，和天下，澤及百姓，明於本數，係於末度，六通四辟，小大精粗，其運無乎不在。其明而在數度者，舊法世傳之史尚多有之。

28. Wang, *Zhuangzi jiaoquan*, v. 3, 33.1297–98. Wang notes (33.1300n8) that Qian Mu 錢穆 argues that the number used in the phrase "Six Classics" indicates "Tianxia" is a Han text.

29. Here, the connotation of *yi* 藝, usually translated as "Classics," is sliding toward the sense of "attainments," although the version of the arts here is rather different from charioteering, etc. Note the absence of the *Changes* classic, which "guides them in transformation" 道化 in the *Shiji* version of this passage.

30. Wang, *Zhuangzi jiaoquan*, v. 3, 33.1311: 俱誦《墨經》，而倍譎不同，相謂別墨。

31. Wang, *Zhuangzi jiaoquan*, v. 3, 33.1298.

32. The view of history as going through distinct stages, or even explicitly dispensational schemes, is not uncommon even in pre-Buddhist Chinese texts. On this phenomenon, see "The *Mengzi*'s transtemporal sage" in Csikszentmihalyi, *Material Virtue*, 191–200; "Équilibre cosmique et logique du déclin" in Espesset, *Cosmologie et trifonctionnalité*, 379–89; and "Hanshi fuxing de zhengzhi wenhua yiyi—Chenwei he *Gongyang* xue dui DongHan zhengzhi de yingxiang" 漢室復興的政治文化意義—讖緯和《公羊》學對東漢政治的影響 in Chen Suzhen, *Chunqiu yu Handao*, 379–484.

33. Burton Watson's treatment of this chapter in the 1958 *Ssu-ma Ch'ien, Grand Historian of China* is titled "The Biography of Ssu-ma Ch'ien." The rationale for not calling it "autobiography" is that it additionally includes the "Letter to Ren An" (Bao Ren An shu 報任安書) from the *Hanshu* adaptation of the *Shiji* chapter; hence Watson's title is of a translation of the *Hanshu*'s "Sima Qian zhuan" 司馬遷傳. The "Letter" was retranslated in a volume of essays by Stephen Durrant, Li Wai-yee, Hans van Ess, and Michael Nylan called *The Letter to Ren An and Sima Qian's Legacy*.

34. For a fuller treatment of many relevant facets of this work, see Csikszentmihalyi and Nylan, "Constructing Lineages and Inventing Traditions."

35. *Shiji* 130.3288–89.

36. John Major, et al., *The Huainanzi: A Guide to the Theory and Practice of Government in Early Han*, 484.

37. *Shiji* 130.3288. Chen Guying 陳鼓應 (*Yi zhuan yu Daojia sixiang*, 179–181) argues that the *Changes* transmission from Tian He 田何 to Yang He is particularly important for understanding Sima Tan's development as a Han exponent

of "HuangLao Daoism" 黃老道家. While we contend that this use of "Daoism" is anachronistic, Chen's survey is important in that it highlights the role of the *Changes* in the intellectual life of the late second-century BCE, a time when Sima Qian outlined a lineage of practices associated with Huangdi and Laozi. See "The *Shiji*'s Counterpoint to the Ru: HuangLao Learning" in Csikszentmihalyi and Nylan, "Constructing Lineages and Inventing Traditions," 80–87.

38. Contrast this with the description of Zhuangzi in the "Tianxia": his connection to the Dao never grows obsolete; however, neither the form nor the content of the chapter implies he incorporated the insights of the other figures described before him.

39. *Shiji* 130.3290: 以六藝為法 . . . 累世不能通其學, 當年不能究其禮. The latter phrase is also spoken by Yan Ying 晏嬰 in a "Kongzi shijia" dialog (*Shiji* 47.1911).

40. *Shiji* 130.3297. Here we have translated the titles of the Classics using italics, but the context indicates that the text itself is just the visible part of larger bodies of practice that lurk below the surface.

41. Qi, *Jiazi Xinshu jiaoshi*, 919. For a more general discussion of this relationship, see Csikszentmihalyi, "Chia I's 'Techniques of the Tao.'"

42. *Hanshu* 30.1767–68.

43. *Hanshu* 30.1734: 歷象日月星辰. This is a quotation of the *Shangshu* 尚書 "Yaodian": 乃命羲和, 欽若昊天, 歷象日月星辰, 敬授人時. Li Ling points out that the connection being made between the historical official position of astronomer and these texts is anachronistic; see *Lantai wanjuan: du* Hanshu *"Yiwen zhi,"* 97.

44. *Hanshu* 30.1768.

45. Zhang identifies Dingzi as the Chu general Ding Gu 丁固 in *Hanshu Yiwen zhi tongshi*, 244–45, and King Xiang is likely Xiang Yu 項羽 of Chu.

46. There are disagreements about how to read Da You, but Marc Winter makes a convincing case that for the "Military" section text *Sunzi bingfa* 孫子兵法, *wuxing* refers to the five visible planets; see Winter, "Suggestions for a Re-interpretation of the Concept of *Wuxing* in the *Sunzi bingfa*." For the conjunction of the five planets as one of the omens of Liu Bang's eventual rise, see *Shiji* 27.1348 and 89.2581.

47. *Hanshu* 30.1734: 及拘者為之, 則牽於禁忌, 泥於小數, 舍人事而任鬼神.

48. *Hanshu* 30.1769: 而小數家因此以為吉凶, 而行於世, 浸以相亂. Similarly, *Hanshu* 30.1743 explains the misapplication of agriculture-related texts in the "Many Masters" (Zhuzi 諸子) section: "However, when the debased performed them, they thought there was nothing that they owed the sage kings, and wanted the ruler and ministers to plow the fields, disrupting the proper sequence of superior and inferior" 及鄙者為之, 以為無所事聖王, 欲使君臣並耕, 誖上下之序.

49. Li, *Lantai wanjuan: du* Hanshu *"Yiwen zhi,"* 212.

50. *Hanshu* 30.1780. The Kongzi quotation here appears not in the *Analects*, but in the "Zhongyong" chapter of the *Liji*. Here we read *su* 素 for *suo* 索, consistent with the *Liji* version. The fourth-century Jiangnan promoter of methods of

transcendence Ge Hong 葛洪 may have had the *Hanshu*'s criticism in mind when he put these very words of Kongzi in the mouths of "people of this generation" (*shiren* 世人) that disparaged Liu Xiang as "always reclusive and acting atypically." Ge writes that Liu's work on spirit transcendence was unfairly dismissed by such persons: "If a work does not come from the gateway of the Duke of Zhou, or if matters do not square with the hand of Zhong Ni (i.e., Kongzi,) then people of this generation end up regarding them with disbelief" 書不出周公之門, 事不經仲尼之手, 世人終於不信. See *Baopuzi neipian* 2.20 and 2.15. Here, Ge seems to also have in mind the *Huainanzi* quotation about texts being attributed to the Shen Nong and the Yellow Emperor examined above.

51. These methods are likely similar to the ones found in some excavated texts such as the *Daoyin tu* 導引圖 uncovered at Mawangdui. Donald Harper translates the title of this text as "Drawings of Guiding and Pulling"; see Harper, *Early Chinese Medical Literature*, 314n4 and 316n2.

52. Yao Zhenzong believed that Ge Hong's discussion of the five categories of *zhi* 芝 fungus (stone, wood, grass, meat, and *jun* 菌 fungus) was based on the eighteen-fascicle text from this section of the catalogue; Yao, *Hanshu Yiwen zhi tiaoli*, 449–50.

53. While the "smelting of gold" sounds like alchemy, in common with later Daoist practices, the phrase is also connected with longevity and immortality. After detailing the Qin emperor's strong interest in spirit immortality practices, the *Hanshu* "Jiaosi zhi" 郊祀志 describes the continuing practices of natives of the region of Qi in the Western Han, who were richly rewarded for their *huangye* 黃冶 "smelting of gold" and "journeys on the seas in order to seek out spirits and harvest the drugs of immortality" 入海求神采藥; see *Hanshu* 25b.1260.

54. This quotation appears in Yan Yan, *Zizhi tongjian bu*, 119.23b. Ye Changqing (*Hanshu Yiwen zhi wenda*, 187) argued that the reason for the name "Recipes and Arts" was that "Spirit Immortality" techniques originated from the "Recipe Masters" (*fangshi* 方士) from the area of Qi.

55. *Hanshu* 30.1728.

56. *Zhouli zhushu* 周禮注疏 9.138a: 使帥其屬而掌邦教, 以佐王擾邦國.

57. *Hanshu* 30.1708 and 30.1715.

58. See the entry for "Liu Xin" 劉歆 in Loewe, *A Biographical Dictionary*, 383–87.

59. Michael Nylan has pointed out to us that we know this from other writings attributed to Ban Gu, including his "Liang du *fu*" 兩都賦.

60. The theory tracing the various masters to ancient offices is often called *zhuzi chu yu wangguan* 諸子出於王官. During the early twentieth century many scholars heatedly debated the historicity of this theory. More recently, scholars have begun to approach this theory from a new perspective: how the theory, regardless of its accuracy, may have provided the rhetorical framework needed to gather together disparate strands of Western and Eastern Han thinking in a single system. For a

recent review of the debates and the new interpretation of this theory, see Deng, "'Zhuzi chu yu wangguan shuo' yu Hanjia xueshu huayu."

61. *Suishu* 隋書 34.1051: 儒、道、小說，聖人之教也，而有所偏。兵及醫方，聖人之政也，所施各異。

Asian-language Bibliography

Chen Guoning 陳國寧. *Hanshu Yiwen zhi zhushi huibian* 漢書藝文志注釋彙編. Beijing: Zhonghua shuju, 1983.

Chen Guying 陳鼓應. *Yi zhuan yu Daojia sixiang* 易傳與道家思想. Taiwan: Shangwu yinshuguan, 1994.

Chen Suzhen 陳蘇鎮. *Chunqiu yu Handao: LiangHan zhengzhi yu zhengzhi wenhua yanjiu* 《春秋》與「漢道」：兩漢政治與政治文化研究. Beijing: Zhonghua, 2011.

Deng Junjie 鄧駿捷. "'Zhuzi chu yu wangguan suo' yu Hanjia xueshu huayu" '諸子出於王官說'與漢家學術話語. *Zhongguo shehui kexue* 中國社會科學 (2017): 184–204.

Fang Yong 方勇. *Zhuangzi* 莊子. Beijing: Zhonghua shuju, 2015.

Ge Hong 葛洪. *Baopuzi* 抱朴子. In *Baopuzi neipian jiaoshi* 抱朴子內篇校釋, Annotated by Wang Ming 王明. Beijing: Zhonghua shuju, 1980.

Gu Jiegang 顧頡剛. *Handai xueshu shilue* 漢代學術史略. Taipei: Qiye shuju, 1975.

Gu Shi 顧實 (1898–1956). *Hanshu Yiwen zhi jiangshu* 漢書藝文志講疏. Taipei: Taiwan shangwu yinshuguan, 1980.

Hanshu 漢書. Compiled by Ban Gu 班固 (32–92). Annotated by Yan Shigu 顏師古 (581–645). 12 vols. Beijing: Zhonghua shuju, 1962. Reprint, 1996.

Jiang Boqian 蔣伯潛. *Zhuzi tongkao* 諸子通考. Hangzhou: Zhejiang guji chubanshe, 1985.

Li Ling 李零. *Lantai wanjuan: du* Hanshu *"Yiwen zhi"* 蘭台萬卷：讀漢書・藝文志. Beijing: Sanlian shudian, 2013.

Li Shuhua 李叔華. "*Zhuangzi* 'Tianxia pian' de zhuzhi he chengwen niandai"《莊子・天下篇》的主旨和成文年代新探. *Zhexue yanjiu* 哲學研究 (May 1995): 72–81.

Lin Xiyi 林希逸 (1193–1271). *Nanhua zhenjing kouyi* 南華真經口義. Kunming: Yunnan Renmin chubanshe, 2002.

Liu Xiaogan 劉笑敢. *Zhuangzi zhexue ji qi yanbian* 莊子哲學及其演變. Beijing: Zhongguo shehui kexue chubanshe, 1988.

Qi Yuzhang 祁玉章. *Jiazi Xinshu jiaoshi* 賈子新書校釋. Taipei: Zhongguo wenhua zazhishe, 1975.

Shiji 史記 [*Records of the Senior Archivist*, aka *Historical Records*]. Compiled by Sima Qian 司馬遷 (?145–?86 BCE). 10 vols. Beijing: Zhonghua shuju, 1959. Reprint, 1982.

Wang Shumin 王叔岷. *Zhuangzi jiaoquan* 莊子校詮. Taipei: Zhongyang yanjiuyuan lishi yuyan yanjiusuo, 1988.

Yao Zhenzong 姚振宗 (1842–1906). *Hanshu Yiwen zhi tiaoli* 漢書藝文志條理. In *Ershiwu shi kanxing weiyuan huibianji* 二十五史刊行委員會编集, *Ershiwu shi* 二十五史. Shanghai: Kaiming shudian, 1935.

Yan Yan 嚴衍 (c. 1575–1645). *Zizhi tongjian bu* 資治通鑑補. N.p.: Shengshi Sibulou, 1876.

Ye Changqing 葉長青. *Hanshu Yiwen zhi wenda* 漢書藝文志問答. Wuhan: Huazhong shifan daxue chubanshe, 2015.

Zhang Shunhui 張舜徽 (1911–1992). *Hanshu Yiwen zhi tongshi* 漢書藝文志通釋. Wuhan: Huazhong shifan daxue chubanshe, 2004.

Western-language Bibliography

Aque, Stuart V. "Pi Xirui and *Jingxue lishi*." PhD diss., University of Washington, 2004.

Cai Liang. *Witchcraft and the Rise of the First Confucian Empire*. Albany: State University of New York Press, 2014.

Csikszentmihalyi, Mark. "Chia I's 'Techniques of the Tao' and the Han Confucian Appropriation of Technical Discourse." *Asia Major* series III 10, nos. 1–2 (1997): 49–67.

———. *Material Virtue*. Leiden: Brill, 2004.

———. "The Social Roles of the *Annals* Classic in Late Western Han." In *Chang'an 26 BCE: An Augustan Age in China*, edited by Michael Nylan and Griet Vankeerberghen, 461–76. Seattle: University of Washington Press, 2015.

Csikszentmihalyi, Mark, and Michael Nylan. "Constructing Lineages and Inventing Traditions in the *Shiji*." *T'oung Pao* 89 (2003): 59–99.

Drège, Jean-Pierre. *Les bibliothèques en Chine au temps des manuscrits (jusqu'au Xe siècle*. Paris: École Française d'Extême-Orient, 1991.

Durrant, Stephen, Wai-yee Li, Michael Nylan, and Hans van Ess. *The Letter to Ren An and Sima Qian's Legacy*. Seattle and London: University of Washington Press, 2016.

Espesset, Grégoire. *Cosmologie et trifonctionnalité dans l'idéologie du Livre de la Grande paix (Taiping jing* 太平經*)*. PhD diss., Université Paris-Diderot–Paris VII, 2002.

Harper, Donald J. *Early Chinese Medical Literature*. New York: Routledge, 1998.

Hunter, Michael. "The 'Yiwen zhi' 藝文志 (Treatise on Arts and Letters) Bibliography in its Own Context." *Journal of the American Oriental Society* 138, no. 4 (October–December 2018): 763–80.

Loewe, Michael. *A Biographical Dictionary of the Qin, Former Han and Xin Periods (221 BC–AD 24)*. Leiden: Brill, 2000.

———. "Liu Xin, Creator and Critic." *Rao Zongyi Guoxueyuan yuankan* 饒宗頤國學院院刊 (April 2014): 297–323.

Major, John, Sarah A. Queen, Andrew Seth Meyer, and Harold D. Roth. *The Huainanzi: A Guide to the Theory and Practice of Government in Early Han.* New York: Columbia University Press, 2010.

Nylan, Michael. "Han Views of the Qin Legacy and the Late Western Han 'Classical Turn.'" *Bulletin of the Museum of Far Eastern Antiquities* 79/80 (December 2013 [published 2020]): 51–98.

———. *Yang Xiong and the Pleasures of Reading and Classical Learning in Han China.* New Haven: American Oriental Society, 2011.

Watson, Burton. *Ssu-ma Ch'ien, Grand Historian of China.* New York: Columbia University Press, 1958.

Winter, Marc. "Suggestions for a Re-interpretation of the Concept of *Wuxing* in the *Sunzi bingfa*." *Bulletin of the Museum of Far Eastern Antiquities* 76 (2004): 147–80.

Contributors

Miranda Brown is Professor of Chinese Studies at the University of Michigan, Ann Arbor.

Jesse Chapman received his PhD from East Asian Languages and Cultures at the University of California, Berkeley.

Karine Chemla is Principal Investigator for the European Research Council's Advanced Research Grant "Mathematical Sciences in the Ancient World" and Senior Researcher at the French National Centre for Scientific Research & University of Paris—Paris Diderot.

Mark Csikszentmihalyi is Marjorie Meyer Eliaser Professor of International Studies, University of California, Berkeley.

Luke Habberstad is an Associate Professor in the Departments of East Asian Languages and Literatures and Religious Studies at the University of Oregon.

Lee Chi-Hsiang is a Professor, Department of History, Foguang University, Taiwan.

Michael Loewe is retired from his position as University Lecturer in Chinese studies at the University of Cambridge.

Michael Nylan is Sather Professor of History at the University of California, Berkeley.

Tian Tian is an Associate Professor in the School of Archaeology and Museology at Peking University (北京大學), Beijing.

Zheng Yifan is completing his PhD in East Asian Languages and Cultures at the University of California, Berkeley.

Index

Analects, 15, 103, 109, 116–19, 131n57, 371, 397n4
Annals (aka *Spring and Autumn Annals*), 9, 201; "King's City," 148, 151–59, 161, 165; and *kongyan*, 311–13; transmissions, 372; and Treatise on the *Wuxing*, 182, 186–90, 214, 248, 252–53. *See also* ChengZhou; Luo City; Luoyang
Annals of Lü Buwei, 102–10, 118–19; "choosing the best course," 102, 109; collaboration, 27, 104, 110, 118; the "masses," 118; "Taking Pleasure in Success," 102–103

Ban Biao, 4, 34, 222, 232, 248
Ban Gu, choosing the right men, 119–20; compiler, 9–10, 109, 115, 223, 237, 242–44, 248; and framing, 384–85; and omen readings, 216–17, 241; and rulers' legitimacy, 232; and Sima Qian, 13, 18–21; tweaking, 241–43; "two capitals" theory, 154. *See also Hanshu*
Ban Zhao, 182, 193, 201. *See also* Treatise on Celestial Patterns
Bibliographic Treatise. *See* Treatise on the Classics and Writings
bing jian, 62, 64
"Biography of the Granary Master" (*Shiji*), 31, 339

bureaucrats. *See* officials

Chang'an, 106, 148, 156, 284–98
Changes, 186–88, 195–96, 214–18; Eight Trigrams, 185–86, 228; transmissions, 371
Chengdi, 3; in omen treatise, 234–35; palace library, 139; and water control, 111–12, 125
ChengZhou, 148–61
Chunqiu. *See Annals*
Classic of the Pulse, 358
classicists: in contrast to functionaries, 18; critique of Wudi, 21, 220, 297; as described by Ouyang Xiu, 33; expertise disseminated, 380; fallible, 251; filling pragmatic needs, 8–9; on portent theories, 248–49; Wudi's patronage of, 247
Confucius, 3, 22, 31, 186, 241; and *Documents* Postface, 136–41, 162–63; loss of teachings of, 369

"Discourse on the Five Resources Tradition of the 'Great Plan,'" 182, 201
Divine Farmer, 374
Documents: conflation of graphs in, 141–42, 144; Fu Sheng version of, 137–39, 144, 167n28; so-called Kong Anguo (Mei Ze) version,

409

410 | Index

Documents (continued)
144, 162–63; Kong Yingda subcommentary, 149–50, 164–65, 393; pre-Han versions, 139; "pseudo-Kong" "forgery," 136, 139–40; and *Shiji*, 17–18; transmissions, 371. See also ChengZhou; Luo City; Luoyang; "New City;" and *names of individual chapters* (e.g., "Great Plan," "Luo gao," "Proclamation for Kang")
Dong Zhongshu, 62, 190, 214–16, 220–21, 238, 244–50, 312
Duke of Zhou. See Zhougong
"Duo shi" chapter (*Documents*), 142, 146, 149–50
Du Qin, 111–17, 130n38

Earth, 5, 183, 231, 288, 293–94
Earthly Unity, 283–84
experts, 112–19, 236, 240, 264n120, 377

fangshi, 283–87, 297–98
Feng Qun, 110–13, 125
Fenyin, 285–86, 293–94
Five Classics, 28, 221, 297, 371–81
Five Hegemons, 184
Five Resources, 187–89, 203–204, 225–29

gong tian, 65–69
Gongyang Tradition, 182, 252, 372
"Great Plan" chapter (*Documents*), 18, 182; and omen treatise, 225–30, 238–40, 246, 249; Nine Divisions of, 186–87; and *Wuxing zhi*, 185–87, 195, 201–202
Great Tang Record of the Suburban Sacrifices, 288
Gu Yong, 67, 192–93, 297–98

Hanshu: biography of King Yuan of Chu, 138; Map of the Purple Pavilion, 286; memorials and water control proposals, 102–103, 118–19; proper decision-making, 117; *shan* ("best" proposal), 103; tables, list of, 20; treatises, list of, 14–17; technical expertise, 117–20. See also Ban Gu; Ban Zhao; *and names of individual tables and treatises* (e.g., "Treatise on the *Wuxing*," "Treatise on Celestial Patterns," "Treatise on Geography," "Treatise on the Classics and Writings")
Hao, 148–61
hao jia, 62
hao zu, 62
Heaven. See Taiyi
Historical Records. See *Shiji*
History of Han. See *Hanshu*
houfan (medicine), 351, 354
Huan Tan, 64, 114–17
Huo Guang, 192, 219–20

"In the Realm," 375

jiao 郊 ("suburban"), 295–98, 302n62
jiaohua ("civilizing influences"), 4, 273n213, 392
Jia Rang, 113, 126
jue ("orders of honor"), 25, 55–70
jue shan ("choose the best course"). See officials; rulers

King Cheng, 145–50, 157–61
King's City, 151. See also ChengZhou
Kong "wall find," 137–38
kongyan ("abstract expressions"). See *Annals*; mathematics; *Shiji*
Kong Yingda, 149–50, 160, 393. See also *Shangshu zhengyi*

kong yu ("empty theories"), 115
Kongzi. *See* Confucius
Kuang Heng, 294

land tenure: and bookkeeping, 61–62; estates, 62–65; government provisions, 51–52; and inheritance, 68–69; land reclamation, 52–53; and population, 75–80; and produce, 53–55; and ritual purpose, 82; salaries, 74–75; and vagrants. See also *bing jian; gong tian; hao jia; hao zu; jue; liu min; ming tian; quan; yu*
Lin Yi, 350–56
liu min ("vagrants"), 59, 61, 67–68
Liu Xiang: on Confucius, 138–39, 141; and Liu Xin, 138–41, 225, 237–42, 370; and omens, 190–93; 244–48; use of classical literature, 371, 395
Liu Xin: and Academicians, 138; on revelation texts, 228; and *Rites of Zhou*, 395; and Sima Qian, 299. *See also* Liu Xiang; Treatise on the *Wuxing*
Liu Zhiji, 22, 29, 223, 238
lü 律. *See* mathematics
Luo City, 141–62, 174n79
"Luo gao" chapter (*Documents*), 146, 149, 151
Luo, radicals used for transcription, 141–45
Luo River, 141, 145, 157–58
Luo Writing. *See* "Treatise on the *Wuxing*"
Luoyang, 27–28, 135–45, 150–59, 172n62

Map of the Purple Pavilion, 286
mathematics: abstract expressions in *Nine Chapters*, 314–28; abstract expressions in *Mathematical Procedures*, 323–25; history of, 307; *lü*, 315–18, 321; *shi* ("action, event, situation"), 312–14, 319–22, 326–27; *shu* ("procedures"), 30, 308–309, 319, 323; *shu* ("quantities"), 315. See also *Nine Chapters on Mathematical Procedures*
medical commentaries: as established practice, 361–62; as explanations for students, 339–40; as glosses, 341–42; scarcity of, 361, visual distinction of, 348–49
ming tian, 65–69
Miu Ji, 283–84, 287, 291

"New City," 145–50, 159, 170n39
Nine Chapters on Mathematical Procedures, The: "abstraction" sense of, 320–21; integral values in, 316–17; *li* ("paradigm"), 333n43; Li Chunfeng subcommentary, 310, 325–26, 330n11, 331n22; Liu Hui commentary, 310, 326, 330n11, 333n43; textual *dispositif* in, 318, 327
Nine Tripods, 146–47, 152, 162
Nishijima Sadao, 50

Odes, 154, 195, 218, 372, 379, 383
officials: 358–60, 390–91; loyal (*zhong*), 102–10, 118; salaries, 74–75; who "choose the best course" (*jue shan*), 102, 104, 118
omen treatise. *See* Treatise on the *Wuxing*

population, 76–80
"Proclamation for Kang" chapter (*Documents*), 145–46

quan ("deeds"), 69–74, 85–88

Record of Rites: "Monthly Ordinances," 193, 197; transmissions, 372
risk, 112, 119, 222, 236, 243–44, 248
Rites of Zhou, 391–95
River Chart. *See* Treatise on the *Wuxing*
ruler: active role of, 118; and "choosing the best course" (*jue shan*), 102–104, 107–109, 118–19; collaboration, 118; loyalty to, 84, 110; master-selector, 109; promising to complete a project, 103–104; securing divine sanction, 102

Seven Kingdoms rebellion, 185, 191, 234
Shangshu. *See Documents*
Shangshu zhengyi, 160, 167n16. *See also* Kong Yingda
"Shao gao" chapter (*Documents*), 145–46, 149, 160
Shi. *See Odes*
shi 事 ("event"), 312. *See also* mathematics
shi 史 ("archivist"). *See Shiji*
Shiji: collaboration, 27; on compiler of *Documents*, 136–37; on emperor's active role, 118; on *kongyan* and *shi*, 312; medical practice in, 339; "Personal Narrative of the Senior Director of the Archives," 381–83; tables, list of, 19–20; treatises, list of, 11–13. *See also* Sima Qian; Sima Tan; *and names of individual tables and treatises*
Shijing. *See Odes*
Shi Qi, 103–104, 107–10, 115, 123
shu 術, 2–11, 37n29. *See also* technical arts; *zhi*
shu 術 ("procedures"). *See* mathematics
shu 數 ("quantities"). *See* mathematics
shu 書 ("treatises"), 10–13, 39n43

Sima Qian: and Ban Gu, 13, 18–21; compiler, 9–10; as dutiful subject, 108; "Forest of Classicists," 167n15; and political fortunes, 184–85; on sacrifices, 296; as Senior Archivist, 182–83. *See also Shiji*
Sima Tan, 182–83, 368, 382–83
si tian, 65–69
Spring and Autumn Annals. *See Annals*
suo ming you ("named as the owner"), 65

Table of Figures, Past and Present (*Hanshu*), 20–23, 251
Table of the Three Dynasties (*Shiji*), 19–21
Table of the Twelve Local Lords (*Shiji*), 10, 19
Tables, 19–23
Tai, 282
Taishi gong shu. *See Shiji*
Taiyi: acquiring human properties, 284–85; as astral deity, 283–86, 298; altar to, at Chang'an, 283; altar to, at Sweet Springs (Ganquan), 285–91; and curing Wudi, 284; equated with Dao, 283; and Fenyin, 285–86, 293–94; as Great Lord, 286; and Heaven 292–99; and stars, 283; three simultaneous aspects of, 283. *See also* Wudi
Tang Mingtang, 288
technical arts: and modern categories of science and technology, 2–3, 374; and *shu*, 2–8; and *yi*, 2; and *zhi*, 5–8
Ten Canons of Mathematics, 310
"Tiandi," 293
Tian Fen, 105–108, 122, 124
"Tianxia," 369, 373–84, 399n18. *See also Zhuangzi*
Treatise on Celestial Patterns (*Hanshu*), 16, 28; and "Celestial Offices," 185, 193–97, 200–201; and the Classics, 195–97; compilers, 182; and dynastic history, 197–200; Jupiter/

Year Star, 198–99, 202–204, 252; Mars/Dazzling Deluder, 197–98, 202–204; Mercury/Watch Star, 202–204; Saturn/Quelling Star, 198, 202–204; Venus/Great Bright, 202–204; Wang Mang in, 199–200; and *Wuxing zhi*, 200–201. *See also* Ban Zhao; *Hanshu*

Treatise on Ditches and Canals (*Hanshu*), 17, 102, 109–20, 123–26, 131n57

Treatise on Five Phases. *See* Treatise on the *Wuxing*

Treatise on Geography (*Hanshu*), 17; and archaeology, 156–61; and *Documents* Postface, 136–40, 143–44, 148, 151; influence of *Annals* and *Documents* on, 151; and two capitals model, 148, 151–59. *See also* ChengZhou; Luo City; Luoyang; "New City"

Treatise on the Celestial Offices (*Shiji*), 12, 28; governance in, 182–83; and historical precedent, 185; lineage of observers, 183–84; not relying on classical texts in, 185; and political crisis, 184–85; and solar eclipse, 185; on Tang Du, Wang Shuo, and Wei Xian, 183–84. *See also* Treatise on Celestial Patterns

Treatise on the Classics and Writings (*Hanshu*), 17; complete taxonomy of, 367–69; on fragmentation and decline, 369–70, 384, 390; on offices and masters, 392–93; on offices and sections, 394; the Way and statecraft, 385–88, 396

Treatise on the *Feng* and *Shan* Sacrifices (*Shiji*), 12; first mention of Great Lord in, 285–86; Qin altars, 290; and "suburban" (*jiao*), 294–98; Wudi in, 291. *See also* Taiyi

Treatise on the Origins of the Various Illnesses, 357

Treatise on the Suburban Sacrifices (*Hanshu*), 15–16; and the *fangshi*, 297–98; and "suburban" (*jiao*), 294–98. *See also* Taiyi

Treatise on the *Wuxing* (*Hanshu*), 16; and *Annals* 182, 186–90, 214, 248, 252–53; author/compiler of, 223, 238–44; chronicle of comets, 189–93; compared with *Hou Hanshu*, 236; complexities of, 248–51; criticizing rebellions, 234–36; and decentralized forms of power, 236–37; diverging from "Great Plan," 225–31; on Dong and Liu, 244–48; initial compilation, 248; lightning strike in, 237; multiple readings of, 242; narrative arc, 189; and an older tradition, 234; omission of ominous events, 237–38; purposes, 240–42; as revelation text, 251; River Chart and Luo Writings, 228, 238; ruler responsible for signs, 189; scope of, 232–34; and *shuo*, 233; silence on Liu Xiang, 247; on *Tian jie ruo yue* in, 246–47; unoriginal aspects of 232; and the word *wuxing*, 227–32; and *zhuan*, 233; *zhuan yue*, 224, 239, 263n104

Treatise on the Yellow River and Canals (*Shiji*), 12–13, 104–109, 122–23

Tunshi River, 125

Wang Bing, 350–57
Wang Feng, 111–12, 117, 125, 235
Wang Mang, 182, 190–201. See also *bing jian*; *gong tian*; *jue*
Wang Yanshi, 111–12, 125
water: control and currying favor, 118–19; control and ruler, 118; managers (*dushui*), 101
Wei, king of, 103–104, 107–109, 115
Wei River canal, 106, 112, 122, 124
"Wei Taiyuan," 293

Writings by the Senior Archivist. See Shiji
Writings on Mathematical Procedures. See mathematics
Wudi 武帝: and Chaori xiyue ceremony, 291; critique of, 297; cured of illness, 284; and the Fenyin bronze, 285–86; and Gongsun Qing, 16, 285–87; *Hanshu* critique of, 297–98; and immortality, 281, 285–87; and *leici* offering, 292; patronage of classicists, 247; reorganizing cult practices, 281–82; rhetorical treatment of, 21; temple for, 220; and water control, 105–109, 122–24; and worship of Taiyi, 283–87, 298; and Yellow Emperor, 285. *See also* Taiyi
Wudi 五帝 ("Five Lords"), 283
wugu ("witchcraft"), 138, 192
Wuwei manuscripts: circulation of commentaries, 358–59; and commentary, 341–50, 361; compared with Wang Bing and Lin Yi, 350–56; content similar to other documents, 359; explanatory material in, 341–42, 357; on illnesses, 344–51; layout, 348–49; on masters, 358; structure of, 359–60
Wuxing zhi. See Treatise on the *Wuxing*

Xia Heliang, 199
Xiang Yu, 189–91
xiaoshu ("lesser computational arts"), 387–88
Ximen Bao, 103, 110, 115, 122–23, 128n11
Xuanfang Temple, 107
Xu Shang, 110–11, 125, 217, 239

Yan Shigu, 138, 143

Yellow Emperor's Classic of Eighty-One Difficulties, 348
Yellow Emperor's Inner Classic of Medicine, 340, 343, 348, 350–51, 358
Yellow River, 122–26; anxiety about, 121; canal, 106–107; damage to area of, 110–12, 120; dike breach, 105–106; flooding reduced post-Han, 121; sacrifices to, 107, 118
Yellow River Chart. *See* Treatise on the *Wuxing*
yi 藝 ("arts"). *See* technical arts
Yi. *See Changes*
Yijing. See Changes
Yin men ("men of Yin"), 147, 171, 176n94
"Yiwen zhi." *See* Treatise on the Classics and Writings
yu ("convey"), 69
Yu, sage-king, 105, 116

Zhangjiashan Statutes and Ordinances of 186 BCE: 53, 56–57; on heir of holder of *jue*, 70–71; *yu* as temporary provision, 69
Zhang River, 103–105, 109–10, 118, 122–23
Zhang Rong, 113–14, 126
Zhang Shoujie, 155, 283
zhan tian, 65
Zheng Dangshi, 106, 122, 124
Zhengguo canal, 105, 122–24
Zheng Xuan, commentary ascribed to, 167n17; contra Sima Qian, 173n73; and "two capitals" model, 154–56
zhi 治 ("ordering"), 5–8. *See also shu* 術; technical arts
zhi 志 ("treatises"), 14–17, 39n43
Zhougong, 3, 145–61
Zhuangzi, 31, 375–80, 384. *See also* "Tianxia"
Zhu Xi, 139

www.ingramcontent.com/pod-product-compliance
Lightning Source LLC
Chambersburg PA
CBHW021239240426
43673CB00057B/630